Indian Agricultural Economy During Pre-Green Revolution Era (1940 to 1965)

VOLUME 1

INDIAN SOCIETY OF AGRICULTURAL ECONOMICS

ACADEMIC FOUNDATION
NEW DELHI

www.academicfoundation.org

First published in 2016
by

ACADEMIC FOUNDATION
4772-73 / 23 Bharat Ram Road, (23 Ansari Road),
Darya Ganj, New Delhi - 110 002 (India).
Phones : 23245001 / 02 / 03 / 04.
Fax : +91-11-23245005.
E-mail : books@academicfoundation.com
www.academicfoundation.com

Disclaimer:
The findings/views/opinions expressed in this book are solely those of the authors and do not necessarily reflect the views of the publisher.

Cataloging in Publication Data--DK
Courtesy: D.K. Agencies (P) Ltd. <docinfo@dkagencies.com>

Indian agricultural economy during pre-Green Revolution era (1940-1965) / Indian Society of Agricultural Economics.
pages cm
ISBN 9789332703339

1. Agriculture--Economic aspects--India--History--20th century. 2. Agriculture and state--India--History--20th century. 3. Green Revolution--India. I. Indian Society of Agricultural

Economics.issuing body.

HD2072.I53 2016 DDC 338.10954 23

Typeset by Italics India, New Delhi.

Printed and bound by The Book Mint, New Delhi.
www.thebookmint.in

CONTENTS

Preface

THE year (2015) is the Platinum Jubilee year of the Indian Society of Agricultural Economics (ISAE). The Society has grown up to this stage because of its dynamism which encompasses a series of activities such as conducting annual conferences, seminars, workshops and trainings, and bringing out a number of publications besides publishing a world class quarterly journal, *Indian Journal of Agricultural Economics* (IJAE) which carries articles based on research and policy analysis. The Society had also embarked in policy advocacy from time to time. ISAE emerging as one of the oldest and reputed professional societies could not have happened overnight. The leaders comprising Presidents, Conference Presidents, Secretaries, Treasurers and Other Office Bearers chartered the Society in a professional manner adding value at different stages of its growth. However, the Members are the building blocks of the Society and their unstinted cooperation in running the Society in a democratic manner added great strength, respect and credibility to the Society.

Wider discussion at various points in time during the past three years among the Office Bearers and Members of the Society led to the decision to bring out a publication of the Society's "Conference Presidents" addresses which forms, of course, one of the outcomes of Platinum Jubilee celebration. It was felt that though these addresses are published in IJAE regularly, a compilation of addresses will help readers to reach them under one publication. Further, it is common knowledge that access to the addresses of earlier years is cumbersome and for some years it may not be available anywhere else except Society's office. Only the time will tell the value of bringing out this publication. And it will be a great asset to the posterity.

While the idea of publishing the addresses shaped up, it was felt that putting all the 74 presidential addresses (one address less of 75 because no conference was held in 1949) in single volume seemed unwieldy and decision was made to bring out the addresses in three volumes which were christened as: Volume I—*Indian Agriculture Before Green Revolution: Concerns and Approaches*, Volume II—*Indian Agriculture During Green Revolution Period—Concerns and Approaches* and Volume III—*Indian Agriculture Post Green Revolution: Concerns and Approaches*. Volume I carries 24 addresses, and the other two volumes contains 25 addresses each. To have a bird's eye view of the addresses in each volume before delving into it, it was conceived to have an introduction to the volume. Accordingly, three eminent and highly accomplished economists who have contributed so much to the field of agricultural economics were approached by the Society to write the introduction. They are: Prof V.S. Vyas (currently Professor Emeritus, Institute of Development Studies, Jaipur), Prof S.S. Johl (currently Chancellor, Central University of Punjab, Bathinda), and Prof. Abhijit Sen (currently Professor of Economics at Jawaharlal Nehru University, New Delhi) who were generous enough to write the "Introduction" to the volumes I, II, and III respectively. These introductions have added tremendous value to the volumes and one can have a brief understanding of what is in the volumes and the writers have taken pains to integrate and summarise the thoughts from addresses which reflected various developmental issues.

In the first period (1940-1965), that is the pre-Green Revolution era, the addresses were concerned with issues of immediate post-Independence situation and policies arisen thereof. The major themes covered in these speeches were: peasant agriculture, agrarian structure, land reforms, role of agriculture in economic development, rural indebtedness, agricultural cooperatives, agricultural organisation and finance, rural development, planning for agricultural development, Five Year Plans and improvements in agriculture, agricultural production and productivity, agricultural price mechanism for agricultural growth, etc. One address was on agricultural economics and agriculturist. The first address was delivered by Sir Malcolm Lyall Darling, an ICS officer. Interestingly, the speakers included the President of India, economists and administrators holding positions in Planning Commission and Reserve Bank, Ministers from the Central and State governments, ICS officers, Vice-Chancellors and Chairman/Directors of leading research and development Institutes.

The addresses during Green Revolution era (1966-1990) focused on: agricultural planning, determinants of and imbalances in agricultural growth, structural change in agriculture, unemployment, resources and its distribution, irrigation, ecological degradation, nutrition and food security, trends in agricultural technology, agribusiness management, and prices, costs and terms of trade, agricultural subsidies, agricultural marketing and taxation and risks in agriculture. These lectures were delivered again by men of eminence such as University Professors, Directors of Institutions, Deans of Colleges, Members of National Commissions, development administrators, experts from FAO, Vice-chancellors, but most of them however are economists/agricultural economists.

Third volume covering the period (1991-2015) encompassed the issues relating to dry land agriculture, mountain agriculture, secondary agriculture, dairying, forest resources and natural resource management, inequality and poverty, risks and technology challenges in the new millennium and human response to them, price and non-price reforms and globalisation and agricultural policy. The addresses were mostly delivered by eminent professors of economics from various Universities, Directors and Vice-Chancellors of leading institutions and personalities from University Grants Commission and International Agricultural Research System.

The knowledge system covered in the addresses, their interpretation, understanding of issues and profound policy analyses and recommendations made would be found very valuable by the students, teachers, and researchers in agricultural economics besides other stakeholders in agriculture. These volumes will be referred and used by future generations to be engaged in the study of agricultural economics.

The Society is grateful to a number of people who provided help to bring out these volumes. The idea originated three years ago and its key proponent was late Dr N.A. Mujumdar, the Editor, IJAE, whose idea was taken forward to the Office Bearers and Members of the Society by Dr C.L. Dadhich, Hony. Secretary and Treasurer of ISAE. Prof Dinesh Marothia, President, National Institute of Ecology, New Delhi, concretised the idea further and helped to consolidate the volumes with technical and organisational support from Dr Dadhich and Mrs Vijaya Venkatesh, Hon. Joint Secretary, ISAE. The most difficult task of collecting all the 74 addresses was ably carried out by Mrs Vijaya. Prof R.S. Deshpande, Former Director, Institute for Social and Economic Change, Bengaluru, was instrumental in making to understand the importance of having an introduction to each volume. The Society greatly acknowledges service rendered by all these individuals. I strongly feel that these three volumes will be very useful reference to understand the trajectory of thoughts of great people in the field of agricultural economics over the past three-fourth of a century

Mumbai

—C. Ramasamy
President
Indian Society of Agricultural Economics

V.S. VYAS

Introduction

THE Indian Society of Agricultural Economics (ISAE) was established in 1940. It was a tumultuous time for the country. War was raging in Europe, Asia, and Africa. Rice supplies from Burma had stopped; the country had to feed large number of troops; the transport system was preempted for military movements, resulting in added difficulties in food-deficit regions. Domestic foodgrain production was less than that in 1880 (Nanavati, 1945).[1] Food prices were skyrocketing, and rationing was introduced. As one of the conference presidents remarked, "The times were out of joint." Soon after Second World War (WWII) ended, there was the partition of the country, which caused a huge influx of refugees. Major part of the wheat-producing areas went to Pakistan. Thus, the supply of two important foodgrain, rice and wheat, was seriously disrupted.

Usually, in such a time, it is difficult to take a long-term, comprehensive view on the issues of growth and equity. Since 1940, annual ISAE conferences have provided a forum to do precisely this. The credit goes to the pioneers who established and nurtured this institution, most notably to Sir Manilal Nanavati and the illustrious people who succeeded him as presidents of ISAE, as well as to eminent scholars and administrators who successively presided over the annual conferences of the society. As one of the conference presidents (Ashok Mehta, 1964) had remarked, in this land of high infant mortality, ISAE had not only continued functioning, but was also going from strength to strength!

1. Names of the conference presidents and the year in which they presided over the conference are given in the parenthesis.

Right from the beginning, the conference presidents set high standards of discussion on the issues confronting India's vast and impoverished rural sector. This tradition continued, as the enclosed first 24 conferences addresses—from 1940 to 1965,[2] i.e., preceding what might be called pre-Green Revolution era—amply bring out. The speakers discussed issues of relevance to the agricultural economy of the country in a forthright manner, bringing their vast experience as well as deep insight to bear on their presentations. This will be obvious from the brief review of the major themes discussed in the presidential addresses during that period. These themes and the depth in which these were discussed have made them relevant to our own time.

The major preoccupation of the presidential speeches during that period centred around the issues of the role of agriculture in India's economic development; strategies for augmenting agricultural output; required reforms in agrarian structure; cooperatives and rural reconstruction; the role of the government in agricultural development; and organisation of research in agricultural economics. There were diverse views on all these subjects and the speakers candidly enunciated their views.

The Role of Agriculture in India's Economic Development

A few of the conference presidents touched upon the broader issue of agriculture's role in economic development (Anjaria, 1955; Madan, 1958; Sen, 1959). They were emphatic that agriculture was at the heart of the problem of development of the economy. All of them were in agreement that labour had to be shifted from agriculture to non-agricultural sectors, for which surplus had to be generated in agriculture. They differed on the modus operandi for achieving this objective. While some favoured extracting surplus from agriculture and investing it in industry, as in the Soviet Union and Eastern European countries (though our methods for surplus mobilisation could be different), others were in favour of demand-induced industrialisation to generate and invest agricultural

2. No conference was held in 1949.

surpluses, as in Japan and most of the Western developed economies. Sen (1959) summed up the various methods of generating and transferring surplus from agriculture as: compulsion, inducement, collectivisation, and strategic stimulant. He, as well as several other speakers, suggested that in our circumstances, only inducement and strategic stimulant could succeed.

Some of the conference-presidents were in favour of 'big push' to get agriculture moving, while others favoured balanced growth of agriculture and industry. Although no one took the 'agricultural fundamentalist' position, there was a unanimity of views on the continually expanding output of agriculture lest it became a bottleneck for overall growth of the economy. It is interesting to note that one of the speakers (Anjaria, 1955) explicitly stated that agricultural productivity should be improved with more intensive use of labour and by organisational changes such as cooperative farming, and that the industrial sector should have a claim on the public investment.

Practically all the speakers who dwelt on the theme of agricultural development emphasised the need for shifting a part of the workforce from agriculture to non-agricultural sector. Most of the speakers recognised the role of rural industrialisation in this context. However, it was pointed out that not all types of industrialisation would help; the industries with linkages to agriculture alone would meet the objective.

Strategies for Augmenting Agricultural Output

In the initial years, the speakers mainly focused on augmenting food supply. Only in the later years was attention given to high-value crops and animal husbandry. This was natural as the food problem was becoming progressively more difficult with rapidly growing population, low yields, and difficulties in importing foodgrain. However, the speakers were not overly pessimist on reaching the goal of food self-sufficiency, as they found that the climate for agricultural growth was favourable, with high agricultural prices, and the state and Central governments' keenness to increase foodgrain production.

It was suggested that prevailing low yields meant that the scope existed for improving productivity in agriculture.

The enormity of the task was not underestimated. For example, after careful screening of the available data, V.K.R.V. Rao (1961) pointed out that not only were the yields low, but there also was no significant improvement during the first two Five-Year Plans. Nor was there any remarkable shift to high-value crops, and the regional differences in yields of principal crops were quite glaring. He also pointed out that the yields were particularly low in the states that were major producers of specific foodgrain.

For improving productivity in agriculture, earlier speakers had placed emphasis on the extension of irrigation. Some of the speakers pleaded for small irrigation projects and construction of wells on farmers' fields (V.L. Mehta, 1947; Rajendra Prasad, 1950). They suggested that the small irrigation projects, apart from other benefits, ensured an early outcome and did not involve the problems of resettlement of evicted people. There were others who were more enthusiastic about major irrigation projects (Ramamurthy, 1948). Even the votaries of large projects were aware of the lack of proper utilisation of irrigation water, and made suggestions for improvement on this score, which are relevant even today. In the later years, the emphasis was more on improved seeds and improved agricultural practices. Those who spoke on this subject were of the view that it was the responsibility of the state to ensure the supply of improved seeds to the farmers, especially for food crops.

Issues relating to farm mechanisation were also highlighted. Predictably, those who advocated small irrigation projects were not in favour of farm mechanisation, and the reverse was true of the votaries of large projects who pleaded for farm mechanisation for improving efficiency in cultivation and other farm chores.

Agricultural research was one area which attracted the attention of many speakers. It was pointed out that most research was done for commercial crops, while food crops were neglected. Also, the appropriate research which could be of use to small farmers, was missing. A number of speakers advocated that serious attention be given to the research on dry land agriculture. They also observed that the available research findings were not extended to the farmers. A word of caution was put forward by Bhattacharjee (1965) in this context. He maintained that many a time in our enthusiasm to propagate a new technology, the element of risk entailed and level of skill needed are not properly appreciated.

Most of the speakers emphasised the role of extension in boosting agricultural productivity. It was suggested that there should be a village-level worker who could keep in contact with all relevant departments of the government that are working for agricultural development (Patil, 1951).[3] One of the conference presidents (Krishnamachari, 1953) reflected the views of several others in stating that the gaps in research as well as the deficiencies in delivery of services could be taken care of by well-organised National Extension Services.

There were interesting comments on the price policy. Ganguli (1963) made it the main theme of his presidential address. The speakers dwelling on the subject of price policies in socialist as well as capitalist economies observed that a feature common to most of the economies was that productivity in agriculture was lower than in non-agricultural sectors. In a country like ours, the differences widened due to unfavourable agrarian relations and defective marketing, which weakened the incentive of the agricultural producers to produce more and produce efficiently. Because of such differences in productivity between agriculture and non-agricultural sectors, in the natural course, the terms of trade would be against agriculture. Any policy to artificially raise agricultural prices to improve the terms of trade for agriculture would be injurious to the whole economy. The importance of non-price factors, for example, investment in infrastructure and application of technology, adjusting the taxation, cheap credit, etc., were emphasised in this connection. These measures were likely to level up the productivity in agriculture and non-agricultural sectors, although the process of leveling up would have to be continuous.

3. This idea later materialised in the institution of the Village Level Worker.

The role of prices in influencing land allocation for different crops was supposed to be positive, though doubts were raised whether this would apply to all sections of producers, especially small farmers (Madan, 1957). It was suggested that the non-price factors would also play a significant role in the allocation of land for different crops, especially in a traditional society where a large section of the agricultural population was characterised by subsistence farming (Bhattacharjee, 1965). Apart from prices, self-consumption, marketed surplus and de-stocking and 'restocking' of the primary producers have an important role in the allocation of resources between different crops (Ganguli, 1963).

Several speakers emphasised the need for price stability. However, it was maintained that "stabilisation of agricultural prices cannot be obviously divorced from the wider question of the stabilisation of the entire price level" (Krishanamachari, 1953). A speaker warned, "Panicky and unwise interventions in the determination of prices and quantity would clearly be neither necessary nor desirable" (Ganguli, 1963).

The speakers were generally not in favour of 'high' prices to stimulate agricultural growth in a poor country like ours, and would like greater emphasis on non-price factors, for example, research and its extension, infrastructure support to agriculture, etc. At the same time, it was pointed out that, paradoxically, while rich countries were making deficiency payment to agricultural producers, poor countries were fixing prices, which were lower than the international prices.

Necessary Reforms in Agrarian Structure

The period preceding WWII had witnessed momentous political changes in the country, culminating in popularly elected ministries in several provinces. The changed political scenario influenced the peasants' relationship with the landlords, moneylenders and the government agencies (Darling, 1940). More important among these were the changes brought about in land relations.

Even before Independence, national leaders voiced concern over the faulty agrarian structure, progressively diminishing holding size, and increase in the number of marginal farmers and landless labourers. The popular ministries initiated the process of abolition of intermediaries and tenancy reforms. The subsequent period saw the continuation of these reforms with measures such as protection of tenants by ensuring security of lease and fair rent, right of certain class of tenants to purchase the land they were tilling, and ceiling on land holdings. The presidents of the conferences generally endorsed these moves and applauded the progress in that direction. As one of the speakers noted, "The land reform legislation in India is impressive judged by the far-reaching character of the changes, the area and population involved, and the time and manner in which it has been carried" (Krishnamachari, 1953).

It was pointed out in one of the presidential addresses that in European countries, land reforms were given precedence over technical improvements. (Nanavati, 1945). However, some speakers wanted to give equal weight to institutional change and development activities; they should be made complementary (Anjaria, 1955). Even those in favour of land reforms expressed their disappointment over the fact that the objectives to be achieved by each land reform measure for overall development of agriculture were not spelled out clearly, and more importantly, there was a hiatus between the promise and performance (Madan, 1957).

A few suggested that a minimum size of holding should be specified, and no cultivation should be permitted below such area. Nanavati (1946) advocated the view that the agrarian structure in the country should be identified with peasant proprietors having economically viable holdings. Patil (1951) introduced the idea of "registered farms" i.e., the farms above the economic holding that were not members of the cooperative farming societies. He, as well as Gadgil, were not in favour of ceiling on land holdings as they argued that ceiling on holdings would not solve the problem of low productivity nor would contribute in a measurable way to social justice. In this context, Gadgil distinguished land reforms from land management and concluded that the land reform legislation, as well as *Bhoodan*, had not contributed to the emergence of economically viable holdings. There were opposite views in favour of land ceiling as well as *Bhoodan*. As an alternative, Madan (1957) advocated steeply graded tax on holdings.

Even those who were not very enthusiastic about the land reforms conceded that due to these measures, the situation for the poor was better than what it would have been. However, the extent of success achieved was not inconsequential. It was argued that the abolition of intermediaries and security of tenure were by and large successful. Intermediate tenures accounted for 45 per cent of land and the overall impact of intermediary abolition was most welcome. For the remaining measures, particularly ceiling on land holding, the reason for less-than-expected success, it was pointed out, was excessive reliance on revenue department officials; public participation was conspicuously absent.

Some of the conference presidents raised the issue of agricultural labourers. V.L. Mehta (1947) remarked, "Now that in most provinces there has been an attempt made to protect the tenant and to afford him security, the protection and the promotion of the labourers on the land must claim the attention of provincial governments." Later, Ashok Mehta (1964) commented that despite all the land reforms, landless labourers and marginal farmers were the single weakest spot in Indian agriculture.

Another important point made was that the beneficiaries of land reforms, small farmers and tenants, did not receive any support for improving their agriculture. As Rajendra Prasad (1950) put it, "The effect of the absence of such provisions was to replace the *zamindar's* agency for collecting rent with a government agency without in any way improving the position of the cultivator".

Major differences surfaced in the stand taken on post-reform agrarian structure. The views ranged from progressively moving towards joint village management of agriculture (somewhat akin to China's commune system), collective farming of small and marginal farmers who benefited from the ceiling legislation, joint farming by the same groups of beneficiaries, and individual holdings supported by cooperative action in different areas.

There was serious concern for small and marginal farmers. It was recognised that there was overpopulation and rampant disguised unemployment, especially on small farms. The problem, according to the votaries of cooperative farming, could be solved by joint farming. However, they wanted to take a cautious approach to cooperative farming even on small holdings or holdings created by sequestering surplus land. The arguments were mainly based on the peasant's attachment to his/her patch of land, and also the difficulty in the management of cooperative farms due to the problem of evaluating the labour input by the members of the cooperative farms. One of the speakers (Nanavati) maintained that the decay of cottage industries was responsible for large number of artisans taking up agriculture on tiny plots of land. They would be better off if they could get employment in their traditional crafts, and 'genuine' farmers get the land to cultivate.

A note of caution was introduced by Dantwala (1960). He wanted the consequences of any major change in agrarian structure to be anticipated and provided for. He was of the view that the defects in our agrarian structure were reflections of the backward economy and could not be corrected by instituting land reforms as isolated measures.

Institutional Support

The conference presidents were concerned with the pitiable conditions of the villages, and as one of them remarked, "(I)t is sadly true that the slums of India are not urban but rural". (Ramamurthy, 1948). In order to redress the situation, most of them took a comprehensive view of rural development, encompassing economic as well as social development. Some of them advocated that the objective of comprehensive development of the villages would be possible only by mustering the resources of the people. The importance of *panchayats* was recognised in this context.

Soon after Independence, several provinces had enacted comprehensive legislation on the organisation and functioning of the *panchayats*. It was stated that the *panchayats* had the potential of mobilising manpower for the development of villages. However, some of the earlier speakers had reservations on the effectiveness of this institution because of the lack of expertise at the local level and the likely political bias of the elected representatives. Many of them suggested a single organisation for the village, encompassing the main civic and economic functions. (Nanavati, 1945; Patil, 1951) Patil wanted an all-embracing Village Production Council, which should work towards joint management of the village's total resources.

Cooperatives attracted a lot more attention. Many speakers wanted to give cooperatives a pivotal role in the rural economy. One of the earlier conference presidents (Vijayraghavachrya, 1941) supported the statement of the Royal Commission on Agriculture that "the modern conception of the cooperative movement differs markedly from that commonly held at the close of the last century; the term 'agricultural organisation' or better still 'rural reconstruction' expresses more accurately the nature of the activities included within the movement." Other speakers also made a plea to extend the fold of cooperation beyond credit, to include labour, housing, marketing, processing and even hire-purchase of farm equipment as parts of cooperative activities. They also recognised the role of cooperatives to increase the 'countervailing power' of the farmers. But the main attention was centred on credit delivery, and the cooperatives were considered the most effective organisations for this purpose.

There were differences—which persist even now—on the size of the cooperatives (large or small), and on the multi-purpose *versus* single-purpose societies. But the main point of contention was the role of the government in the cooperatives. The question was posed whether state partnership, which was advocated and spelled out in detail in the Rural Credit Survey Report, was the desired direction for the growth of cooperatives or was it likely to result in a 'dependency syndrome' and what now is called 'free riders problem'? Both Anjaria (1955) and Karve (1956) supported the stand of the authors of the Rural Credit Survey and endorsed state partnership. However, another speaker, Dantwala (1960), severely criticised government partnership in cooperatives. Quoting the example of the Sugarcane Processing Cooperatives, he demonstrated that a small section of rural population with some political clout could take undue advantage of the state support. One of the conference presidents was keen that even with the state partnership, the cooperative should generate their own resources. The state should play, essentially, a catalytic role. (Madan, 1957). Many others also took note of the fact that because of the assured government support, members of cooperatives were not putting enough effort to generate their own resources. Doubts were raised whether cooperatives provided panacea for all economic ills of rural society, especially when their entrepreneurial talent as well as risk-taking capacity were limited.

Government's Role

Overall, the role of the state in agricultural development was considered extremely important. As one speaker put it, "In a deficit economy like ours, any attempt on the part of the state to allow matters to drift without any planning of production, procurement and distribution is a dereliction of duty and is playing with the lives of millions of men" (Ramamurthy, 1948). It was pointed out that just five per cent deficit of food supply in the early 1940s resulted in huge increase in prices, largely due to an inefficient state machinery, and produced the Bengal Famine, resulting in deaths on a terrible scale. There were also voices which were cautious in putting too much reliance on government (Rajendra Prasad, 1950). Their reservation was due to the factors such as, poor human material, corruption, lack of coordination and lack of accountability.

It was generally felt that in the developing phase of economies, the government has to pay an active role. The government's intervention can be in terms of legislation, regulation, and active involvement through projects and programmes. The scholars who addressed the conferences in the initial post-Independence years commented on all three roles. In India, since agriculture is a state subject, it is difficult to examine policies for the country as a whole. However, because of the overriding role of the central government and political homogeneity with one party dominating the political scene in large parts of the country, the Centre's writ was followed in practically all states. Therefore, concentrating on the Centre's policies gave a fair picture of the country as a whole. In fact, many speakers were in favour of a strong central government department of agriculture taking a lead role in agricultural development in the country.

On the general area of planning, speakers were satisfied with the importance attached to agriculture in the Five-Year plans. However, it was felt that adequate investment was not provided for the agricultural sector, and for this and other reasons, performance of agriculture was not up to the mark. Ashok Mehta (1964) maintained that lack of success in agriculture was as much due to slow growth of industry, particularly those providing agricultural inputs and tardy progress in irrigation projects, as due to some weaknesses in agriculture. The latter were identified as slow pace of implementing land reforms, inadequate deployment of extension services, and absence of proper price policy.

Most of the speakers highlighted the importance of increasing investment in agriculture and allied sectors. It was pointed out that as arable land was limited and dependence on agriculture by a growing number of workers was increasing, the balance could be struck by increasing investment in agriculture at a corresponding scale. Low growth in agricultural output translated in the low income of the people; as a result, the rural urban differences widened. Keeping in view the paucity of resources for investment, some speakers argued for concentrating on high-potential areas with assured rainfall or reliable irrigation.

The government's major initiative was in the area of land relations, which were sought to be influenced by the successive land reform legislation. Another area where the government intervened both through legislation and regulatory measures was access to credit by the farmers, and the relief from huge debt incurred by the agriculturists. On the seriousness of the issue of rural debt, one of the conference presidents quoted with approval the Royal Commission on Agriculture, which had noted that "the worst policy towards agricultural debt is to ignore it and do nothing" (Vijayraghavacharya, 1941).

Even before Independence, some of the provincial governments, e.g., Punjab, Bombay, Madras, had enacted legislation to prevent alienation of the land of the defaulting farmers by moneylenders.

A number of conference presidents commented on the measures to regulate moneylenders' operations. Most did not recommend their total abolition but were in favour of registering them and monitoring their activities. They suggested a number of measures, such as defining the maximum interest to be charged, prohibiting land alienation in lieu of overdues, reconciliation of old debts, compulsory scaling down of rural debt, restriction on the force to recover debt, etc.

The speakers also commented on marketing reforms under Agricultural Produce Marketing Committees, and generally favoured the provisions of the legislation. Absence of an effective programme for cattle insurance was pointed out as a big gap.

A significant number of the past presidents were associated with the Planning Commission, as chairman, member or high official (for e.g., V.T. Krishnamachari, Ashok Mehta, R.K. Patil, Tarlok Singh, S.R. Sen). It was natural that they would be commenting on the planning process. Some of them criticised overambitious targets and weak implementation. Many were keen that the planning should be more comprehensive, and due emphasis should be placed on social aspects as well. Although they were satisfied with the priorities set out in the Plans, they were not satisfied with the outcome of the planned programmes. In the first four Plans, the planned targets of the growth in agricultural output could not be met. The performance of the agricultural sector was found not up to the mark. Blame was laid, as being done now, on the weak delivery system.

It was hoped that the National Extension System (NES) would not only make the farmers aware of the advanced methods of cultivation, but would also facilitate in meeting their requirement of necessary inputs such as improved seeds and fertilisers. Additionally, one of the speakers (Bhatttacharjee, 1965) wanted the NES machinery to train cultivators in modern methods of cultivation. High hopes were pinned on the National Extension Service and the Community Development Programme. It became clear over a period of time, as later speakers admitted, that NES did not succeed in its objectives despite deployment of resources and manpower.

Directions for Research in Agricultural Economics

A recurring theme in a large number of addresses was the paucity and inadequacy of the agricultural statistics. The speakers underlined the need to study various facets of agricultural economy in the context of overall development of the economy. The contribution of various committees appointed by the government to look into the rural issues such as inquiries conducted by the National Income Unit, National Sample Survey, Survey of the Conditions of Agricultural Labour, and the comprehensive inquiry on Rural Debt and Finance by the Reserve Bank of India (RBI) were appreciated. However, it was asserted that they were no substitute for detailed, systematic and regular information on different aspects of rural economy.

The speakers were not indifferent to the difficulties in collecting statistics from large number of smallholdings in lakhs of villages. They also appreciated that many aspects of rural change were difficult to measure. Even after recognising these difficulties, they were not satisfied with the coverage, quality and relevance of research being conducted by various agencies.

In the initial years, ISAE itself organised a number of studies and wanted to establish regional centres for the same (Nanavati, 1944). The need to strengthen statistical base in collaboration with universities and research institutions was emphasised. Apart from the absence of necessary data, the gaps in holistic understanding of rural society in general and the agriculture sector in particular was highlighted and a plea was made to undertake a comprehensive rural survey of the country and, also, village surveys in large numbers.

One of the conference presidents, Vakil (1952), devoted most of his address to the need for proper data collection and appropriate agencies to do this task. He was against ad hoc inquiries conducted by the Central and state governments through their departments, or by the agencies created only for such purpose, and wanted such research to be entrusted to the universities. He pointed out that apart from being more economic, such an association would benefit government functionaries as well as the academics. On the other hand, Karvey (1956) and a few others were appreciative of the contribution of official studies.

While discussing these gaps, one of the speakers (Bhattacharjee, 1965) wanted to look afresh at the theoretical formulation of the development process and relationships in agriculture that were largely based on Western concepts and methodologies. He categorically stated that available economic theories do not explain sufficiently usefully the practice of agriculture in the so-called traditional societies. For example, he questioned whether, in traditional societies such as India, agricultural producers actually and invariably seek to maximise profit. The conventional theory does not take into account social constraints and structural rigidity. He pleaded for an interdisciplinary approach, especially taking advantage of the understanding provided by sociological and psychological research.

Conclusion

Not only in identifying the problems of the rural economy but also in proposing solutions, the conference presidents sound contemporary and relevant; in fact, ahead of their time. Their insistence on treating agricultural problems of low productivity and over-dependence of the workforce on this sector as the heart of the development problem of the country is as true today as it was in the beginning years of planned development. Similarly, insights provided by some of the speakers on augmenting surpluses from agriculture through 'inducement' and strategic intervention has a lot to commend itself. A plea for forging strong linkages between agriculture and industry while planning for the development of the economy is as valid today as it was when proposed in the 1950s and 1960s.

There are several lessons that can be drawn from these addresses to design strategies of agricultural development. Two of these need particular mention. In many of the conference presidents' speeches, the importance to non-price factors in incentivising agriculturists to produce more was highlighted. They also gave due importance to the simultaneous development of the supportive infrastructure (e.g., irrigation) and industries (e.g., fertilisers) to ensure steady growth in agriculture. The speakers' emphasis on intensive efforts in some areas, use of superior technology in agricultural production, and the importance attached to improved seeds sounds like precursors of the Green Revolution.

An overwhelming impression left after reading these forceful speeches is that the learned speakers were disappointed by the outcome of the two major initiatives of that period, land reforms and community development, and national extension, in terms of their impact not only on agricultural output but also on the agriculturists, particularly the small and marginal farmers and agricultural labour. They were also getting disillusioned with the cooperatives as institutions for imparting collective strength and meeting the immediate requirements of the poor.

In the addresses of the conference presidents of that period, the ground was prepared for fresh thinking on the vital aspects of agriculture and rural institutions. It is not surprising that when new technology and HYVs (high-yielding varieties) appeared on the scene in the mid-1960s, public leaders and officials did not show any hesitation in accepting it, rather welcoming it.

1

SIR MALCOLM LYALL DARLING

Peasant and Politics

Life is infinitely complex, and even for oneself it is not always easy to say what is good and what is bad. I therefore feel certain diffidence in prescribing for others, and in dealing with the peasant, my chief study, and my object has been less to apply a spur than to hold up a mirror in which he might perhaps see his life as it is and judge for himself how far it should be changed. Finally, I have always sought contact with the individual rather than with the mass; the individual is more human and understandable, and ultimately the problem of human life is the problem of the individual. I mention these personal details as they colour the whole of my paper.

The subject of this paper is the peasant and politics and how we as a society can help the one to fit the other. I make no apology for speaking of the peasant, since India is predominantly a country of peasants. In certain tracts the nomad, the hunter or the shepherd prevails. But the man who is most typical of the 300 millions who live by the soil is the cultivator who has one or at most two pairs of plough bullocks and who, more often than not, owns most of the land he cultivates. The peasant therefore must provide the very pith and marrow of our studies. And could there be a more interesting type for a society like ours which brings together East and West in a happy collaboration of effort? The peasant is the true link between East and West, and no one who has observed his ways in the countries that lie between us and Western Europe—in Turkey, Bulgaria, Yugoslavia, Austria and Italy—will feel that there is an impassable gulf between the two and that "never the twain shall meet". In Turkey, to take a homely instance, the first thing I saw two years ago on entering a peasant's courtyard was a spread of dung cakes. In Bulgaria, one of the villages I visited had just consolidated its scattered holdings, and the headman of the village told me with pride that it was the first village in Bulgaria to do so. It was also with pride that I told him that Punjab had already consolidated nearly a million acres. In Italy, I was constantly reminded of Punjab and when an Italian read my book, *The Punjab Peasant*, he said that he was constantly reminded of Italy. In all those countries, although there is great variation of temperament, custom and circumstance, the cultivation of the soil presents the peasant with much the same problems, and close contact with nature develops in him a simplicity, a dignity, and a contented acceptance of life rare amongst the more sophisticated.

But, the peasant is not only the link between East and West, he is also the strongest link on the chain which joins our age to the remote past. In Homer's *Odyssey* you will find descriptions of peasant life which might have been written of the Mediterranean peasant today. Similarly, the Old Testament illustrates almost every aspect of Indian peasant life; for instance, "The sluggard will not plough by reason of the cold", and "He that is hasty to give credit is light minded". I emphasise these two points because they provide the background for my theme. In the name of constitutional advance, we are pouring new wine into very old bottles and in undertaking this difficult operation we shall be wise to look for guidance or warning to other countries that are dealing with the same problem. In 1938 when I was in Bulgaria, I found that the constitution was under suspension because, for the time being at least, it did not suit the country, which is essentially a country of peasants with perhaps the simplest and sturdiest peasantry in Europe. This constitution was of the usual democratic type with ministers responsible to a Parliament elected mainly by peasants. It is this type of Government which I shall have in mind throughout this address, since it is the type that is being planted in India today.

Man, said Aristotle, is a political animal. This may be true of urban man, perhaps even of western man, but it certainly is not true of the Indian peasant. On the contrary, he is essentially a non-political animal. He has always "let the legions thunder past" and plunged in work again, in the work of wrestling for daily bread with a despotic and capricious nature. But great changes are taking place. His life is no longer bounded by the distance that a horse or a bullock cart, or a small donkey, will take him in a day. Nor are his numbers any longer severally limited by drought, pestilence and war. The world with all its hope and stir is at his gates. The lorry is there to take him a hundred miles or more in a day to waste time and money in Kacherry, to visit a cinema or vote in an election, and he has so increased in numbers—the last 20 years have probably added 80 millions to the population—that it is a question whether or not 20 years hence, when the babes of today are full-grown men and women, they will be able to maintain the very modest standard of living of their fathers. Last of all, we are involved in a war that threatens to crack our globe in two.

The Great War released forces which are spreading all over India, and their effect upon the peasant has been to set him thinking about his rights instead of his obligations. Three years ago, too, he was given the vote. This may well prove to be a momentous change in his political history. It is certainly a hazardous experiment to give the vote to over 20 million peasants most of whom are illiterate and who moreover represent more than two-third of the total electorate. Never yet has parliamentary democracy been built on such foundations.

If, however, the Indian peasant is illiterate, he is certainly not deaf, and not always dumb. His ear is now being assailed by journalist and propagandist, by politician and patriot, by lawyer, trader and moneylender, and by the officials of half-a-dozen different departments, all competing for his attention, custom or cash. Greatly bewildered and sometimes dazzled, he listens to each in turn, but ever thinking of crops, prices and payments, and the limited but inescapable necessities of daily life. In the past, when approached by men of the outside world, he has usually listened in silence, a silence taught to him by centuries of dealing with strangers, and it is now fashionable to call him and his kind "the dumb millions". But he has always been ready to talk to those whom he trusted and to talk even loudly when too much is asked of him. When 10 years ago prices crashed, the murmur of his discontent passed from one end of India to the other.

In the last 80 years, there has been nothing in village life comparable to this fall in prices. Flood and famine have hit, and still hit, particular tracts with greater intensity, but the rupee famine which following the fall in prices has affected the whole of India and profoundly disturbed the peasant's triple relationship with landlord, moneylender and the government. The adjustment of these three relationships has taxed the energies and resources of provincial governments to the full and what has been done is full of significance for the future. The difficulties between landlord and tenant have, owing to the widespread prevalence of peasant proprietorship, affected the country as a whole the least, but where they have arisen they have been the most difficult to deal with and in one province at least have led to the passing of a drastic measure in favour of the tenant. How dangerous a situation of this kind can be when the peasant asserts his rights and the landlord relies for the defence of his interests upon hereditary right, rather than upon service to his tenants and the community, is shown by what happened in Europe after the war. In Romania, the landlord was deprived of nearly 12 million acres in return for an almost nominal compensation. In Poland, he also lost much of his land and was only partially compensated, and in Russia, he lost everything. In all three countries, this was done by the party in power to gain the political allegiance of the peasant, in the first two to avoid revolution, in the last to confirm it. In India, the rapid increase in the population is already putting the tenant at a disadvantage, and if we are to avoid serious disturbance, of which there are already ominous signs, he will have to be increasingly protected. But it is not easy say how this can best be done and the problem is essentially one for politician and economist to consider together.

The disturbance in the relation of debtor and creditor has been felt throughout India and there is, I believe, not a single province which has not passed some new Act to protect the debtor, weighed down as he was by a burden of debt that the fall in prices doubled in a few months. One of the most interesting studies that await the agricultural economist is to ascertain the final result of these measures and to gauge whether on balance they do more good than harm. Meanwhile, we may note one result which has shown itself in different degrees throughout British India. I refer to the acute restriction of credit. In my own province of Punjab, I doubt whether until war came more than two annas could be borrowed for every rupee borrowed before the fall in prices. One result has been a marked tendency to substitute the buffalo for the bullock, and in the poorer tracts even the purchase of a buffalo has often been a matter of difficulty. If this process were to continue, it could hardly fail to have a deteriorating effect upon agriculture, and any deterioration in agriculture must, sooner or later, have serious political consequences in a country with a rapidly increasing population. Now that prices have gone up again owing to the war, the moneylender may start lending again. If on the other hand, he remembers the biblical warning quoted above and refuses to lend on the principle of once bitten twice shy, governments will be faced with the problem of how to supply the cultivator with the credit which all agree he needs. Some will suggest the cooperative movement as the best way out of the difficulty. I wish I could endorse those, but at present this is only possible to a very limited extent, since there is no province where the movement touches more than 15 per cent of the population. Will not governments then be forced by political pressure to take the moneylender's place? If so, they will find themselves involved in all the difficulties that moneylending in India brings, and these difficulties are not likely to be mitigated by a democratic constitution with periodic elections. In Yugoslavia, for example, the peasant has used his vote to get 50 per cent of his debt cancelled. Any government which wishes to avoid losses on this scale should be careful, in controlling the moneylender, not to drive him out of business. As long as the peasant is illiterate and stupid about the use of money, he will need protection, but only those who know him well can say how far this protection should go if it is not to harm him. Here again is a question in which the assistance of the economist might be invoked with advantage.

Finally, there is a disturbance in the relationship between the peasant and government. Until 10 years ago, this had been the least disturbed of the three. It is true that there have been cases like Bardoli where the peasant has resisted payment temporarily, but speaking broadly he has never seriously questioned the government's right to a share in the produce of the soil. The fall in prices suddenly doubled this share and made it very difficult for him to pay it. Like the landlord and the moneylender, the government has been forced to forgo a substantial part of its dues, in Punjab to the tune of many crores of rupees. But this has not everywhere been done without pressure from the peasant, and now that he has got the vote, he may easily develop this pressure into a general attack upon the whole land revenue system of the government. This has happened in Bulgaria. There the land revenue, which under Turkish rule absorbed 10 per cent of the gross produce, has been reduced under political pressure accentuated by the spread of demagogy, to little more than a local rate and has been replaced by a number of indirect taxes upon what the peasant consumes. Whether this is to his benefit is a moot point, and some say that he pays more now than he did before. The land revenue system is so well suited to India—if anyone doubts this let him read the report of the Punjab Land Revenue Committee[1]—that I hope that the example of Bulgaria will not be followed. But, whether it is followed or not, we may be certain that government dues are not likely to be paid by the peasant in the future with as little question as in the past. This is one of the danger points for the governments of the future. The peasant's sense of obligation towards both moneylender and landlord has been definitely weakened, and this can hardly fail to react upon his attitude towards government.

It is clear that the last 10 years have set in motion a number of currents which may be expected to gather force as the peasant realises the power put into his hands by the vote. The autonomous governments of the last two years have certainly not been indifferent to this. Whatever their party, they have devoted earnest attention to improving the peasant's lot and my own province has been foremost in this. The interesting question is whether the very varied measures taken have on the whole been wise. Here the politician must ultimately give way to the economist, but it is too soon yet to pass judgment. One comment, however, may be made. Taken in the name of progress, these measures are in danger of making progress difficult if not impossible. Their immediate effect has been to deprive the

1. See more specially paras 151-154.

peasant of credit, to make it difficult for the enterprising to obtain capital to develop their land and easy for the inefficient to remain on it however badly they cultivate it and as I have said, to undermine the moral sense of obligation which makes a man pay his dues and repay his loans. All of which is more consistent with a static than with a progressive peasantry.

In saying this, I am far from condemning these measures. I am merely pointing out one of their probable effects. If they make progress difficult, it is the price that India has to pay for the contentment of her peasantry. Whether the price is worth paying depends upon the point of view. The pure economist, with his mind fixed upon the production and distribution of wealth, will be tempted to say—no. But I confess that to me contentment has long seemed of greater value than material progress, never more so than today when so many people are eating the bread of affliction. I belong to a generation which was brought up to believe in progress as part of the divine order of things, and even to regard discontent as divine order of things as well as divine. But I have lived to see a great part of the wealth gained in the pursuit of progress destroyed in its wake. The latest example of this is Russia. Here we have a country which for 20 years has been the admiration, almost the gospel, of those who believed that the progress of the peasant could be rapidly achieved by the State when determined to enforce its will. And what is the result? In the sacred name of national development, lakhs of peasants are being flung, without a thought for their individual welfare, upon another peasant country, whose only crime is that they hold liberty dearer than life. And the Russian peasant cannot resist because in the name of progress his individual will have been crushed.

It was the proud boast of those who have done this that they were doing more for the prosperity of the peasant that had ever been done before. Whether the boast is justified or not I cannot say, but the twofold moral is—respect the individual will and set not your heart upon riches. The politician should remember the first, and the economist the second. The totalitarian states have always ignored the first and as to the second, it may almost be said that when the pursuit of wealth becomes the chief aim of society, it leads to its ultimate destruction. That is one reason why the peasant survives when empires perish, as any traveller may observe who journeys overland from Constantinople to Baghdad: he will see many villages, but of the towns that once adorned the empire of Babylon, Nineveh and Rome, hardly a trace. To the peasant as a class, the accumulation of wealth is impossible, and nowhere in the world will you find a peasantry which according to urban standards is wealthy. For brief periods and in restricted areas, a boom in some money crop may set wealth flowing into the village, but like those rivers which start in flood and end in sand, the flow often does more harm than good and ends as suddenly as it began. In India, we have the case of jute in Eastern Bengal and of cotton in Berar and in the canal colonies of Punjab. In all three areas, the last war produced a boom followed by a slump, and psychologically at least, the last state of the tract was worse than the first.

There is another point to be considered in judging these measures: they all mean more officials. It was a common charge against the old regime that it was bureaucratic, but if legislation proceeds at the pace of last two years, the bureaucracy of the past will be nothing to that of the future. I am reminded of what a well-known socialist said at the end of his first visit to Russia when asked what the country needed the most. "A hundred thousand highly trained officials," he replied. "And what is our greatest need now?" He was asked at the end of his second visit some years later when planning was in full force. "Two hundred thousand highly trained officials." The lesson applies to India, and its importance for the governments of the future is that the best men, though not necessarily the cleverest, should be chosen for the public service irrespective of community or caste, men who understand the peasant and who will serve him instead of quietly scorning or exploiting him.

It is a common charge against all bureaucracies that they find it difficult to maintain personal touch with anything but their files. When the mass of people are educated this matters less, for they can ventilate their grievances in a number of different ways. But when they are illiterate, grievances tend to smoulder and sometimes erupt as suddenly as a slumbering volcano. Personal touch is the only safeguard for a trouble of this kind, and the motto of all governments should be, in the words of Confucius, "be in advance of the people". Even in the past when the peasant was much less concerned with his rights than he is today, the volcano has not always slumbered—witness the Moplah rebellion in the south and the Red Shirt movement in the north and the no-rent campaigns in Oudh and Bihar. What then will happen now that the peasant is awake and had got the vote? Of one thing we may be certain; in the new world which will emerge from the war, he will demand a larger share of the fruits of the earth than he has had in the past. Could our society then start at a more pregnant moment for the future of the peasant?

I must now consider very briefly what part we, as students and economists, should play in determining this future. In our country so overwhelmingly rural as India, voting power must lie more in the village than in the town. The politician must therefore pay increasing attention to the demands of the peasant, and the election of three years ago show that this process has already begun. Further, since democracy involves some system of party government, parties will compete with each other for his vote, each trying to outbid the other, some even trading on his ignorance, and promises will be made which cannot be kept. This process has also begun and the shrewder amongst the peasants are beginning to ask themselves whether the party that made them should be supported again. Since, however, support is given for social as well as political reasons, many will continue their support. But others will transfer their allegiance seduced by the fairer promises of some other party, only to be disillusioned once more. Disillusionment is almost inevitable with the type of promise commonly made by the politicians today for all such promises involve conferring some substantial blessing before the next election, that is, within at most five years, and such shortcut blessings are either very costly to the State—for example, a substantial reduction in land revenue—or costly to others—for example, to the landlord and moneylender; and in either case their number is extremely limited. What then will happen when the peasant finds that his hopes are being continually raised and as continually dashed? In Bulgaria, having some education, he has come to distrust the politician and to depend more upon himself. In India, being ignorant and uneducated, he may be slower to put two and two together, and in the meantime he may well be led into supporting policies entirely subversive of the constitution.

There is clearly a great opportunity in all this for a society whose stated object is "to promote the investigation and study of economic, social and statistical problems relating to the industry of agriculture and animal husbandry and to communities of persons concerned therewith". If we investigate the more pressing problems and do this in the completely dispassionate yet human spirit proper to investigations of this kind, if too we collaborate closely with kindred bodies working in other provinces, such as the Boards of Economic Enquiry in the Punjab and Bengal and the Gokhale Institute of Politics and Economics at Poona, we may hope in time to be able to advise parties in regard to their agrarian programmes and even, though this is perhaps optimistic, to exercise a moderating influence on their electioneering promises. Already one government is considering the appointment of an Economic Adviser, and in the last 12 months, two governments have consulted bodies of experts in regard to the development of the cooperative movement. These are promising signs, and if democracy is to survive, governments will be more and more compelled by the realities of politics and village life to turn to those with experience and knowledge for advice and guidance; for under no democratic system will an awakened peasant tolerate his life being manipulated, still less exploited, by those who do not understand him.

Our first duty then is to understand him. There is a great gulf fixed between the highly artificial life of the intelligentsia and the natural primitive life of the village. The bridging of this gulf is one of the great problems of democracy in India, and we must now consider it from the point of view of the economist. First of all, if I may be so bold, I would warn those of you who like myself have been trained in the English school of Economics, against the urban and industrial outlook of this school. The outlook of the English economist is naturally dominated by the vast commercial and industrial enterprises of Great Britain and by the infinitely complex problems of international trade and finance, and in my time at least little attention was paid to the problems of peasant life, although they affect more than half of the world's population. Agriculturally, too, England is dominated by large-scale farming, and the peasant survives there only as a museum piece. It was not until I came across books like *Carver's Principles of Agricultural Economics* that I began to realise that agricultural problems had a special character of their own and that for their proper understanding they require students steeped in agricultural life. This alone would justify the foundation of our Society.

Accordingly, when we approach the village, if we have been brought up in the town or in the school of commercial and industrial economics, we must divest ourselves of our urban notions and realise that, in India at least, the village is of greater importance than the town and far more acceptable to the bulk of its people. As a Punjabi puts it:

> In the village there is simplicity and good fellowship, and one lives with one's brotherhood. In the town only the police and the fire brigade look after you, neighbours do not come to your marriages, and when you die, only the doctors knows.

In the town money is all important and it is forever drawing money into its arteries, so much so that it has obtained far more than its fair share of the wealth that has flowed into India during the last 50 years. What is worse, the larger towns exercise a demoralising influence on many of those who migrate to them and markedly upon the villages round them. Let no one therefore approach the village in a spirit of superiority, as so many do. Let us rather approach it in the spirit of those who feel that they have at least as much to learn as to teach, and for my part I believe that the town has more to learn from the village than the village from the town.

My second point follows from the first. Since we are dealing with human psychology as well as with the operations of nature, we must make up our minds on which we think is the more important—agriculture or the agriculturist. I have myself never hesitated between the two. I believe that the land was made for man and not man for the land. This sounds a truism, yet many have acted as if they thought the contrary. England in the 18th century is a case in point. Under the stress of war and the need for food, it took to enclosures and large-scale farming, and the peasant disappeared to the great loss of the country. In India, the opposite policy is well illustrated by the Punjab Land Alienation Act, the main object of which was to prevent the peasant losing his land to the moneylender. The two policies may be seen in opposition in the cultivation of sugarcane. In some tracts the small-scale peasant system prevails, in other, the large-scale capitalist system. Under one, the cultivator farms his land as owner or tenant; under the other, he is a mere labourer. The factory-owner and the capitalist prefer the latter as it increases the sum of wealth; but the former increases the sum of welfare, for it is better for the cultivator to be an owner or tenant than a mere labourer.

To those who agree with me that the agriculturist is more important than agriculture, my third point will present no difficulty. It is that in trying to improve village conditions and heaven knows there is room for improvement—we should remember what I have called in one of my book, *The Old Light and the New in the Punjab Village*. So ancient and developed an institution as the Indian village community, with the delicate balance of rights and obligations could not have held together for many hundred years without some illumination from the spirit of man. But now, if I may be allowed to quote from this book, the peasant:

> Stands between the old light and the new. Like a man walking at night with a lantern who suddenly finds himself confronted with an approaching car, he is dazzled by the headlights of the new age and can scarcely see the modest lamp at his side. Yet dim as it is, this lamp has for him a special value, because it has guided him through the centuries and it does not shut out the light of the stars. Through it he has attained a simple, if primitive, relationship with both God and his neighbour; and if the result is poverty and ignorance, it is because superstition has come between him and God, and custom between him and some of his neighbours. The value of the new light lies in the importance it attaches to knowledge, freedom and effort, and its dazzling attraction in the escape it offers from poverty. Its danger is that knowledge, freedom and effort may be prostituted by the few to the exploitation of the many and that the escape from poverty may lead to materialism.[2]

2. Rusticus Loquitur, p.376.

This is perhaps the greater danger that threatens the peasant. I have asked many who know him well and who can look back 30-40 years to know whether he is better off than he was. As far as material things are concerned, the answer will vary with the tract; but if it is a case of spiritual values, almost all agree that there has been deterioration. The peasant is less truthful than he was, less moral and more inclined to faction and litigation, and the natural culture born of such books as the *Ramayana* and the *Mahabharata*, and of such poets as Tukaram and Eknath is waning. The only improvement conceded is a growth in self-respect. I was discussing this the other day with an unusually intelligent peasant in the Deccan and his comment was: "The older generation knew morality and honesty; it has gone. The new generation receives no training in these things. Fathers issue notices against sons and sons against fathers". Someone standing by mentioned as an antidote the new laws that were being made. "The more laws that are made, the more people try to escape from them" was the caustic comment. And it is a comment that a politician should ponder. It illustrates a point I made in discussing the trend of recent legislation, namely that it was undermining the moral sense of obligation in regard to debt. Now you cannot weaken the moral sense at one point without weakening it at another, and if this process goes beyond a certain length village life may revert to the conditions of the jungle from which it has arisen by centuries of effort, and the well-spring of India's spiritual heritage will dry up at the source. Accordingly, in spreading the new light of science and democracy, let us remember that the village has its own knowledge and its own democracy, both rooted in age-long needs that still exist.

Of the three foundations of modern democracy—liberty, fraternity and equality—fraternity is the one which matters most to the village. Without it, there can be no unity, and without unity there can be no real life. *Ikatth lohe di latt*—unity is the strength of iron—is a Punjabi proverb, and the village has needed an iron strength to survive all these years with a culture that lifts it far above the level of barbarism. The basis of modern culture is independence—independence of speech, action and thought. But the basis of this ancient village culture is interdependence, the interdependence of those who live together with common walls, common needs, and common ties, who depend for the means of life as much upon each other as upon themselves, and who feel a common dependence upon the mysterious and overpowering forces of nature. What is needed is that this interdependence should be harmonious, that the strong should not prey upon the weak, the young upon the old, the knaves upon the fools and that all should combine to this end, whether the attack comes from within or without the village.

Unfortunately, this harmony, this sense of fraternity, or *biraderi* as the Punjabi call is, has been seriously disturbed, and all over British India the village has lost much of its former strength. There are many reasons for this, but prominent amongst them is the acquisitive spirit, which modern life has greatly developed. In this respect, no age has been more inimical to village life than the present, and one of its most unwholesome products is the man who tries to grab power for himself or his party by stirring up strife. If faction is increasing and communalism is spreading, the blame is largely his; and if, led by agitator and demagogue, he grows in authority and number, it is not impossible that village life may gradually disintegrate and that the chaotic conditions of the 18th century may then return.

This brings me to my final point. Unity is so important to the Indian village and so seriously threatened by present political and economic conditions that all possible means should be taken to strengthen it. It is now not so much a case of achieving material progress as of arresting a manifest spiritual decline. Two means suggest themselves, both consistent with a democratic constitution: one is to revive the *panchayat*, and the other to develop the cooperative movement. Some governments have already taken steps to revive the *panchayat*, and the results achieved in the Punjab in the north and in Mysore in the south, to take but two examples, encourage the hope that with patience and sympathy the *panchayat* may in time become a valuable bulwark to the village.

As to the cooperative movement, so much has been written about it that all I need say is that, though entirely unpolitical—and long may it remain so—it performs the great political service of binding all classes together irrespective of caste, community or creed, and through its motto of "each for all and all for each", it gives the village that sense of harmonious interdependence which, as I have said, is its fundamental political and economic need. It is to the great credit of all the autonomous governments, whatever the colour of their flags, that they have realised the importance of the movement. The danger is that more may be expected of it than is possible with a peasantry that has not yet learnt the proper use of money, which cannot read an account or write a receipt, which regards social and religious obligations as more important than the satisfaction of economic wants, which tries to make living out of tiny holdings often badly fragmented, and which uses the great blessing of peace for blind multiplication of the species; which moreover, has to contend with a climate that saps energy and with a nature that mocks the energetic with the fury of its caprices. So considered that the Indian peasant rouses both sympathy and respect; sympathy, because he is the living expression of fragility and uncertainty which we must all feel in life, and respect because we see in him the beauty that comes from prolonged struggle and suffering uncomplainingly borne.

This peasant, if I may briefly recapitulate, is illiterate and for centuries he has been working in his fields isolated from the outside world and entirely ignorant of its doings. This isolation is now disappearing and the modern world is at his gates. The sudden fall in prices made him realise how dependent he is upon this world and profoundly disturbed his traditional relationship with landlord, moneylender and the government. In the midst of this disturbance, democracy appeared and gave him the vote. Now in her wake come propagandist, politician and patriot, agitator and demagogue, and an ever-increasing number of advisers and well-wishers. Roused by their exhortations and promises, he has begun for the first time to think more of his rights than his obligations. But in his mind, these rights hardly extend beyond the circuit of his village, and he still cares nothing for the higher verities of politics, nor, with rare exceptions, is he capable of understanding them. He therefore falls an easy prey to all who appeal to his hopes or to his passions. Disillusionment must follow an appeal to the one and faction an appeal to the other, and democracy and the village can thrive on neither. The ultimate remedy is to educate in a day, perhaps not even in a generation. Meanwhile, if democracy is to take healthy root, all political leaders should do their best to curb the agitator and the demagogue, and resolutely eschew all promises not based upon the realities of village life. This is where the student and the economist can render valuable service to the politician and the official, by giving them dispassionate advice after careful enquiry and, if necessary, timely warning of possible strains. But whoever it is that approaches the peasant; whether politician, official or economist, let him set himself first to understand the peasant before attempting to prescribe for him. Let him remember that he has as much to learn as to teach, and that ultimately it must be the peasant who must decide his fate for himself. A month or two ago, I asked a fine old Brahmin farmer in the Deccan how we could best help the peasant. "We must mix with them (he said) study his conditions and teach him by sweet words to improve. Up to now there has been no attempt". And he went on: "We must win his confidence and teach him by slow degrees; each improvement must be taken up one by one, not all at the same time." Who are the best teachers (I asked), officials or non-officials? "Whoever it is", he replied, "whether officials or non-officials or even one of our own kin, he must feel for the peasant. So only shall we win his confidence".

That is the crux of the matter: we must feel for the peasant, for without feeling there can be no understanding, and if we do not understand him, we shall certainly mislead him and also ourselves.

Indian Journal of Agricultural Economics, 24-25 February 1940.

2

T. VIJAYARAGHAVACHARYA

Cooperation as Applied to Rural Reconstruction

RURAL reconstruction is a term which can bear wider and narrower interpretations. In the widest sense, it includes all that a government can do to ensure prosperity in rural areas, and particularly every effort it can make to redress the balance between town and country. In a slightly narrower sense, it includes all those activities that tend to increase the health and the income of those who live in villages and isolated homesteads, whether through better facilities like education, communications, irrigation, credit and justice, or through the provision of particular assets like wells and school-houses, or through subsidy of one kind or another, or measures to check the operations of adverse forces, as in legislation for the relief of debt. If we look at these wider aspects of rural reconstruction, we find cooperation playing an important part in improving the conditions of rural credit and of agricultural marketing—and in fact as the Royal Commission on Agriculture in India (RCAI) has stated, "rural reconstruction" expresses more accurately the nature of activities included within the cooperative movement. The passage in the Royal Commission's report is so illuminating that I quote it below:

> The modern conception of the cooperative movement differs markedly from that commonly held at the close of the last century; the term 'agricultural organisation' or better still 'rural reconstruction' expresses more accurately the nature of the activities included within the movement. It is now accepted that cooperative principles can be used in overcoming most of the obstacles to progress in rural communities. Wherever agriculture is the predominant industry, cooperation is coming to be regarded as the natural basis of economic,

> social and educational development and India is no exception. Whatever view is taken of the failures and disappointments, it cannot be denied that, in almost all provinces, there are to be found outstanding examples of the successful application of cooperation to the problems of rural life; and, where success has been obtained, it should be within the power of the people to repeat it.

In this paper, I concern myself with rural reconstruction in the narrower sense of particular efforts to improve the economic and social condition of particular villages. I think the subject may be divided into two parts—what the departments of government do through their officials and what the people do for themselves. Some are more interested in the first, some in the second, and there are others who hold a view which range all the way between the extremes, one extreme being that cooperative societies are to be regarded as mere agencies through which the government will reach the villagers with advice and material help, and the other extreme being that they are organs of self-help, which will stultify themselves if they accept anything more than applause from outside.

The best work seems to have been done in countries where the cooperative society served as a link between the will to self-help through honesty and industry of the individual and the special knowledge and ability of outside experts. In many provinces, the outside experts until recently were to be found in the government services, but an increasing number of private citizens are equipping themselves to participate in the arduous task of instructing, organising and helping the villagers. The well-known words of the RCAI will bear quoting again:

> Where the problems of half a million villages are in question it becomes at once evident that no official organisation can possibly hope to reach every individual in those villages. To do this, the people must be organised to help themselves and their local organisations must be grouped into larger unions, until a machinery has been built up to convey to every village whatever the different expert departments have to send it. It is by such a system and such a system alone that the ground can be covered. Only through the medium of cooperative associations can the teaching of the expert be brought to multitudes who could never be reached individually.

In spite of the general acceptance of the view of the Royal Commission, the expert departments have not yet succeeded in this country in reaching the rural population through cooperative societies. Not only the credit societies who own readiness to forget their secondary objects might serve as an excuse, but even agricultural improvement societies and agricultural demonstration societies, societies for the execution of public works of utility to the whole village, joint cultivation societies, and better living societies have been living an unprofitable existence, where they could have been spearheads of the campaign for improved cultivation and less ignorant villages. Past experience makes it all the more necessary to insist on the need for change in the attitude and methods of both government departments on one side and cooperative societies on the other.

I have examined with some care the programme of the Punjab United Provinces and Bombay governments. All three aim at bringing the efforts of specialist departments to a focus in the village society by a twofold drive to coordinate the policies of departments at the centre and the activities of their officers in the districts, and to multiply village societies of a suitable constitution and adequate enthusiasm. The schemes differ in detail; the Punjab uses a single Rural Development Commissioner, the United Provinces and Bombay, Boards of Rural Development. The United Provinces government has set up a rather larger Board with rather more non-officials and have appointed a Rural Development Officer without other function; Bombay has preferred to make the Registrar of Cooperative Societies the Director of Rural Development. In the districts, Punjab relies on the voluntary Dehat Sudhar Committees and the initiative of Deputy Commissioners in calling occasional meetings of informal Officer's Board. The United Provinces scheme includes District Rural Development Associations in every district with non-official chairmen, Deputy Collectors as Secretaries, and a membership fairly equally distributed between official and non-official. There are also rural development officer for his charge. These two provinces also have Inspectors for smaller territorial units; Punjab seems to rely on the ordinary staff of various departments.

These experiments are interesting and should throw light upon the most economical and effective means of using resources of the government and energies of the people. I am in favour of a less ambitious scheme for a beginning in provinces where there is no organised official rural reconstruction scheme. I am against taking specialised staff from a specialist department as the Bombay Scheme does in transferring agricultural overseers and sub-overseers to the control of the cooperative department in its new form. Agricultural demonstration, propaganda and research are interrelated and go together and should not, in my opinion, be separated as in the Bombay Scheme. On the protest of the Indian Central Cotton Committee, the Government of Bombay have recently gone back on the separation policy.

There is need in many provinces for a greater degree of coordination between the expert departments—particularly those "development departments" which are immediately concerned with welfare of the agriculturists—the Cooperative, Agriculture, Industries, Veterinary, Public Health and other allied departments. Although the need has long been felt and expression has often been given to it, I find that in actual practice the departments work rather in water-tight compartments. As things stand, it is difficult to secure teamwork between the several departments.

In some provinces, a controversy has arisen as to whether a Development or Rural Development Commissioner may not be appointed to ensure teamwork between the "nation building departments". On theoretical grounds such an appointment seems to be an excellent method of coordinating them. But objections have been raised that this method of coordination is likely to lead to friction and to the suppression of the spirit of initiative in enterprising heads of departments. It has been urged that the Commissioner is more likely to interfere with individual discretion and judgement of the heads of departments, to control rather than coordinate, to dictate rather than suggest; instead of being a coordinating officer, he may tend to become an administrator and reduce the heads of departments to a position of subordination. It is contended that this arrangement may rather impede and delay the working of the departments than promote coordination.

The advocates of the view consider that the method best adapted to secure formulation of programme and harmonious cooperation of the departmental heads at the provincial headquarters and judicious allocation and coordination of work in the districts is through the constitution of coordinating committees. They recognise an objection frequently levelled against Committee as part of the administrative machinery of the government.

> That Committees are apt to be wasteful of time anyone who has had long experience of them can confirm; many varied hours rapidly spent in much talk and in few definite results with much nibbling at issues and little thorough consideration of any; committees are apt also to be relatively inefficient largely because responsibility is too thinly spread like a little butter over a long piece of bread.

They reply to this criticism by saying that what they contemplate is not a committee for the purpose of administration, but one charged with the formulation of measures and plans for the rural and economic development of the country in the form of Provincial Standing Committee of the heads of departments, with whom a certain number of non-officials should be associated. The provincial Standing Committee should have as its Chairman the Minister most closely concerned with rural development and as its secretary, an Under Secretary to the government. Whereas in come provinces all the development departments are not under one Minister, there should be a redistribution of portfolios among the ministers so that the Minister who is to be the Chairman of the Provincial Standing Committee may be in charge of all the development departments that are intimately concerned with rural welfare. The job of the Standing Committee will be to initiate measures of rural uplift, to watch the execution of measures and schemes, and to review the progress made; it will supply points of view and comments on, and impart cohesion and unity of purpose to, the several schemes of rural development proposed by the heads of departments and make them related parts of a comprehensive programme of rural betterment. In any big schemes of the government, there must be some Committees for the purpose of coordination; and provided the Committee is a well-planned, active and live organisation, the advocates of this view see no reason why it should not prove an effective agency for securing teamwork between the departments and for ensuring the maximum efficiency with a minimum expenditure of the government's resources. The Committee they propose will be composed of administrators and specialists and representatives of the people with the Minister at the head. They think that such a Standing Committee will be better adapted to modern conditions than a single permanent official and may prove to be a very helpful adjunct to the Government machine. They also suggest that in the districts, similar committees comprising officers of various development departments should be constituted with the Collector or Deputy Commissioner as its Executive Chairman. At present in some provinces, there are periodical meetings of District Officers and non-officials for rural betterment. They consider that this arrangement should be placed on a permanent basis and the bodies they suggest should be designed more as committees of coordination than merely as advisory bodies. The District Committee will execute the policy and the schemes suggested by the Provincial Committee. They may in addition suggest schemes on their initiative for the consideration of the Provincial Standing Committee. In this controversy as to whether the coordination should be performed by an officer bearing the designation of Development Commissioner or by a Committee composed of officials and non-officials, it is experience that will decide which is the more efficient plan; perhaps different provinces may try different methods so that the final verdict, whichever it is, may be brought in on the evidence of the results of practical working.

The appointment of 'village guides' to serve as a link between the agricultural population and the officers of development department executing plans of village uplift has been advocated in some provinces. The work of rural reconstruction will require the services of an increasing number of well-educated and imaginative young men who can understand the general principles of village and district administration and agricultural organisation. They need not be specialists, but should be men with hand and brain as well as be able to give a lead in those little things which may make all the difference between success and failure of the efforts to fit modern technical progress to the needs of an economically backward agricultural population. They will render advisory services and put the rural population in touch with the specialised services of development departments. I consider that it will be at once desirable and economical to have one common agency of rural guides for village *panchayats* and cooperative societies for efficient supervision of societies and for the provision of guidance, direction and supervision of rural work. They need to be properly trained, paid and controlled.

What is the proper organisation to awaken the enthusiasm and mobilise energies of the villagers? Some people favour the statutory village *panchayats* that have been set up in some provinces with powers of taxation and compulsion. Others, while recognising the usefulness of *panchayats* for certain purposes, point to the recent history of *panchayats* and stress their limitations. They opine that these are apt to arouse faction rather than enthusiasm and to neglect even sanitation, their simplest duty. It is contended that they are not adapted to promote agricultural improvement, better marketing or social change. I feel that when a village has progressed in cooperation to the point where mutual assistance comes more readily to their minds than backbiting, there is scope both for the cooperative society and the *panchayat*, the former for voluntary activity and the latter for whatever involves the compulsion of a backward minority or the interests of villagers as a whole and not cooperators only. If in due course an elected *panchayat* should take the place of village officers as the executive of the village and the main organ of administration—the time is not yet—the possibility and necessity of combining the efforts of the *panchayat* and the cooperative society would be greatly increased.

The question whether there should be one kind of cooperative society or several has recently come to the front. The controversy about multiple purpose societies is familiar to all cooperators. I think that agricultural credit societies in several provinces should be re-organised so as to give more weight to the secondary objects and make them instruments of 'better farming, better business and better living'. There is no need however to discourage the formation of agricultural improvement societies, land reclamation and colonisation societies, seed societies and so forth. But the main effort should be to re-organise existing societies and constitute new one in such a way that a single organisation will serve as far as possible all economic needs of a villager. I see no need to have separate better living societies, arbitration societies or supply societies in a village, which have a satisfactory primary credit society. Joint work, joint responsibility and joint security are the contributions of cooperation to village welfare, and they can be secured for valid purposes in a single organisation.

The danger in single societies with many purposes is that the business ability of the villager, which is sometimes unequal to the conduct of even a single enterprise, may be overstrained in attempting several. I think it is essential that efficient supervision should be provided. Not only paid secretaries for the larger societies, but village guides for every village are in my opinion necessary. Regarding village re-construction as a threefold process, where the stages of education, association and better living overlap, I think cooperative societies should undertake the education of the villager in the principles and practice of cooperation with the aid of departmental officers, promotion of literacy by adult schools, reading rooms and other devices, and instilling of the principles of health and economics, enabling those who know nothing to learn from those who know something. They should at the same time promote association in activities and enterprises useful to each and to all, putting into practice what they have learnt about health and economics, clearing was maternity care and child welfare and first-aid training, and establishing centres of recreation both outdoor and indoor. At the same time, besides giving credit and collecting savings, suppling good seed and improved implements, they should take an interest in the conservation of grazing grounds and the planting of trees, encourage the castration of scrub bulls and the control of breeding, experiment in the rotation of crops and increase in fodder acreage, instruct their members in cottage industries, occupy the excess of their spare time and supplement their income, form connections with sale societies and enable their members to secure a better return for their work; hence, by paying attention to the biological, economic and social needs of their members give them a new hope and a new interest in life.

I do not think that any society can make an advance all along the line all the time. Every society will have to make a choice between the possible enterprises and prefer steady but real progress to a sudden efflorescence without roots. But if officers of the various departments concentrate their efforts on the cooperative society, stay long enough to discuss the debate instead of hurrying through a lecture, carefully supervise actual enterprises like demonstration plots and village libraries, and concert their tours beforehand with each other so that every department takes a hand in forwarding each enterprise as part of a well-considered programme, there is no reason why a society should not make a wise choice of activities and gradually develop them into enduring improvements.

I am not here suggesting definite lines of coordination of rural services. I think they should form the subject matter of study and investigation by the proposed district and provincial standing committees; they should formulate measures in the light of the results of investigations made in consultation with or through specialist departments and voluntary associations. There are however certain aspects of rural problems which have been borne in upon me by my study of agricultural conditions and I should like specifically to mention them.

An outstanding fact that emerges from these studies is that several provinces suffer from a deficit of food crops. Every schoolboy knows that Madras is one of the largest rice-eating provinces in India and that the staple food crop of that province is rice. But how many know that Madras does not produce all the rice it eats? In spite of an increase in the rice crop in Madras this year by 12.5 per cent, there is now an acute shortage of food in that province, largely due to a slump in the imports of rice from Burma in order to supply this deficiency. I had the same difficulty in 1919 when I was the Diwan of Cochin. We had to adopt special measures to import Burma rice to the extent of one crore of rupees, to start rice depots in every considerable village in the State and adopt a system of ration tickets to prevent starvation.

The problem is not peculiar to Madras, Bengal, Bihar and Bombay share it too. India as a whole does not even produce enough rice to feed those parts of the country where it is the principal grain eaten. Ten years ago, Madras imported 16,000 tonnes of rice in addition to what it produced to feed its population; now it imports 818,000 tonnes of rice. The main cause for the escalation in imports is the increase in population. This phenomenon is not peculiar to Madras, but is found in other provinces as well.

The minimum subsistence family holding in India may, in general terms, be put at 4-6 acres, although differences in soil, productivity water supply, crop rotation and agricultural practice may alter the size of the subsistence family holding. Five acres may fairly be accepted as the minimum subsistence family holding. It may also be stated that one acre of land per capita is indispensible for human nourishment, as 4.2 persons is the average size of family in India. But the crop area per capita for the whole of India is 0.78 acres, while that for Madras is 0.74 acres. The Punjab, the Central Provinces and Bombay have 1.12, 1.58 and 1.61 acres respectively.

Considering the present world conditions, conditions which are likely to last long after the war is over, it is necessary to make the rice-eating provinces self-sufficient in respect of their food supplies without giving up industrial or money crops. This ideal seems to me possible of attainment through scientific research, intensive farming of land under cultivation, introduction of improved seed and increased supplies of manure and extension of area under cultivation, by land reclamation and colonisation wherever possible, by undertaking new irrigation projects and a liberal and prompt extension of irrigated areas in deltaic tracts.

In my opinion one, of the most important tasks for the Agricultural Departments and the governments in the provinces mentioned is to increase the output of food crops through the adoption of a well-planned scheme of distribution of improved seed. The responsibility for producing the seed should rest on the Agricultural Department and the responsibility for distribution should ordinarily be entrusted to non-official agencies. Cooperative organisations are particularly well suited to this task. Where no such agencies are available, the government should make the necessary arrangements. The technical supervision necessary should be supplied free by the government. There is a danger that in the increased attention now being paid to the production of seed of commercial crops, that of seed of food crops may be neglected. It is doubtless true that a varied cropping protects the farmer in seasons of depression more efficiently than the production of a single crop. But I wish to emphasise the fact, not so well known as it ought to be, that in spite of the enormous acreage in rice—it is somewhere about eight million acres—the country is not self-sufficient in her rice production, and that this is a potential danger in an unsettled world.

Items of census news published in the newspapers indicate that the population of India at the present census will in all probability jump up from 352 millions to 400 millions. Whether this rate of increase should cause anxiety or be regarded as normal is a matter of controversy among economists. Broadly speaking, there are two schools of thought. One holds, with the Census Commissioner for 1931, "that of all the nations that we know, it is India that has the largest population." The Census Commissioner considered the 10.6 per cent increase in population disclosed at the 1931 census "a cause for alarm rather than satisfaction." We do not know what the Census Commissioner of 1941 is going to think. The second school thinks that there is nothing in the population increase in this country which suggests any inherent abnormality. It points to the fact that during the first three decades of this century, the population of England increased by 23 per cent, while that of British India by only 17 per cent. This school of thought approaches the matter from the point of view of what is called the "net reproduction rate" and not from that of births, deaths and increments. From the available statistical evidence, there seem to indications of a lower specific fertility rate and considering the very low number of female survivors, a low net reproduction rate; it holds that there is *prima facie* proof that the net reproduction rate is low and indicative, at the most, of a normal stable population growth. Apart from this controversy, however, I am immediately concerned with one result of the population increase and that is the effect it has had by way of increasing pressure on land for cultivation purposes. I have refrained from examining its effect on curtailing land available for pasture and for fodder to cattle, resulting in a loss of efficiency in working bulls and a loss of productivity in milking animals.

Whether the pressure is such that the land cannot support all the people dependent on it, or in other words, whether there is real agricultural overpopulation as such is a matter of economic and statistical enquiry. I would, however, content myself with pointing out that modern economic theory does not suggest any reason for regarding overpopulation as abnormal. For overpopulation only means that the capital resources of the population fail to increase fast enough to maintain the output per head: it would be equally accurate to speak of a shortage of capital. If, as population increases, the area of land and capital equipment remain the same, the output per head must fall, but it need not follow that labour is unemployed. But the limits of labour intensification are more quickly reached when no addition to capital can be made. Consequently, land shortage may easily mean unemployment if the qualities of land do not make more intensive cultivation practicable. Physical conditions may set a fairly clear limit to the amount of additional labour which can be employed. Where climatic and soil conditions are poor, an increase in population means redundant labour. Thus, in terms of modern economic theory, to say that a region is overpopulated means that the working population is excessive with reference to a certain technique of cultivation, which must be taken as fixed as things are.

Whatever economic theory may suggest, we find that in fact overpopulation appears in such crude form that it is easily recognisable. Recent investigations in certain European regions have shown that even in the absence of satisfactory statistical data, the existence of overpopulation can be indicated through certain phenomena generally recognised as symptomatic. A common feature of regions in which there is a pressure on agricultural land is the great reduction in size of the typical holdings by successive sub-divisions, resulting in the so-called 'dwarf holdings'. Another feature that is common to many of these regions is fragmentation of holding, the dispersion of individual holdings in small parcels—the holdings are generally too small not only for economic operation, but for subsistence also. Another symptom of agricultural overpopulation is the high percentage of food and cereal crops in the total arable area. Again, a low unit yield is common to practically all the regions which are overpopulated. The small size of agricultural holdings, the high proportion of cereal crops to cultivated area, the low unit yields—whether or not they can be related to agricultural overpopulation—are of significance in that they all lead to undernutrition. Many of the agriculturally over-populated regions have, as one of their main results, a state of chronic undernutrition which, in turn, means disease and general loss of vitality. Together these conditions eat away what remains of the spirit of enterprise.

The conclusion to which we may fairly come seems to be that most provinces in India suffer from agricultural overpopulation. The remedies suggested for this are many—emigration, industralisation to bring about a more balanced national economy, pursuit of subsidiary industries, a rationalisation of agriculture and agricultural colonisation. The Royal Commission on Agriculture has recommended schemes of colonisation for relieving pressure on land and settling on disafforested or wastelands the surplus agricultural population.

Schemes have been tried in Madras of putting University men to work on the land. There schemes aroused very little enthusiasm and have not been attended with success. Schemes of colonisation can be divided into two classes: those for putting on the soil townbred graduates, and those are for settling landless agricultural labourers on land. The chief difficulty in this matter is want of definite knowledge in every province about the extent and nature of the land fit for agriculture but still unoccupied. The figures under 'cultivable waste' in official statistics are no guide, including as various investigations have revealed much of land which is fit for no agricultural use. If there is useful land to be had, the ordinary cultivator who already owns land will take it. If the land needs such reclamation as an individual cultivator can achieve, the revenue rules in most provinces afford facilities for it. The real difficulty come in where a large amount of money has to be spent in reclaiming the land before it can be brought under the plough. The Madras Committee on cooperation has suggested that in such cases, before the land is handed over to the colonist, the state should undertake all the preliminary measures to make the locality fit for habitation and the land fit for cultivation, such as clearing the land of jungle, the sinking of wells for irrigation, the provision of roads, the building of dispensaries in malarial tracts, and in all cases the provision of free technical advice through the agricultural veterinary and public health departments. In this connection, the experiment which the Madras Government has just started for financing cooperative colonisation schemes in two districts of that province will be watched with interest. The difficulty that the government had to meet was that in the past, a good deal of the land that they had assigned to the landless labourers had found its way into the possession of rich influential neighbours. In order to ensure that land is assigned to the really poor and is retained in their enjoyment, the government propose to make assignments hereafter under the scheme not to single individuals but to groups of colonies formed into cooperative societies. It is to the society that the government will look for the payment of its taxes and for the repayments of loans for reclamation raised from the Central Cooperative Banks (CCBs) or other sources.

Another aspect of the rural problem to which I want to invite your attention is that of cattle insurance. Either from the point of view of the income from the sale of cattle products (i.e., milk and milk products, hides and skins) or from that of cattle is of great importance to the *ryots* of the country. Nor can we overlook the contribution of cattle to the fertility of the land through supply of manure.

It is well known that the amount of capital invested by the *ryots* in livestock is always considerable, and the small *ryot* who loses his cattle by disease or other misfortune is frequently unable to replace them out of his own resources. Of all, the disasters to which he is liable, there is none which he has more reason to fear than the loss of his livestock. This is one of the sources of his indebtedness. It is true that he can borrow from a cooperative credit society to repair his losses and to buy fresh cattle, but even so, his credit for other purposes will be impaired. One of the ways to provide against cattle mortality is by insurance.

An attempt to form Cattle Insurance Cooperative Societies (CICS) was made in Madras as long ago as 1919. The main features of the scheme as formulated by the Madras Government of the time were briefly:

(a) That all cattle above one year and below eight years should be insured;

(b) That a member could insure 10 animals at the rate of one animal per share;

(c) That the animal must be examined by the Veterinary Assistant Surgeon and certified by him to be healthy;

(d) That the insurance covered only two-third of the actual value of the animal;

(e) That the premium was one anna per rupee on the sum insured for male animals, and one and a half for female animals;

(f) That the bones and hides of the insured animal which died were to be the property of the society; and

(g) That the income derived by the way of premium and by the sale of bones and hides was intended to be utilised for the payment of the insured amount to the cattle owners who were members of the societies.

Five societies were registered under the scheme. Four of them never started work and the fifth one was short-lived. All of them were eventually liquidated.

Cattle insurance societies were organised on a large scale in Burma, but out of 219 such societies on the list, only six societies were working in 1938. It is understood that cattle insurance has not proved popular in Burma because the people considered the rate of premium was high, that the indemnities paid to them on the death of the cattle to the extent of two-third of the insured value were too small, and the premium paid by them was not a good investment as they did not get back even a part of it if no cattle died.

Cattle insurance is one of the most difficult forms of cooperative effort. First, statistics relating to mortality among cattle are necessary to assess the actual premium rates. They are not at present available, and it is essential that suitable arrangements is be made for the collection and collation of accurate statistics showing the incidence of mortality among cattle from all causes in each age-group. Secondly, it is necessary that the *ryot* is keep the animals well and in good condition. As it is, most of the *ryots* often allow their cattle to sustain themselves on communal pastures, which leads to under-nourishment and contraction of contagious diseases. The average quality of cattle is poor and insurance of cattle is not worth while except in the case of pedigree stock, but in their case too the risk of losses through diseases is at present great. In western countries, cattle are generally allowed to roam about on communal pastures. Thirdly, there should be an adequate organisation for the control and prevention of contagious disease and veterinary aid should be easily accessible and properly regulated. The existing lack of proper and timely veterinary aid and the ignorance and prejudice of the *ryots* to avail themselves of such little aid as may be available are obstacles to experiments being made.

The number of qualified Veterinary Assistant Surgeons available for inspections and production of documents, treatment etc., when necessary in rural areas is too small, and the question of identity of animals, when false or doubtful claims are put forth, is too difficult to decide. I am afraid that insurance of cattle cannot be worked successfully through private enterprise in the existing conditions of the country and that the State alone can attempt a venture of this kind even as an experimental measure.

The Indian *ryot* is thus under a double disadvantage. He is at the mercy of the seasons and at the mercy of cattle diseases.

Indian Journal of Agricultural Economics, 12-13 February 1941.

3

T. VIJAYARAGHAVACHARYA

Study of Rural Indebtedness in the Province

THE Madras Committee on cooperation of which I had the honour of being Chairman has made an interesting study of rural indebtedness in the province. The results of that study are so valuable in themselves and are so capable of application to other provinces that I need make no apology for making them the theme of my address. The problem of agricultural debt is intimately bound up with welfare of the countryside, and perhaps a study of it from the provincial field may yield more fruitful results than an all-India study, which is bound to be vague and general and consequently of less practical assistance in division measures to deal with a situation, which is rapidly becoming acute and which may if not solved in time develop into a menace to the public peace. Even in so conservative a state as Udaipur where the cultivators still maintain the traditional attitude of blaming Providence for economic ills, which in more advanced parts of India will be laid on the shoulders of the government, the problem is now insistently calling for solution and the State Council is considering what action would be appropriate. The truth of what the Royal Commission on Agriculture said in 1928 is now being realised all over India: "The worst policy towards agricultural debt is to ignore it and do nothing".

The first difficulty that we meet with is one with which we are confronted in almost every agricultural inquiry in India. That is the absence of reliable data. In the Madras committee's inquiry, it was assisted by three documents, namely:

(a) The first document is Sir Frederick Nicholson's book on land and agricultural banks published in 1895. He was the founder of the cooperative movement in India and his book is now an acknowledged classic. He drew from his inquires the conclusions that three-fourth of the rural population of Madras borrowed for one purpose or another and that the borrowings had resulted in little economic profit to the *ryot*. He estimated the total rural debt of the Province at about ₹ 45 crores at the beginning of any given harvest. An amount of 18-20 crores of this sum were covered by mortgage of land, the balance being in the form of grain and non-mortgage and cash debts. The most common rate of interest was 18 per cent.

(b) The second document was the report of the Madras Provincial Banking Inquiry Committee published in 1930. It computed what may be called the total standing debts at about ₹ 150 crores and the debt from year to year at ₹ 70 crores.

(c) The third document was the report made in 1934 by an officer of the Indian Civil Service (ICS), Mr W.R.S. Sathianathan, who was appointed to make a special inquiry into the question of rural indebtedness on a motion by a non-official member of the Madras Legislature. The great depression had led to what is usually called a 'catastrophic' fall in agricultural prices and in the land and the inquiry was specially directed to the conditions produced in rural areas in consequence. The method adopted was to make economic surveys of 141 selected villages. This produced a great deal of interesting information, but the method suffered from a defect that was common to all such inquiries in the past, namely, that no expert statistician had been called in to assist, and the villages where the investigation was made were not selected on the random sample method. This method was advocated by Prof Bowley and Prof Robertson in their scheme for an Economic Census of India published in 1934 and has been independently adopted with brilliant success by Prof P.C. Mahalanobis in his survey of the jute area for the Bengal Government. Mr Sathianathan's inquiries reveal many interesting facts, nevertheless. The total agricultural debts of Madras was estimated at ₹ 200 crores, of which 47 per cent was mortgage debt, 50 per cent unsecured debt and 3 per cent debt was on the pledge of moveables. The detailed surveys made in the 141 selected villages disclosed that the purposes for which the debt was incurred might be classified as follows:

	Per cent
Payment of prior debts	25
Purchase of land	14
Trade	13
Marriage, funerals and other ceremonies	11
Agricultural expenses	10
Relief of distress	6
Building of houses	6
Improvement of land	4
Payment of land tax	3
Education of children	2
Other purposes	6
Total	100

According to the above classification, at least 42 per cent of the debt may be classified as unproductive.

The inquires show that the agricultural population in the selected villages consist roughly of 54 per cent of registered landholders, 18 per cent of tenants and 28 per cent of farm labourers. Only 21 per cent of the first class, 23 per cent of the second class and 36 per cent of the third are free from debt.

The average debt per head of farm labourers was ₹ 15, while the debt per head of land owners and tenants paying land tax or land rent of ₹ 1-100 was ₹ 211, and that of those paying above ₹ 100 was ₹ 918. According to revenue returns, 97 per cent of the holdings in the province pay an assessment of ₹ 50 or less and 77 per cent ₹ 0 or less. That is, the small and medium landholders and occupancy tenants constitute the bulk of the rural population and are the foundations in which the agricultural industry is built in Madras. According to Mr Sathianathan's inquires while the debt per head of landowners and occupancy tenants paying tax varying from ₹ 1-100 was ₹ 211, the debt per acre of land owned or cultivated by the class was ₹ 72 and the debt per rupee of assessment was ₹ 29.

These figures are eloquent evidence of the serious position to which the *ryot* has been reduced. No word of comment seems necessary. What was the action that the Madras government took to meet the situation?

State action in relation to agriculturist debtors may be broadly divided into three main categories. The first consists of measures designed to give farmers relief from the burden of their debts by means of either reduction of the capital owing or conversion of their liabilities. The second category also consists of measures of direct intervention, but of a nature less harsh upon creditors, such as the repayment of loans by installments, the suspension of coercive action and measure of distraint against debtors, the formulation of a special procedure to bring about direct agreement between debtors and creditors, provision of adequate credit facility, etc. We now proceed to examine how far these forms of debts relief were granted in Madras.

The Madras Debtors' Protection Act (VII of 1935) is the first legislative enactment for the protection of small debtors who have borrowed loans below ₹ 500. It prescribes maximum rates of interest for secured and unsecured debts and requires creditors' rates of interest for the same, and requires creditors to maintain proper accounts. This was found inadequate as it was not devised specially for the solution of the agricultural problem.

The Agriculturalist Loans Act, 1884, was amended in 1935 and the government undertook the distribution of state loans to agriculturists for the redemption of prior debts through a special staff of loan officers appointed for the purpose. Under the Act, the government advanced ₹ 1.68 lakh during 1935-36 and ₹ 16.85 lakh in 1936-37 and ₹ 7.41 lakh in 1937-38. The relief afforded by the grant of loans for the discharge of prior debts in this manner was infinitesimal.

A practical manifestation of the public's serious concern at the agrarian crisis was the placing on the Statute Book of the Madras Debt Conciliation Act (1936) for voluntary and amicable settlement of debts by bringing together agricultural debtors and their creditors through the medium of Debt Conciliation Boards (DCBs) constituted for the purpose. Six such boards were constituted in some districts of the province, but they did not achieve anything substantial by way of affording relief to the indebted *ryots*. The boards were expected to mediate between the creditor and the debtor, and bring about a composition arrangement with the creditor. The hope that by virtue of the Act rural indebtedness would be scaled down appreciably did not materialise.

The next stage was reached when the government, realising the acuteness of the problem, got through the Legislature the Madras Agriculturists' Relief Act, 1938. With the passing of this Act, we proceed from the stage of voluntary settlement to a compulsory scaling down of rural debts in accordance with the provisions of the Act. Act registers the first bold legislative attempt at compulsory reduction of rural debts and embodies some of the common features of debt relief legislation undertaken in Central and Eastern Europe —reduction of the capital amount of the debt, conversion by way of reduction of indebtedness through a lowering of the rate of interest payable on loans, fixation of a maximum rate of interest, etc. The Act came into force in March 1938 and intended to give relief to indebted agriculturalists by compulsorily scaling down their existing debts through ordinary civil courts, by fixing the maximum rate of interest, and by writing off arrears of rent due to *zamindars* and other landholders. The Act is also intended to ensure a more satisfactory working of the Debt Conciliation Act. The main principles of the Debt Relief Act may be simply stated.

As regards debts incurred before 1st October 1932:

(a) All interest outstanding on 1st October 1937 shall be deemed to have been discharged and only the principal will be payable; or

(b) If the payment on account of principal or interest or both is twice the amount of the original principal, the whole amount shall be deemed to have been discharged and if the payment is less than double the principal, only the balance shall be paid.

In the case of debts contracted after 1st October 1932, the amount of the whole transaction shall be made out at 5 per cent, simple interest on the original loan and only the balance after deducting all payments shall be payable. The rates of interest on debt so scaled down and on loans incurred after the coming into force of this Act may not exceed 6.25 per cent. The relief afforded by the courts under this Act during the period from March 1938 to September 1939 may be summed up as follow:

(a) Applications to the extent of four crores of rupees were filed in the courts during the period.
(b) Two crores and five lakhs were scaled down
(c) Forty-eight per cent of the original debt was remitted.

The committee on cooperation has recommended that the existing debt relief legislation in Madras should be effectively supplemented and strengthened so that the agricultural industry may be rehabilitated. The present procedure under the Debt Conciliation Act is that an agricultural debtor or any of his creditors may make an application for the settlement of debts giving the particulars mentioned in Section 6 of that Act. On receipt of the application, the board fixes a date and place for hearing the application and the necessary notices are sent by registered post to the debtor and each of his creditors. If, after examining the debtor, the board considers it desirable to attempt to effect a settlement, a notice is issued calling upon all his creditors to submit statements with full particulars of the debts owed. The debtor and each creditor are then required to explain their case regarding each debt due to any creditor or the assets of any debtor, the board decides the matter after taking evidence and such decision is binding on all parties in all proceedings before the board. The provisions of Agriculturalists' Relief Act by which the amount due as under the ordinary law is scaled down is taken into account by the board. The Board then prepares a complete schedule of the creditors and of the assets and liabilities of the debtor. If the creditors to whom more than 50 per cent of the total amount of the debtor's debts is owing come to an amicable settlement with the debtor, the settlement is reduced to writing and, signed by all the parties concerned, and is registered within 30 days by the Chairman of the board. It then operates as a decree of the Civil Court. If no amicable settlement is arrived at within 12 months from the date of application, the application is dismissed. The Committee on Cooperation recommends that the settlement of debts under this Act should not be conditional on the consent of the creditors to whom more than 50 per cent of the total amount of the debtor's debt is owing. This consent is difficult to secure in practice. They therefore recommended that this provision may be dispensed with and the compulsory conciliation of debts may be introduced. To accelerate the procedure, the Committee also recommends that whenever an application is made by a debtor or a creditor, all the creditors of the debtors should, on receipt of a notice, disclose the debts due to them within a prescribed time and that claims not so presented should be deemed to have been extinguished.

The next recommendation of the Committee is that a simple Rural Insolvency Act should be enacted and insolvency tribunals established for villages to administer the law on the spot promptly. Every honest agriculturalist debtor who is unable to meet his creditors' claim should be given speedy discharge, when he unreservedly places his estate at the disposal of the insolvency tribunal or commissioner. The reason for this recommendation is that there is a fairly good number of rural debtors who are either actual or potential bankrupts and whose liabilities far exceed their assets. Even if their debts are scaled down in accordance with the provisions of the Madras Agriculturists' Relief Act, it is impossible for them to raise credit to pay the reduced debts. It is not desirable that the debt should be allowed to hang round their necks throughout their lifetime and should be passed on from father to son. The Committee relied upon the recommendation of the Royal Commission on Agriculture which stated as follow:

> Just as creditors have a right to insist that all debtor's assets should be impounded and applied towards the payment of the debts, so also a debtor who has given up all his assets should have the clear right to be allowed to earn his living if he can, and to be free to make a new start in life.

The Committee emphasises that the procedure for these Rural Insolvency Courts should be simple and cheap and within the reach of any ordinary indebted *ryot* who may choose to take advantage of the law.

Another measure in the same direction urged by the Committee is the addition of a legislative provision to the those already existing in the Civil Procedure Code and in other laws under which the homestead and the minimum agricultural holding should be exempted from liability to attachment and sale. What the minimum holding is should be determined by the State with reference to the local agricultural and economic factors. The object of this provision is to prevent the on-going process by which a small landholder is reduced to the position of a tenant or a labourer. The small holders impart a certain degree of stability to the countryside and constitute the backbone of the agricultural population. It is necessary to protect them for extinction.

The Committee considered the advisability of an enactment prohibiting or restricting the transfer of agricultural land in order to check the tendency for the land passing from agriculturists to non-agriculturists. After considerable discussion, the Committee decided that the principle of the Punjab Land Alienation Act should not be extended to Madras. One of the main reasons which influenced the Committee in this decision was the fact that the chief moneylender in the rural economy of the province was the agriculturist himself, and even though the land is constantly changing hands, most of it is from one *ryot* to another. Another reason was that the bulk of the agriculturists in the province are small landholders and are more or less continuously in need of credit. Any restriction on the alienation of land would reduce its saleability and in consequence its value, and would either dry up the source or increase the cost of credit much needed by him. Further whatever may be the case in northern India, in the social and agricultural conditions prevailing in Madras, it is difficult to draw a sharp line between the cultivators and non-cultivators. The townsman in Madras is more often than not an emigrant from the village and his heart still hankers after the village which to him is his home. As an alternative to the Punjab restriction plan, the Committee recommends that legislation may be undertaken for providing:

(a) That the existing mortgages of agricultural land in which the mortagagee has enjoyed the fruits of the land for not less than 25 years shall be deemed to be extinguished; and

(b) That, in future, no usufructuary mortgage of agricultural land should be permitted unless provision is made for automatic redemption within a fixed period of years of which 20 should be the maximum, after which the land must be re-delivered to the mortgagor free from all encumbrances.

The basic idea of such provision is that the profits of a land during the period of mortgage are equivalent to the interest and sinking fund on the debt, and in the absence of such provision, the proceeds from the land are usually taken as equivalent only to interest or fraction of interest on the debt, and in such circumstances, the mortgage may continue indefinitely. It is true that such legislative action may carry with it the risk of evasion by creditors taking sale deeds instead of mortgages, but the Committee is inclined to think that with progress of time and spread of literacy, these practices will be rare. The Committee considers that such a measure will on the whole be beneficial to the *ryot* and the Committee recommends to the government the need for legislation in this direction.

Yet another measure recommended by the committee was the statutory control of the professional moneylender. The Committee recognises that the interest of the creditor should receive due consideration in all schemes of debt redemption not only on grounds of equity, but also on that of economic necessity. The moneylender occupies a very important and influential place in the provision of rural credit. He is a major provider of capital to agriculturists and whatever may be the malpractices indulged in by many of his tribe in their dealing with the *ryot*, he has continued and will continue to be a necessary evil in the rural economy of the province—indeed of all India—until and unless he can be replaced by an efficient and widespread cooperative credit system. At present, it is a fact that cooperative credit has only touched the fringe of the problem of rural indebtedness. The moneylender cannot therefore be eliminated, but power of doing evil should be curtailed. The Committee pointed out that though the maximum rate of interest on agricultural loans is now governed by the Debt Relief Act, the malpractices associated with the bulk of moneylenders still continue to prevail, perhaps in a subtler form, and other new practices have come into play. Execution of pronotes for amounts larger than those actually advanced, collection in advance of the difference of interest between the higher rate usually charged by the moneylender and that permissible under the Act from the amount of loan, and the disbursement only of the balance of it, the taking of usufructuary mortgages on a large scale than before and the execution of sale deeds on condition of reconveyance of land on the repayment of loan according to stipulated conditions, absence of accounts and the failure to give receipts—these are the practices indulged in by the dishonest amongst the moneylenders. Usury manifest itself not only in the form of charging high or excessive rates on loans but also under other guises referred to such as the deduction in advance of a fixed amount from the loan, or delivery of produce for a disproportionately low price fixed before hand. The result is that in an industry which usually leaves a small margin of profit and which is, therefore, in need of a reduction in the cost of production, these practices have a demoralising effect on the agriculturist debtor; there is no inducement for him to make serious attempts to improve his holding or to grow a bigger crop as another person will get the lion's share of the produce. This state of affairs is most prejudicial to agriculture. To see the evil in its most lurid form, one has to visit the cultivators on the hills, usually timid, ignorant and illiterate, exploited in the most unscrupulous fashion by the sharp-witted moneylenders from the plains below. To remedy these evils, the Committee suggests that legislation should provide for:

(a) Compulsory licensing of moneylenders,
(b) Proper maintenance of accounts by moneylenders and periodical inspection and audit of their accounts,
(c) Fixation of a maximum rate of interest,
(d) Imposition of an obligation on moneylenders to furnish periodical statements of accounts to debtor,
(e) Prohibition of compound interest, and
(f) Signing of the contract by borrowers, the contract being in the language of the district to which borrowers belong.

The Committee considered the suggestions made that the agriculturists should be prevented from borrowing for unnecessary purposes and that this should be done by restricting the sources or means of credit. Left to himself, it is argued, the agriculturist will soon get heavily into debt again and if this credit is improved by the removal of his old debts, the downward process is likely to be rapid unless it is countered by suitable checks, and his future borrowings are restricted to what is required for productive purposes. To that end, the Reserve Bank of India (RBI) has suggested that the total future liability (after the liquidation of previous debts) of agriculturists might be limited, whether by fixing it in terms of a suitable multiple of land revenue or on the basis of the average value of land held in proprietary or occupancy right, so as to enable the debt to be liquidated after providing for the bare necessities of life of the owner (or tenant) and his family within a period, say, 30 years. A second safeguard against excessive credit and evasion of restrictive measures suggested is to prohibit borrowing from more than one source.

The Committee has considered these suggestions and feel skeptical of the feasibility of enforcement of the measures suggested. While the Committee entirely agrees with the view that unproductive borrowing should be prevented and credit should be restricted in the interests of agriculturists themselves, it is not sure that a statutory prohibition or restriction of the kind suggested will be either practicable or desirable. A mathematical fixation of future liability of the agriculturist is difficult to ensure; nor will it foster his economic freedom or independence. On the other hand, it is more than likely that the sense of absolute security engendered by the proposed restraint might destroy prudence and regularity in the activities of most agriculturists as well as the spirit of saving and economy. The Committee is, therefore, opposed to statutory restriction of credit by way of the fixation of the total future liability of agriculturists. This economic tutelage is not on the whole desirable. The Committee would rather rely on the corrective influence of education, the spread of which, however slow, may offer a lasting remedy.

The Committee is also not certain whether the sterilisation of credit implied in the second suggestion is possible or desirable. In this connection, the Committee's attention has been drawn to provisions of the Bombay Agricultural Debtors' Relief Act, 1939, according to which a debt adjustment board make an award call on the debtor, if he is not a member of a resource society, to become one within a prescribed time. If he defaults, the board is required to dismiss the debtor's application for adjustment of his debts. The proposal would in effect mean the control of all the activities of the debtor in relation to his credit needs and is probably designed to ensure the sterilisation of all private credit which competes with institutional credit. For the reasons already explained, the Committee is not inclined to support this proposal either.

The Committee closed their investigation into the question of rural indebtedness by the following paragraph which deserves reproduction.

> We are conscious that many of the measures already taken by the Government or recommended by us involve departures from the principles of freedom of contract and of the sanctity of formal undertakings. But we must place the prosperity of agriculture—which means the general prosperity of the country—above all other considerations. Agriculture as a producer of wealth, as an employer of labour, as a fine mode of life, as a provider of homes for many, as an industry which develops some of the finest qualities in those engaged in it, deserves substantial and reasoned support from the Government, the legislature and the public. It is possible that the general effect of the measures proposed may be the curtailment in some measure of the supply of credit by private agencies. But then it is known that a portion of the present borrowings by agriculturists is unproductive and avoidable, and the curtailment of such unproductive borrowing should prove really healthy. The moneylender will continue to ply his trade—though it may be restricted—and to supply the requirements of many of his old clients. With the extension of village credit cooperative societies under the planning which we have suggested elsewhere, it is hoped that reasonable demands for productive credit in the areas covered by them will be adequately met.

Indian Journal of Agricultural Economics, 26-28 December 1941.

4

SIR MANILAL B. NANAVATI

The Problem of Agricultural Reconstruction

WHEN we think of the progress of agriculture and the welfare of the peasantry our mind immediately runs to Bengal which has been passing through an unprecedented famine. The word "famine" in this context is misleading for Bengal's sufferings are not due to any freak of nature; its roots are to be sought elsewhere. It is due not only to a deficiency of supplies but also to a deficiency of purchasing power in relation to the high prices which are an inevitable concomitant of inflation. A variety of causes have combined to bring about this state of affairs as a result of which the economic and social life of the country-side in Bengal has disintegrated. The food situation in India even in normal times is bad enough and the exceptional demands of the war have only aggravated the condition. There has also been a diminution of supply on account of the loss of Burma and there have been difficulties due to the haphazard operation of Provincial and Central controls and their procurements policies.

The 'grow more food' campaign is practically a failure and was bound to be so because no attention was paid to the parity between the prices of non-food crops and those of food crops. So long as the grower of food crops is not assured of a fair price for his produce—'fair' in relation to the price of non-food crops as also with reference to his cost of production and minimum needs—there is no real incentive to the farmer.

The situation in Bengal as also in some parts of Southern India has brought out to the surface the fundamental weaknesses of the Indian economic system. The Famine Commission of 1880 pointed out the growing pressure of population on the soil and indicated the dangers of a lop-sided economic development involving too much reliance on agriculture. Since then, several Committees and Commissions have reported on agricultural conditions and problems, but throughout all these 60 years we have done pretty little to cure this fundamental weakness. Of course, some new lands have been brought under the plough, more irrigational facilities provided and means of communication and transport improved and extended. The cooperative movement too has spread widely and a number of legislative measures have been undertaken to improve the economic condition of the farmer. All this, however, amounts to so little that not only is the productivity of our agriculture low, but it seems even to be diminishing. At the same time, the population has gone on increasing and there has been, during the last several decades, a continuous rise in the man-land ratio. The area under non-food crops has increased relatively faster than the area under food crops so that, the production of food per head of population has been declining. It is true that the agricultural statistics of India have many shortcomings and they have to be used with great caution. It does however appear from these figures that during the last two or three decades there has been a worsening of the food problems quantitatively as well as qualitatively with all the consequences of under-feeding, starvation, liability to diseases and epidemics, premature deaths and in general, economic inefficiency and intellectual stagnation. The Foodgrains Policy Committee's Report discussing the available food supply remarks that these figures must necessarily be underestimates because the natural growth of population that has been recorded would have been impossible if the picture of deficiency of food supply as revealed by the figures was true. A similar line of argument has been advanced in the Russell Report also. But these views are based on the assumption that in ordinary times everybody got the minimum required for the efficient maintenance of life, which is not true. The capacity of our people to stint themselves and to face privations seems to be so great that I would not rule out the possibility of an actual worsening of the food situation over the last 40 or 50 years in spite of this increase in numbers. It is only a devitalised half-starved population that can submit to slow tortures of flesh and inch by inch and minute by minute without protest, without murmur and it is this class that procreates most. In no civilised country of the modern world would such a condition be observed.

The low standard of life in this country and its consequences have been discussed by economists and statesmen, and has been confirmed in studies and enquiries by members of the medical profession and nutrition experts. The investigations of Sir John Megaw and of Dr Aykroyd as also of Dr Mukherji in this direction are well-known.

The basic weakness of our economic system manifests itself in increasing landlordism, an uneconomic land system which includes increasing landlordism, declining size of holdings with their growing sub-division and fragmentation, decreasing productivity of land and low incomes, the chronic indebtedness of the peasantry, the decay of rural industries, the comparative backwardness of large scale industries, lack of heavy capital goods and key industries, etc. Time does not permit one to go into details about the various defects of our rural economy and the causes thereof. I would like to impress upon and I may repeat it here that even apart from the war our problem was serious enough to demand close attention and utmost effort.

The war aggravated these weaknesses. The stoppage of imports, the increase in exports, the larger consumption in India by the troops, prisoners of war, by evacuees and by people engaged in heavy war work, all these increased the pressure on our limited stocks. The continued inflation of currency in spite of warnings and protests from different parts of the country has undermined the purchasing power of the poorer section of the community and aggravated the mal-distribution of wealth, so that the process of the rich getting richer and poor getting poorer has been accelerated by the war. The Government of India (GoI) somehow failed to understand the true significance of all these forces and let matters drift far too long a time.

It is satisfactory to note that the Indian army with the guidance from the present Viceroy has undertaken to distribute supplies to the rural areas. But the problem of prices has not been solved. Gruel kitchens alone would not meet the situation and we are afraid that before normal conditions are restored several million people would have either perished or been so devitalised through disease and starvation that if an epidemic of the type of the last war spreads it will take a still greater toll than in that period and we may have a repetition of the story of the 1900 famine. Let us hope that timely steps will be taken to save us from such a calamity.

So far our administrative machinery has failed and the cooperation of the people with the Government in this crisis is, to say the least, inadequate. India presents before the world a very sorry spectacle of a rich country with poorly developed and inefficiently organised economy. The rise in population as indicated by the last few censuses was at one time hailed as a sign of the fulfillment of the British Rule in India but this self-delusion has been exposed, and at a terrible cost, too. The problem is, therefore, not only to import sufficiently large quantities of food grains to be able to restore confidence among the population but also to make same available to them at prices within their reach in the immediate future and to plan on a comprehensive basis for the long run.

Thus we come to the fundamental long-term aspect, an aspect which must be borne in mind when we consider the vital problem of increasing the productivity of agriculture to provide our teaming millions, if not a high standard of life, at least a reasonably decent standard so that they will be free from the gross evils of under-nourished, diseases and early death.

The question is: What is really wrong with our agriculture today? Are our natural resources inadequate or is there something wrong with the human factor or is something wrong with the psychological and sociological environment within which economic activities take place in this country? I do not want to discuss here the various theories of over-population, but I think there is a common agreement, on the point that India's main problem is the failure to utilise the varied resources of nature with which she is endowed, so as to foster an all-round economic life which would give scope to her people for a stable and diversified economic existence. This can be done in the light of experiences of other countries of the world which have passed through similar conditions, but which have placed their economy beyond want and beyond famines. We cannot blame nature. We have to blame man. In other words, we have to blame the economic, political and social framework of our life, including, of course, the policy or the want of it on the part of the authorities. Several Committees and Commission have reported on this question with all its different aspects and steps have been taken by Provincial Governments and the GoI to improve matters through famine relief, railways, irrigation, cooperation, *takavi* loans, marketing facilities, debt legislation and the like. And yet, the net result of it all has been small for our work has been haphazard and of a piecemeal character.

What is wanted, first of all, is therefore a change in outlook, a new orientation of policy and a new drive to reach the farthest village and the lowest strata of the people. This means that the initiative must come from the GoI who alone can afford to employ the best experts and to find the larger funds necessary for research and reconstruction.

We have found as a result of recent experience that in vital matters such as food supply and price control, provincial autonomy has limitations. The Bengal famine has brought out fully the economic unity of India. Even in a country like the United States of America (US) where the constituent States have been more zealous of their rights, there has been during the last three or four decades an increased control by the Federal Government over economic activities and the Federal Government has undertaken more and more responsibility for rehabilitation of agriculture through financial and other forms of assistance. Their Federal Department of Agriculture, which employs nearly 5,000 men in their service in all sections of the rural problem, is a continuously working body collecting data on every aspect of the problems and applying corrections where needed. The whole problem of modernising agriculture, of extending irrigation, or conserving the soil and of promoting the best possible utilisation of our land resources can be solved only through the initiative of the Central government. I do not believe that the Provincial governments would be averse to such interference if the Central government offers them not only advice, but also grants-in-aid and loans as necessary.

In 1941, the GoI appointed a number of Committees to deal with post-war problems of commerce, industries and agriculture. For nearly two years, these committees did not function. Recently these have been reconstituted with an elaborate arrangement of such committees to deal with the varied post-war problems. But each Committee seems to be working independently of others. The methods pursued appear to be most dilatory. Notes are prepared by Government departments discussing the problems and these are submitted to unwieldly committees for their opinion and comments. And in the case of food problems, Committee after Committee would meet and opinions would be collected and it will take years before decisions are reached. If, in the meanwhile, war is over, new problems of immediate importance would arise and quite possibly, the important issue would be side-tracked, as happened with the recommendations of the numerous Commissions and Committees that were appointed after the last war. My feeling is that the GoI centralised all power in itself for a number of years and pursued their British colonial policy and then in 1919 divested itself of them in favour of the provinces. Since then, they have ceased to be active directors of the country's economic policy particularly of a developmental character, and when they found themselves confronted with larger issues they failed to function. They have not yet recovered from that position.

As a contrast, look at what England has done during the period of war. The British Prime Minister Mr Churchill, has, in his recent Mansion House Address, promised to the British people after the war food, work and homes for all. Sir William Beveridge has worked out a comprehensive plan of social security to cover all possible contingencies in the life of an individual so as to guarantee to every man, woman and child in the country a certain minimum standard of comfort under all circumstances. This report is now, we are told, to be supplemented by another, a report dealing with the problem of full employment. Several Commissions and Committees have been working in Britain even during the war to solve those problems. Recently, we have heard of the appointment of a commission to deal with the problem of population. There is little doubt that Britain in the post-war period will undertake a planning of economic life on an extensive scale.

However, it is refreshing to find that the Central government has arrived at decisions to take energetic action on some important aspects of national life such as education and health. The Sargent Committee has outlined a scheme of compulsory education and laid down the stages by which it could be brought into execution. In respect of health, an important Committee has been set up to chalk out a programme of health reform. Let us hope that the recommendations of these Committee will find support from the centre which will have to come forward with the necessary grants-in-aid and even assist in organising the machinery to put the schemes into operation.

The most urgent problem for India is to increase the productive capacity of the country. For that purpose, we have to deal with the enormous problem of the already existing large surplus of population on the land. This surplus has been estimated at about 30 per cent of the population now subsisting on the land. As our population increases and the process of de-industralisation continues, this problem becomes more and more acute as large scale industries fail to absorb the increasing number and the small scale rural industries continue to decline and decay. The result is too much pressure on the land, uneconomic cultivation, rural indebtedness and a devitalisation of the rural population. In order to solve this problem, it is necessary to bring more land under cultivation by making use of as much as possible of the 110 million acres of areas classified as cultivable waste other than fallow. A large capital expenditure will undoubtedly be necessary before these lands can be reclaimed, cleared and made fit for human habitation. It is for the State to devise a scheme which would answer these needs. Possibly the use of the demobilised soldiers after the war affords the best opportunity to launch land reclamation schemes with army machinery, so that these soldiers and others may be settled on land so recovered. On the whole, however, the scope for extensive cultivation in this country is limited, for much of the area which is at present fallow may not be really suitable for cultivation. It could be possible if all the processing of agricultural products is done on the farm itself and if the purchasing power of the rural masses is increased through rural industries to create a much larger scope for employment in the villages than we have at present. The development of large scale industries can open numerous avenues for employment in the towns for suitable classes and although the total employment we can thus create may appear small as compared to the numbers to be employed, it is improper to ignore the fact that the process of industrial growth is cumulative and as it proceeds, it creates new avenues of employment which were not originally dreamt of. In the main, however, we must rely on small and medium scale industries in the towns and throughout the rural areas to absorb the surplus population on the land and also the new entrants into industry as a result of the increase of population. India's resources are vast. We have hardly begun to exploit them on a systematic scale. There is a vast amount of work to be done by way of engineering, housing, sanitation, social services, etc., all over the country and if only we devise a proper plan we could find employment for our enormous human resources.

As far as agriculture is concerned, there are, I believe three main items of reform on which we need to concentrate: (i) land reform, (ii) tax reform, (iii) administrative reform. As regards land reform, the situation today is, I use the word deliberately, desperate. There are two main defects in this system. Pressure of population is progressively reducing the average size of these holdings and with every inheritance they are getting more and more scattered. At the same time in spite of the fact that the *zamindar's* rights and privileges are controlled in permanently and semi-permanently settled areas, the *zamindars* continue to exercise a depressing influence on agriculture; and this is a significant fact as landlordism is spreading and making the position of tenants everywhere increasingly difficult. The various systems of tenure introduced by the early British officials in this country have failed to provide the farmer with the proper incentives or to create a class of landlord capitalists who would take interest in the work of modernising agriculture. The Floud Commission of Bengal has brought out startling facts such as extensive sub-infeudation of land, the growth of the *Bargadari* system, the growth of a rural proletariat and a decline in the yields of cultivated land. Even in the *Ryotwari* areas, share tenancy has been growing and the functionless landlord has become parasitic on the cultivator. Land has been passing from the cultivating to non-cultivating classes and a considerable part of the actual cultivators is not in a position to look after agriculture in any satisfactory manner. Agriculture is thus falling into the hands of the wrong type of people. The reform of the land system should aim at giving to the cultivator a compact, possibly an economic holding which he can call his own. The solution, I believe, lies in extending peasant proprietorship. Is not *Ryotwari* the same thing as peasant proprietorship, I may be asked. I would say it is not, to the extent that even under this system the landlord is often a non-cultivator and cultivation is done on the basis of crop-sharing. Peasant proprietorship may not appear to be an ideal solution to those who in terms of large mechanised farms like those of Russia and the US. Considering the historical traditions of this country and the large number of population we have to support, no other solution is practicable. Peasant proprietorship is a system very much suited to subsistence farming. The peasant, it is true, is not always very enterprising. He is often poor, illiterate and conservative. But these defects can be cured. It is for the state to make arrangements to see that ample funds and legislative provisions are made available for the reconstruction of the holdings. The first step in this direction should be the abolition of *zamindar* as an intermediary between the government and the cultivator and the stoppage of lands passing into the hands of non-cultivating classes, and the progressive acquisition of land that has already passed into their hands. It is not necessary for the state to purchase all the rights of the *zamindars* by lump sum payment all at once. It would be a very complicated and difficult undertaking. The state may simply take up all the responsibilities of the landlord, establish direct contact with the cultivator and give to the landlord a reasonable rent—not a competitive rent—as may be fixed by a competent authority. We shall need very careful legislation to deal with the colossal problem. It will also require an efficient administrative machinery to deal with these problems with despatch and rectitude. Gradually, the government may have to assist in reconstituting the holdings by purchase and reallotment so as to restore peasant proprietorship. I cannot go into the details at this stage. The problem of land reform is one of making the cultivator the owner of the land and also of promoting consolidation of holdings so as to bring then up to a fair economic size. Measures in this direction have been adopted in various countries with a large measure of success. Local variations and peculiarities of the problem in the different regions of the country would have to be taken into account but it is certain that there is no solution to the problem of rural poverty unless this question of land reform is placed in the forefront of our programme of rural reconstruction. In the *Ryotwari* areas a fixation of rent by the state and a limitation on the right of transferring land unless a proper case has been made out for the same before the right authority, would go a long way towards restoring to the peasant the interest and initiative which he has lost on account of the growth of cash economy and his dependence on an absentee landlord or the moneylender. There are a few large estates in India today on which the results are, broadly speaking, satisfactory and I believe that these may be used for purposes of demonstration and propaganda as to improved methods of cultivation. Similarly in the areas newly reclaimed, collective farms of the Russian type can be established under government supervision and control so as to serve as model for the people roundabout. For the major part of the country, however, there is no solution other than peasant proprietorship coupled with adequate measures for an all-round improvement in the earning capacity of cultivator so that he becomes a more efficient agent of production and also a better man.

The next important item of reform is the reform of taxation. There is at present a great diversity between the land revenue systems in the different parts of the country. The basis of assessment, the rules regarding settlement and the incidence of land revenue vary very widely. In permanently settled areas while rents have gone on increasing as a result of the growth of population and increased competition for land, the state has not shared in this growth because of permanent settlement. In other areas, the procedure for determining the government's share is often empirical and revision has until recently meant only an increase in land revenue. In the case of large number of uneconomic holdings the present burden of taxation is evidently too high; in fact they should be exempted. It is now admitted that, in theory, the taxation of agricultural incomes should be on the same items as the taxation of all income. There should be a certain limit of exemption and the principle of progression should be introduced. Recently, on account of the contingencies of the war, an attempt has been made to introduce progression in the land revenue in the Punjab and a Bill for agricultural income tax was considered some time ago in Bengal. The whole question needs to be gone into carefully at the end of the war so that we have an equitable system—equitable not only as between the different categories of cultivating classes but also as between the agricultural classes on the one had and the business and industrial community on the other.

Thirdly, it is necessary to reform the administrative system. The present system is not capable of bearing the burdens of the new constructive policy that must be taken in hand after the war. There is too much red tape, corruption and aloofness on the part of the present bureaucracy to enable then to put into effect the larger view of the state as a social service agency for an all-round economic development. The villager today is reluctant to receive advice or help from the officials and it has been difficult to work the various rural reconstruction schemes of the provincial governments just because of this fact. The village officials such as the Patel have lost contact with the people. They are no real leaders in the villages and the so-called leaders are just moneylenders or others interested in catering for government officials and creating on them a good impression for their own benefit, but taking no interest in the work of the village as a corporate body.

In this connection may I say that it is necessary on the one hand to change the entire spirit of the administrative system from the top to the bottom and also at the same time to prepare the people themselves for a greater participation in their affairs. Any amount of propaganda is futile unless there is amongst the people a sympathetic understanding of its objects and purposes.

Ultimately, therefore, the question turns on the revitalisation of the village life. This, as I have always insisted, implies the creation of three institutions in every village, the school, the *panchayat* and the cooperative society. I have myself tried during my career several experiments along these lines and although I may not narrate them I may say that on the whole they have produced remarkably satisfactory results. The school should be a centre of physical intellectual and moral training and it should provide facilities for the training up of the younger generation for the new responsibilities they will have to take up in the rural areas. The village school, as I envisage, it is not merely a place where the three R's are taught but a genuine nation-building institution where instruction is imparted in regard to agriculture and the environs of the village and a new culture is instilled among the younger generation whereby the differences of caste and creed are forgotten and a sense of national unity and solidarity is built up.

The *panchayat* was an ancient institution in this country. It was this institution which enabled India to stand up to all the cataclysms of her political history. As a result of the policy of administrative centralisation followed by the British in this country, these institutions have decayed. Attempts have been made recently to revive them. But local bodies including municipalities, district boards and *panchayat* have been characterised by a woeful lack of responsibility and of by genuine desire for collective improvement. All the same, given proper supervision and guidance, *panchayats* can play a very effective role in the improvement of rural life. There must be in every village or group of villages a statutory *panchayat* with an assured income and these should be placed in charge of the right kind of officials who would enlist the cooperation and support of the people of the locality. Much of the rural development work which needs to be done so urgently can be done if these *panchayats* are revived as an expression of corporate unity of the village people.

The cooperative society with a proper supervising banking union is the third of the institutions which I regard as vital to the plan of improving the economic life of the villager. Here again, I am a believer in the multipurpose cooperative society as recommended by the Reserve Bank of India (RBI) in their Bulletins and Reports.

Cooperation cannot succeed if it is confined to credit alone nor can it work if we have different societies for credit, marketing and number of other purposes. Cooperation must take the whole man and must provide for all his needs. It must enable him to buy better and to sell better; it must save him from debt. It must increase his earning power and it must help him in regard to all his needs. Here again, the question is one of right leadership and right guidance. It is a question also of associating the people with the movement. No movement can work if it is merely another department added to the other departments of government. The government must of course advice and assist, but this advice and assistance can and should be so organised that cooperation becomes like democracy a movement of the people by the people and for the people.

One of the most vital issues in this connection is how to raise funds for these developmental activities and reforms. Till now, valuable recommendations of many committees have been shelved and good schemes of reconstruction allowed to rust by the government for want of funds. This policy has set up a vicious circle for our economy, for it implies that if we want more funds, we should exploit our national resources, but if we want to exploit these resources we must first have more funds in our hands. The vicious circle has to be broken up at several points. This can be done by a proper plan of agricultural and industrial advancement—plan which would aim at increasing the amount of employment and raising the level of incomes and taxable capacity in the country so as to enable the people to build up a capital reserve rapidly and to invest it in properly worked up schemes of development. Once a proper plan has been set in operation, it will have a cumulative effect on the purchasing power and taxable capacity of the people and thus we may be able to build up as Russia did a vastly more efficient economic system through such planned development

To bring about these developments as a part of the post-war reconstruction scheme, large sums of money will be necessary. Apart from the imposition of taxes and the raising of loans in a judicious way, I suggest that half the profit of the RBI, which are annually accruing to the GoI, should be earmarked for the development of agriculture as is done in respect of the profits of the Central Bank of Australia. A suitable amendment may be made, if necessary, in the Reserve Bank Act, to give effect to the above suggestion.

In short, we have to reform the land system, reconstitute the administrative machinery and to train people in taking an intelligent interest in their own upliftment before we can hope to make improvements in agriculture really effective on a noticeable scale. Simultaneously, we will have to deal with other equally important problems such as the absorption of the surplus agricultural population into industry and other diversified occupations, particularly in such pursuits as would yield supplementary incomes to small farmers. In transforming the villager into an efficient producer and a useful citizen, we will have to redeem him from orthodox beliefs and old social practices which come in direct conflict with his economic well-being. Legislation will have to be resorted to for rescuing him from tyrannies of caste, customs and social disabilities as was done by the late Maharaja Sayaji Rao in Baroda. Education and enlightment will have to be brought to them to full up the intellectual and social cleavage between the rural and urban communities. A useful method for bringing about these developments is to open rural reconstruction centres of the Martandam type constituted by Dr Hatch in South India, with a number of experimental stations at the back of them. The country as a whole would need at least 15,000 such rural centres for the rationalism of rural life and their establishment, in turn, would need a large amount of funds. But as already said above, finance for these developmental activities should come from a well-planned exploitation of the country's agricultural and industrial resources. But more than anything else, India needs today statesmen with vision and practical grit who can bring all the national forces together, mobilise them for re-orientating the whole economy of the country and save the people from the curse of unemployment and poverty in the midst of plentiful resources. For, unless this is done, India, as in the past, will find herself in the post-war world a mere spectator with the rest of humanity marching on.

Indian Journal of Agricultural Economics, 25-28 December 1943.

5

SIR MANILAL B. NANAVATI

The Fundamentals of Planning for the Developmnent of Agriculture in Our Country

WHEN we are planning for rural life we have to bear in mind a few fundamentals facts. First, that rural population is composed of men of all grades of civilisation from the primitive men to the highly efficient *kunbi* cultivator. While it is necessary to foster all-round development, special efforts are called for to assist the lowest and the most backward to come to the level of the highest. With this background we have to plan to raise the status of nearly 100 million men who are illiterate and who are living under sub-normal conditions.

Secondly, within each class or community we have men with special characteristics which react on economic development. Within the communities we have men of different castes who are influenced by social customs and religious beliefs which lead them on to unproductive expenditure and perpetual debts. In any scheme of economic planning that we adopt, it will be necessary constantly to bear in mind the social environment within which our reform have to work. Economic planning in general or planning of agriculture in particular does not take place within a social vacuum. This aspect has been neglected in the past, but we cannot afford to do it in future. Planning in India should be a sociological basis covering the whole man and every aspect of his life—neither economic alone nor technical alone.

Thirdly, all the territories are not of a uniform type. There are areas which are highly congested with hardly space for expansion while some of them have hardly an organisation or even funds to plan on an adequate scale. Hence, planning must keep in view the needs and possibilities of the different regions.

The experience of the European countries and of the United States of America (US), which have had to do with similar problems has abundantly proved that unless a comprehensive view of life is taken, the lot of the poorer classes in the rural areas cannot be materially elevated. These countries have studied the relevant conditions very carefully and have adopted measures to bring about a permanent change in the living conditions of the people. They pay special attention to social and cultural activities of life. They have come to believe that the human mind does not work in compartments; it is an organic entity and must be so treated. Therefore, along with the economic and technical measures for the development of agriculture, these countries think of the enrichment of the conditions of living.

Ours is an old civilisation which has undergone many changes and some of the strata are lying thick and firm and will require strong and continuous efforts to place people on the road to progress. This is hardly the place where an elaborate analysis could be given of the steps necessary for full planning. I shall, therefore, confine my discussion to some aspects of agricultural planning which appear to me to have been more or less overlooked in the various schemes—official and non-official—till now drawn up towards this end.

Let us first examine the memorandum prepared by the Government of India (GoI). It appears that their original plans for agriculture were based on Dr Burns' report which deals solely with technological possibilities of Indian agriculture. When the Indian delegation to the Hot Springs Conference returned with a new inspiration and broadened outlook, it seemed for the time that there would come the much needed change in the government's policy. But their report on agricology, though a magnificent document if judged by the traditional government standards, fall short of the higher objective of an all round socio-economic improvements of rural life which should be our goal. It deals mainly with technical improvement in agriculture and factors which would contribute to such development. It proposes to deal with tenures and tenancies only so far as these operate as a hindrance to technical development. This may mean much or little as it all depends upon the degree of emphasis put upon the several items in the plan by the various government agencies. They do not contemplate the resetting of the uneconomic holder and the landless tenant, nor do they show any concern over the growing menace of absentee landlordism.

The government's plan does not provide for a detailed study of the land conditions nor of the various problems of rural life—social economic and administrative.

As for the provincial plans that I have come across, the Bombay government's plan does not touch land problems except in respect of the consolidation of holdings. The authorities admit the need for the improvement in every aspect of the Province's social, economic and cultural life but so far these larger plans have not been evolved. Of the popular plans—the Bombay Plan and the Radical Democratic Party's "People Plan"—are the two most important. The former mentions soil erosion, the liquidation of old debts and the consolidation of holdings or cooperative cultivation. Possibly the authors will deal with the other main problem in their more detailed memorandum which it is said they are now preparing. The Peoples' Plan has an appreciably wide programme of land reform practically on the lines of the Floud Commission's recommendations for the restoration of ownership of tenant cultivators. Technical improvement has of course its utility. It would, no doubt, help to increase the yield from land but only up to a point. That does not carry us far enough. We may have better seed, better manure and improved plough. But what will these avail if the man behind the plough has not the incentive to put his best to utilise them? Today, he is being ground down between two evil forces; on one side, there is the primitive land system with a hierarchy of people not interested in the good use of land but having the lion's share in the produce; on the other side, there are social customs and laws which, with every increase in population, aggravate his misery by increasing his indebtedness, by accentuating the sub-division and fragmentation of holding and by driving him to pass on his lands to the non-cultivating classes.

The problem therefore calls for a wider vision and for bolder measures. The first essential is that we must have a vast scheme for reorganising our agrarian economy by reconstituting holding on the basis of peasant proprietorship. In other words, we must pull the cultivator out of this vast tangled web of land tenures and tenancies that we find today and then take adequate measures to set him on his own. In this connection, I wish to draw your attention to the fact that in the European countries, land reforms were given precedence to technical improvements, and most countries have felt the need of establishing peasant proprietorship on a sound and stable basis, in some cases by a process of breaking up large landed estates, in others—often simultaneously—by consolidating fragments which proved uneconomic and ruinous. I am fully convinced that peasant proprietorship, wisely managed, can give us yields in no way inferior to those obtainable from large-sized farms worked on a joint-stock basis. This, I am sure, is the right approach to the solution of our agricultural problem.

But, unfortunately the administrative system which we have inherited has always fought shy of land reform measures. We had a splendid opportunity to investigate into this problem when the Royal Commission on Agriculture was appointed several years ago, but the land system was excluded from its purview and the report, which is otherwise an admirable document, failed to deal with this fundamental defect in our agricultural economy. The unwillingness on the part of the government to face this vital issue is further evidenced by the provision in the Constitutional Act of 1935 which makes it incumbent on the GoI to secure the permission of the Crown before any changes are introduced in the *zamindari* system.

It must be said here in fairness to the popular ministries in the Provinces that when they came to power, they took up with enthusiasm and energy the problem of land reform. Thus came about the appointment of the Floud Commission in Bengal, the Prakasam Committee in Madras and the Darling Committee in the Punjab. Several provinces like Bombay and Central Province (C.P.) enacted laws to secure better working conditions to the cultivating tenants. But for the political deadlock which ensued on the outbreak of war and for the subsequent resignation of these ministries, we should have heard more of land reform legislation. The Indian National Congress in its Faizpur and Karachi resolutions also stressed the need for land reforms. The National Planning Committee set up by the

Congress began very well and did some useful work through its many sub-committees, but its work has also been suspended on account of the war and other developments in the political situations in the country. No wonder, therefore, that almost all the plans consider the establishment of a genuine national government as a pre-requisite to the economic progress of this country.

Another defect in the present planning is that sufficient emphasis has not been laid on the building up of an adequate administrative organisation to study our rural problems in all their aspects. The first essential of planning is that we must have all the basic facts of the rural conditions not only in respect of agricultural technique but also of social and economic conditions. It is only the basis of such facts that a solid foundation can be laid. India is a vast sub-continent with varied groups of people in different stages of civilisation. India may be said to be unrivalled in her diversity of conditions from region in respect of climate, natural resources, density of population, occupation and standard of living. Each region will have to be treated differently in view of the order of priority in which the different lines of development ought to be taken in hand. For some areas, the main problem would be to relieve the pressure of population on the soil; for others it would be a radical change in the land-system; for some again, it might be irrigation and hydro-electricity, for some others transport and marketing.

Here again the government have failed to realise the magnitude of the problem. In the 1000 crore plan prepare by the GoI, they provide for only one economist and one statistician for the whole of India to deal with the complicated problem of rehabilitating India's economy when a score of such experts would be too few for the task. It was because of this failure to consider all the sides of the problem that the Imperial Council of Agricultural Research could not do much to improve the lot of the average cultivator nor even to increase his yields substantially. Government must, therefore, set up a Department of Agriculture at the Centre, not on orthodox lines but one which have, directly or indirectly, a bearing on agriculture. It must have a number of sections with well qualified and trained personnel, administrative and advisory, and each section must be in charge of specific aspect of this manifold task of reconstruction and planning. The government must also provide for the necessary training of technicians, economists, statisticians, social workers and agricultural experts who will take up the work of administering the plan. A plan is not a thing made and executed once and for all. It is necessary continually to watch the results of various measures, to appraise the domestic as well as international trends and to modify policies from time to time so as to get the best results. All this cannot be done unless we had, on the one hand, trained research workers, and on the other, a suitable administrative staff working in the best spirit of social service.

India has a long leeway to make up in this direction. So far as the government are concerned, no further comment is necessary when I state the fact that the Imperial Council of Agricultural Research has not a single economist on its staff. The Economics Adviser too is not concerned with these basic problems as he is generally preoccupied with others, such as tariff, trade and currency policy. Our universities also, except one or two, have completely neglected this side of research. It is, therefore, now necessary to have a network of research organisations under the Centre's guidance for regular collection of precise data on all these aspects of our economy. The Provincial organisations would collect facts and figures, maintain them up-to-date, study the effects of various government measures and recommend suitable changes in them whenever necessary. Government should also come forward with financial aid to the universities for establishing chairs in agricultural economics and for extending sufficient facilities for socio-economic research in the higher stages of post-graduate study.

It is natural that we expect the centre to set up an organisation which should be a model to the Provinces and States. This organisation should be assured of funds enough to get the best men available and to train many more. The men at the Centre should be better than the best in the provinces because the success of the Centre's activities would depend on the ability of its experts to influence and guide the workers in the local administrations. Already, we have several plans before us, official and non-official, central and provincial and some from the States. It should be the function of the new organisation to take the necessary steps to harmonise these different plans so as to reconcile the claims of different regions and sections of the people and also those of agriculture and industry.

This failure to see the economic problem of India in its proper perspective and to see it as a whole is, to a large extent due o the dearth of the necessary facts which would help us to realise the magnitude of the problem and the various implications of the issues involved therein. The statistical and other information available at present shows us no more than what floats on the surface. As a result, our knowledge on the subject is at present only superficial. That is so because the scope of this work of collecting statistics is circumscribed by the narrow end which the government have had in view, *viz.*, the collection of revenue. Even the information which is collected is not made fully available to the public in an intelligent form. Occasionally, an enthusiastic officer digs out these facts and makes good use of them as we find from the studies made by Messrs. Keatings and Jack, Dr Mann and Sir M. Darling and a few others. I, therefore, feel that a great deal would be gained if an attempt is made to put together in proper form all the information which at present lies scattered about in the various administrative and other reports issued by Provincial governments and the Centre as also by the expert Commissions and Committees during the last 20 years or so. It would certainly be worth while to bring together all this scattered material; and to place it in the hands of expert statistician, economists, and sociologist. This in itself, should be a good beginning.

The next step in this direction would be to widen the scope of our statistical information. For planning is a stupendous task. It needs a great deal of preparation, a good deal of preliminary survey, estimating and prospecting. We have to plan the Plan itself. In fact, well-organised research is the foundation of all social and economic planning. But it is to be regretted that even the officially published data are not sufficiently reliable and the government themselves have now come to realise, not to say confess, this fact. With such inadequate and defective data as our basis, we cannot hope to go to the roots of our problem. To take agricultural statistics, for instance, these are collected today through defective agencies and have little reference to the important aspects of our agricultural economy. Each province has its own way of collecting and presenting these data and this makes difficult any comparative study of conditions in the several provinces. For example, many provinces confine their attention to ownership of land under various groups. A few provinces like Bombay distinguished between agriculturist and non-agriculturist holders of lands. For some provinces, however, this information is not available at all. For the purpose of land reforms we would want to know how much of land is owned and cultivated by the several groups of agriculturist and how, and between whom, the produce is shared. This information should be collected in all Provinces and States on a uniform basis. The scope of our land statistics should be widened to show: (i) who owns the land, (ii) who cultivates the land and on what terms, (iii) what is the equipment of the cultivators, (iv) what crops he raises, and (v) what is the approximate value of the produce. This would lay bare the basic framework of our agricultural economy and show how far and in what respect the present land system does damage to our agriculture and what changes are necessary in our agrarian system as a whole.

The scope of our statistical information would have to be expanded also in other directions if we are to get a fuller view of the problem which we have set ourselves to solve. There must be a rural census which would give a complete idea of the strong and weak points in the social structure of our rural society. This would have to be supplemented by typical village surveys dealing not merely with economic conditions of the people but also with rural sociology, rural health and other important subjects which directly or indirectly influence the worker's efficiency and outlook.

The GoI have been contributing funds to the League of Nations and to the International Institute of Agriculture. These two institutions have been carrying on comprehensive studies of rural life in all its aspects. Some time ago, the League of Nations published 17 such studies which are a mine of information on land tenures, tenancies, economics of cultivation, education, culture, amusements and folklore, measures for land reclamation, land reform and promotion of subsidiary industries and various other aspects of rural life. If the GoI has only sought the help and guidance of these institutions we would have had by now a wealth of data throwing light on the problems which we are trying to solve today. It is to be hoped that these institutions would be approached for advice and guidance as soon as conditions permit after the war.

To sum up, it now admitted on all hands that planning is the immediate need of our country, if chronic poverty is to be avoided. But then, we do not want the plan to be half-hearted and experimental measure. What we need is a comprehensive plan which would first go to the foundations of our economic and social structure analyse the root cause of our various levels and then provide an adequate organisation for their removal. Any plan which over-looks these essentials would not have its basis in 'social reality' and would therefore prove sterile. There are several other aspects of the problem which I deliberately refrain from touching upon in this address. There is the question of population planning, for instance, or the question of financial and trade policy in its bearings on agriculture, or, again, the problem of extending and rehabilitating the cooperative movement. Every one of these problems has a technical aspect, an administrative aspect and a social aspect, which must be taken into account. A satisfactory solution of these will demand all our energies, all the goodwill and cooperation we can muster for this noble, idealistic work. I sincerely hope that the suggestions which I have made in the course of what I may call a constructive criticism of Government's plans would receive due consideration while we are trying to evolve the final plan which is to lead our country onward and upward.

Indian Journal of Agricultural Economics, 26-28 December 1944.

6

SIR MANILAL B. NANAVATI

Economic Revival in the Post-War Period

WE are living in highly dynamic times and the whole world is facing new problems which have to be solved in a new way by far-reaching and drastic reforms. It is necessary at this stage to be very clear in our minds as to our objectives and the bearing of international developments on them. Agricultural economics including the broader issues of rural reconstruction is a subject of the most vital importance for us at this stage of our development.

The outstanding event of the year is undoubtedly the cessation of the war. The world has passed through another crisis. The atom bomb which accelerated or perhaps precipitated the end of this titanic struggle has opened up untold possibilities—creative as well as destructive. The war has strained the economic system all the world over and the task of rehabilitation and relief is by no means simple. But the real problem of reconstruction and development lies, not necessarily along the old lines but along new lines as the political developments in the various European and Asiatic countries unmistakably indicate. No one can today outline with confidence the shape of things to come in the course of, say, the next 5 to 10 years. Though the war is over, peace has not been fully established in the world. In the sphere of international politics new spheres of influence are being perpetuated. One already hears whispers about the third World War—more catastrophic and destructive than the one which has ended. The use of the atom bomb by any disgruntled nation may set the whole world ablaze. On the other hand, scientific discoveries are making for vast changes in productive technique, in locomotion and in communication of ideas. It is certain that international economic cooperation in increasing sphere of trade, industry, monetary policy, etc., will be a condition precedent to economic revival all the world over. Trade rivalries as between powerful nations are bound to exist. They are also bound to exercise a far-reaching influence on the economy of backward nations. All these problems will have to be faced. Plenty of goodwill and good sense on the part of the various nations will be necessary for this purpose. Equally necessary, I think, will be the capacity for evolving suitable institutions for cooperation and working them properly in the spirit of give and take.

There were similar problems to be faced after the last war when an attempt was made to make the world safe for democracy and to end all wars through the organisation of the League of Nations. The short-coming of that institution and its failure are now well-worn themes. Perhaps the only lasting benefit of the League has been the setting up of the International Labour Organization and its various publications, periodicals and others, on financial and economic matters. The League of Nation thus gave an impetus to a systematic study of social and economic problems all over the world and imparted a new perspective in economic research. India was an original member of the League organisation. Except one or two reports on Indian problems, the League of Nations hardly devoted any attention to India. Nor did the Government of India (GoI) prepare special studies or memoranda for submission to the League of Nations.

During this war the President of the United States of America (US) and the Premier of Great Britain announced their new idealism for the world by way of the Atlantic Charter. President Roosevelt also enunciated his gospel of Four Freedoms as the basic creed for human happiness. Thereafter, Dumbartn Oaks, San Francisco; Bretten Woods, London, Paris and Quebec, have been the meeting grounds of historic conferences dealing various problems of the postwar period, problems of security, of money and banking, of rehabilitation, of food agriculture, of reconstruction and development, of labour welfare, of education and culture. The objective behind all these has been to stop future wars and to give the common man his due, to place him at the centre of things and to assure him through full employment and social security those essential conditions of decent civilised existence without which self-expression and self-realisation are meaningless catch-phrases. The institution of organisations such as the United Nations Relief and Rehabilitation Administration (UNRRA), the Food and Agriculture Organization of the United Nations (FAO), United Nations Educational, Scientific and Cultural Organization (UNESCO) and the International Monetary Fund (IMF) is prompted by the desire to see that this war to which the common man was called upon to contribute his utmost is followed by peace in the blessing of which he has a reasonable share. The new era that is drawing on us will, I trust, be the era of the common man. This implies not only political but also social and economic democracy. The war has transformed all valuations all over the world. Leftists movements and organisations are coming on top indicating that the people in general expect their leaders to take far-reaching and radical steps to assure them better conditions of life. Every country has now accepted the idea of a socio-economic order in which the common man's standard of life will be raised above want and stint, in which the vested interests will no longer be allowed to have privilege without responsibility and men and women will feel themselves not the tools or instruments of some superior individuals or groups in power but as free agents in a position to think and act on their own. We in this country have hardly any idea of what this standard of life means in advanced countries. And their conception of the standard of life is going up continuously. They utilise every means known to science for creating more goods, comforts and facilities and their Governments are keen on building up organisations, whatever the initial cost, powerful enough to cope with the social problems of the age of science and technology.

This war has brought home to us the weakness of the Indian economy. On account of Bengal famine and the epidemics in other parts of the country in 1943-44, over two million people lost their lives. The influx of destitutes into Calcutta continues even now and papers report from time to time cases of deaths of such destitutes. The Indian Famine Commission has brought out two volumes of a

Report in this question. The commission found that the situation of this country in respect of food is fundamentally unsound. The Grow More Food Campaign, in spite of heavy expenditure, has yielded but small results. Prices of foodgrains are beyond the reach of certain classes of the population. Nutritive and protective foods are not available in desired quantities. Nearly 30 per cent of the population in India is estimated to be suffering from chronic mal- and under-nutrition. The cessation of the war is not going to ease the situation immediately. The war-devasted countries of Europe have to be fed. There is scarcity of shipping which will last for a considerable time to come. Our population in the meanwhile, as during recent years, goes on increasing while the general economic condition is not improving. During the last 75 years, continuous deterioration in the condition of the masses has been taking place. In 1880, India had a surplus of food stuffs to the extent of 5 million tonnes and now we have a deficit of 10 million tonnes. The consumption of food was then estimated at 1½ lbs. per individual and now it is 1 lb. the man-land ratio is steadily rising. In spite of the development of modern industries, de-industrialisation is still continuing. In 1880, industries absorbed 12.3 per cent of the population and now the figure is 9.0 per cent. In 1872, 56.0 per cent of the population was supported by agriculture; this proportion has now increased to 73.0 per cent. In spite of the Government measures to improve yields and qualities of crops, actually the yields are going down while the land systems and land tenures are getting more and more complicated. In other respects also the Indian economy has been overstrained. In spite of men and materials being available in plenty, organisation for production and distribution remains inefficient.

The Indian Famine Commission have in their report given close attention to this and allied question. They have rightly pointed out that it is the Government's responsibility not only to provide famine relief but also to prevent famines and more positively, to see that every man, woman and child is able to get sufficient nutrition in the interests of health, efficiency and a good life. This is a welcome step forward. In this connection the following observations of the Famine Inquiry Commission deserve notice:

> The state should recognise its ultimate responsibility to provide enough food for all. We enunciate this here as a broad principle, the implications of which emerge from the Report as a whole. In India the problems of food supply and nutrition are fundamental and must at all times be one of the primary concerns of Central, Provincial and State Governments. It is abundantly clear that a policy of laissez faire in the matter of food supply and distribution can lead nowhere and would probably end in catastrophe. All the resources of Government must be brought to bear in order to achieve the end in view.

The implication of such a policy have been stated very aptly by Sir John Russell:

> Broadly speaking agriculture should be so organised as to make the maximum contribution to the well-being of the nation and this means that it must: (a) provide food for the people, (b) help in the social problems involved in national health, population strength, unemployment and restoration of broken lives, and (c) be ready to expand in times of emergency.

The question is: How is this to be secured in practice? The immediate policy is clear. We must make use of the machinery that has been developed during the war at the Centre and in the provinces for stimulating food production, for procurement, for rationalisation of transport, for control of prices and for efficient and equitable distribution. Some of the defects which limited the success of the Government's food policy during the war will disappear shortly with the return of more normal conditions and the administrative system will also prove itself more equal to the tasks ahead as a result of the experience gained during the last five years.

The last 25 years have brought home to us the need for maintaining the income of the rural masses. Under our conditions, a decline in agricultural prices and incomes has a deflationary effect on the entire economy. If agricultural prices fall sharply, all attempts at the amelioration of the peasantry through cooperation and other methods are bound to fail, for its loss in income more than compensates any small gain accruing as a result of such beneficent measures. The immediate policy must, therefore, be to stimulate the production of foodgrains by assuring reasonable prices and by giving technical as well as financial aid to the cultivator to rationalise distribution and to inculcate the habits of healthy consumption among the people.

Then there is the question of long-term policy. There is not much difference of opinion, I imagine, about the objective of the food policy in India. Economically as well as politically, a policy of self-sufficiency in regard to the supply of food is highly desirable for the country. In order to attain this, it will be necessary to bring about an increase in the area under cultivation by bringing new lands under the plough and the expansion of irrigation and if necessary, even by diverting land from non-food crops to food crops. In the main, however, we must increase the yield per acre by improving the efficiency of the land, of cattle, of implements and above all, of the human factor which works these all. There is a tendency in certain quarter to consider this problem from the narrow point of view of technological possibilities, but our task is to see how these technological possibilities can become economic realities. The mere technological approach to such problems is, I think, quite sterile. What we want is a human or humane or socio-economic approach which looks upon agriculture not merely as a technical transformation of matter from one form to another but as a way of organising resources—natural and human so as to ensure certain social ends. The need for this wider outlook has been realised in countries like the US.

A proper food policy for India involves the adoption of deliberate planning in respect of the production and distribution of essential items of food, so as to make them available to every individual. This is a far-reaching change. It involves a radical change in the outlook and method of work on the part of the various Governments in the country and a much closer cooperation between the Government and the people. To state that India needs a planned food policy in fact does not solve the problem. The question is: how is this planning to be done? What are to be the targets and by what means are they to be attained? Once again we have got face to face with the fundamental problems: why is Indian agriculture in a backward state? Why is its productivity so low and why is the population unable to produce sufficient food for itself? Why is there so much poverty?

I do not want to reiterate the obvious, but I am convinced that we cannot solve the Indian rural problem unless we change the basis and structure of our rural economy by a frontal attack on the land peasant proprietorship, so that the productivity of land improves and the worker gets the fullest benefit from his holding. The reform of the land system is a measure of which the GoI have always fought shy. It is encouraging, however, that for the first time the land problem has been discussed in fair detail by a Government Commission. The 2nd report of the Famine Inquiry Commission has drawn attention to the seriousness of this problem and has recommended that the Provincial governments should make comprehensive enquiries to find out measures which will be necessary to remedy the defects brought out in the discussions of the subject by the Commission. I trust that this recommendation of the Commission will be given effect to at an early date and a definite policy will be worked out. There is not much dispute about the facts. The holdings are getting smaller and smaller and fragmented, making effective cultivation difficult. This reacts adversely on the standard of life and efficiency of the worker. Land is also continuously passing into the hands of non-cultivating owners. There is evidence all round that the land system under which the landlord does not cultivate but exacts a parasitic toll is economically, socially and morally indefensible. The evils of tenancy have been realised all over Europe and even in a country like America, they have started encouraging family farms on a relatively small scale. The Indian Government's neglect of the land problem is responsible for the poverty of the people and the squalor and misery in the rural areas. Government have deliberately followed the policy of ruralisation as opposed to industrialisation. Having followed this policy they were unable to adopt measures of agrarian reform, for such reforms would mean releasing as part of the surplus rural population and finding employment for them elsewhere. This policy has had a doubly disastrous effect. On the one hand the conditions in the villages have grown progressively worse. The entire village is often a slum—in fact some of the villages are dung-heaps. The bolder spirits of the village naturally leave the countryside, with the result that both agriculture and handicrafts lose their guidance and patronage. On the other hand, new vested interests have been created in the land especially in the *zamindari* areas and this has barred effectively the path of progress.

Let us be clear that any scheme of land reform will involve a displacement of superfluous labour from the land. To some extent—I should say to a considerable extent—the improvement of productivity that will follow the enlargement and consolidation of holdings will stimulate greater production on the farm and employment in the village and thus add to the wealth of the village. There is also scope for a number of small industries in the rural areas if the country side is prosperous. If the authorities have a proper plan, this problem of unemployment can be tackled successfully. But I should like to stress the point that partial employment for a majority of the rural population, which is only another world for widespread disguised unemployment, is not good substitute for visible unemployment. In prosperous countries like the US and United Kingdom (UK), the majority of the people are fully employed, that is to say, their productive capacity is fully utilised. The problem of unemployment is therefore concentrated on a small section of the population which is unable to find work at all and has to be supported from the public funds. I know the evils of the dole and we all agree that the right to work is as fundamental as the right to exist. However, actual unemployment attracts attention. It becomes a matter of public discussion and the Government have to take steps to root it out. It is, so to say, an obvious manifestation of a disease within the body economic. I think it would be better if such a disease is brought out and adequately dealt with rather than suppressed and allowed to poison the vitals of economic life. It is our policy of allowing any number of people to press on the land with or without work and earn meagre and uncertain income that has led us to avoid facing the whole problem of removing rural poverty. It is this endemic disease of the economic system which needs to be treated urgently at the present juncture. Small and fragmented holdings not only reduce productivity directly but they also have a demoralising effect on the human factor. The very spectacle of idleness and superfluity of manpower that we see in the village is a deterrent to any development. In my previous addresses I have stressed this aspect of the question. Here, I will only say that agrarian reform is the fundamental problem in India and that unless steps are taken to give the cultivator the status of owner of at least an economic unit under active State help and guidance, an extension of the other piecemeal reforms that have been tried in the past without conspicuous results will not solve the problem.

While on this subject, it is gratifying to note that a member of this Society Mr Tarlok Singh, has recently published a book: "Poverty and Social Change"—in which he has suggested a new method of dealing with the problem of small and scattered holding and of the evils of rack-renting and tenancy. The scheme of joint village management that he suggests involves a far-reaching change and it is possible that difficulties of an administrative and of similar nature may be discovered when the scheme is put into operation. For example, I do not agree with him in his treatment of absentee landlords and some other appropriate conditions to see how far the scheme is practicable. Mr Tarlok Singh has made a deep study of Indian rural life. As settlement officer, he has come into personal contact with villages. The suggestions he has made are born of practical experience and should certainly be tried on an experimental basis so as to find out under what conditions a further extension or modification of the same would be necessary.

Another interesting development of a similar nature is the Baroda state's experiments in reconstituting holdings in four villages in the Tilak Wada, Peta Mahal of Baroda district in accordance with a scheme approved by the Imperial Council of Agricultural Research. It is a Bhil area with a very backward population where lands have mostly passed into the hands of money-lenders. The state has now decided to reacquire all these lands and rearrange them into separate peasant holdings, into cooperative farms and into collective farms. The object is to test the efficacy of these different methods of farm management and organisation. This is a step in the right direction and I am sure the country will watch this experiment with great interest. It is interesting to note here that the suggestion for a re-arrangement of holding came from a Director of Agriculture, Sir Robert Allen, who is now Commissioner of Development in Baroda. With his experience of over 40 years, he has seen the imperative need of changing the land tenure system before technical improvements could be effectively carried out. I know some other Directors of Agriculture have also come to a similar conclusion. I am sure, if their opinions are brought out fully and discussed, government would be compelled to adopt measures for land reforms.

In order to solve the problem of India's poverty it is necessary that: (i) further sub-division and fragmentation of holdings are stopped, (ii) that holdings are reconstituted into an economic size with due regard to the needs of each locality, (iii) that the passing of land to absentee landlords should be stopped and lands that have already so passed should be restored to those cultivators who can put the lands to the best use, and (iv) the intermediary interest between the cultivator and the Government are eliminated. How best this may be done can be found by experiments along right lines. All I should like to say here is that the country must face this problem boldly if it wants to increase the productivity of agriculture, to open the avenues of full employment to the masses and to raise their standard of living.

This leads me to a consideration of an aspect of the matter which has often been overlooked—I mean of psychological and sociological factors on the efficiency of the rural workers and the productivity of agriculture. Agriculture even on a small scale is, like any other profession, a highly specialised operation. It requires a lot of ability, perseverance and resourcefulness. Under modern conditions, if agriculture is to become progressive, the farmer must be capable of following new trends, of understanding new discoveries and methods and of applying them intelligently so as to secure the best results. Further, he must be able to cooperate effectively with his fellow workers, for there are many tasks connected with agricultural operations which he cannot do by himself unaided. Under our social conditions, the outlook of the cultivator is limited and he has neither the will nor the capacity for effective cooperation. Many of our cultivators, whether tenants or labourers, specially those who have been recruited from non-agricultural pursuits, and their number in large, are thus not qualified for the work that they are doing. Differences of caste, creed, religion and sect encourage fissiparous tendencies and give rise to a certain type of complexes among these people making united action difficult.

Even the old cultural influences which sustained life and vigour of the people are dying out without being replaced by something new or fresh. On the top of these general disabilities, there are the further evils of illiteracy, indebtedness, a continuously diminishing size of holdings, a faculty land system and an unsympathetic administrative machinery. It is owing to these conditions that our yields have been falling and even under the stress of war conditions the "Grow More Food" campaign did not bear commensurate results.

The consequence of all this has been chronic mal- and under-nutrition paralysing effort and undermining efficiency. It is as a result of these conditions that a million and a half of the poor of Bengal fell victims to a famine without a murmur, without a protest. That is the position of the common man in this country. This apathy, indifference and lack of enterprise of the rural masses must be removed before food production drive can succeed.

The seriousness of this sociological problem is not often realised. Often the adoption of appropriate action baffles human thought. Old customs, vested interest, religious bigotry present impregnable barrier. Governments in India have always stood aloof from all these questions and the people have also come to resent any interference from the State. Even some of the advanced thinkers have come to shun the problem. Not unoften again does it happen that although there is a right understanding of the problem and the remedies for it, the action taken is only by way of small driblets in isolated areas which bring no significant results. What we really want is a new vision of life which would make the people homogenous and strong so that the standard of life may be continuously raised. Mere education of the routine type would not achieve this. Special measures of social reform will have to be devised.

All these presuppose a new administrative system with a new outlook and ideology. The Indian administrative system has come in for severe criticism from various sides. It has been pointed out that its primary function has been to maintain law and order, and that it is not concerned with the social, moral and cultural development of the people. It has not taken any active interest in the industrial development of the country. Even in the execution of the limited functions it has adopted slow and dilatory methods. The system lacks vigour and promptitude. Certain departments of life are treated as taboo and the entire machinery functions with divided loyalty. There is not sufficient contact and cooperation between the administrators and the people, they do not work together and in harmony.

Recently, the administrative system of one of the Provincial governments came in for review by a special committee. The report of the Bengal Administration Enquiry Committee has brought out the weaknesses of the system from the point of view of the execution of even the elementary functions of Government which were accepted as a basis for review. The administrative system in Bengal failed to carry out efficiently even the limited functions assumed by it. No wonder under such an administration the people's lives and property are insecure in emergencies. The Indian Famine Inquiry Commission has also noted the weakness of Indian administrations in similar terms: "A great responsibility rests on Government, administrators and Government servants of all grades in organising and stimulating the work of nation building." The duties of modern Governments extend far beyond the maintenance of law and order. They include within their compass social and economic development in all its aspect. To build a new India, dynamic rather than static qualities on the part of administrators are required. Governments permeated by the ideas expressed in the aphorisms "safety first" and "*après moile deluge*" are inevitably stagnant and can achieve little or nothing. If the future pace of progress must be that of a motor car rather than of the bullock cart, the administrators will have to approach their duties in a new spirit so that the people as a whole respond and cooperate in the new work of development.

The first essential is that the GoI should take a broader view of their functions as in other countries. The Beveridge Plan of Social Security in UK which aims at providing security to the individual from the cradle to the grave as the English-man would say or from the womb to the tomb as an American would say, and similar plans in US are instances in point. After the last depression, the US Government treated 8,00,000 cases individually for rehabilitation dealing with every aspect of their lives and set them up economically and socially. Later on, nearly 3½ million cases were treated under the Food Stamp Plan so as to provide nutritive foods to people who could not other wise purchase them. In England during war time every farm was examined with a view to finding out its efficiency. For this, nearly 4,000 men were employed. In India, even big famines and epidemics do not impel action on a big enough scale; such things are treated with indifference as natural calamities or decrees of God.

Secondly, India needs today a complete change in the spirit of the administration in order to secure the fullest cooperation of the people in the new tasks of development. The rural masses will have to be organised, cooperative institutions for better farming, for better business and for better living will have to be set up and worked. The cooperative movement which has hitherto worked along narrow lines has to be expanded into a rehabilitation movement and the village *panchayat* has to be used as the basic unit of corporate life for social and administrative purposes. These institutions well organised would cooperate with Government in carrying out the plans agreed upon. The scope for development in this is limitless, but the results will depend upon the wisdom and the energy of the administrators and their success in getting the cooperation of the people to the fullest extent.

In contrast to this, one marvels at the results achieved by the food drive in the UK during the war where the State and the people work together. The arable area increased from 11 to 17 million acres and the net output of foodstuffs measured in terms of calories increased by about 70 per cent. The producers, the traders and the members of the professional classes cooperated among themselves and with the

Government in the task of attaining new targets. The women's land army of 60,000 and the Young Farmer's Club providing 400 camps with 25,000 boys contributed their own share in the production of food. These are achievements any country could be proud of.

Part of our difficulties in this country is the lack of scientific studies of our social and economic problems and where they are investigated, our inability to appreciate them and to take appropriate action. Whatever information or statistics that are available are not properly published, and if they are, in several cases not in a readable form. Even where they are available they are not given out to the public. In the absence of these primary materials, which can only be collected and published by Governments, our social and economic studies are incomplete. These defects must be remedied.

The task we are facing are so many and so complicated that Government departments, however well equipped, cannot solve them by themselves. They will have to seek the cooperation of universities and other research organisations in the country. This is the practice in US and in the UK research is not only necessary measures but also for testing and comparing the results of the various reform measures under different conditions in different areas. The principle of modern State action is that the Government machinery must be able to reach every class and group and even individual in need of some kind of assistance or the other. For such purposes, it is necessary to have complete data which in turn should be continuously analysed by experts in the various fields.

In this connection, I must note one or two developments which you will welcome. The Reserve Bank of India (RBI) has recently opened a new Division of rural economics and finance, cooperation and agricultural marketing and a new staff has been recruited for this purpose. This will be in addition to the Agricultural Credit Department now working as laid down in the RBI Act. This is a welcome development. From my personal knowledge of the Reserve Bank, I can say that it has always taken a broad view of its functions and responsibilities in such matters and I am really happy that it has now taken this further step of expanding its research activities. We have always received full cooperation and help from the Reserve Bank in all our studies.

The second development which I have in mind is the establishment of a section in Agricultural Economics at the School of Economics of the Bombay University. A private donation has made it possible this year to make a beginning. The Government of Bombay, I am glad to say, have agreed to finance two more units consisting of a Reader and a Lecturer in this Department as from next year so that the University will thus have a decent department dealing with Agricultural Economics. I have referred in my previous address to the need for a Chair in Agricultural Economics in every university with a view to placing on sound footing education and research in this subject which is perhaps the most important in a country like ours. I trust that this example of the Government of Bombay will be emulated by other Governments and chairs in Agricultural Economics will be set up in the other universities also.

In conclusion, I must emphasise the need for bold and comprehensive action. Small and spasmodic efforts have been made by Government and non-official bodies in the past. They have produced negligible results and the situation has gone on worsening in the course of years. If the common man is to be put at the centre of things and if he is to live and work as an intelligent and efficient member of society, it is necessary that we lift him out of the morass by large scale and comprehensive action which covers all aspects of his life—physical, mental economic and social. India has in her all capacities for growth and development if we adopt a comprehensive plan. If the high and the low all combine and organise, India will enter upon a new era of prosperity and progress. May I conclude with a significant paragraph on this point from the Report of the Indian Famine Commission?

> In our report we have insisted that India does not lack the material resources necessary for advancement and prosperity, but these can be developed only by the effort of human beings—by the governments and people of India—and success depends on the spirit which sustains the country in the tremendous task with which it is

faced. In the past, apathy and defeatism have been all too prevalent. Doubts have been expressed whether the basic economics and social problems are soluble at all. Poverty and hunger have been too often accepted as art of the nature of things and much of the countryside may almost be described as a rural slum where the hopelessness engendered by slum conditions prevails. Such an attitude of mind, on the part of either rulers or ruled, is incompatible with progress. Without vision and faith in the future little can be achieved.

Indian Journal of Agricultural Economics, Vol. 1, No. 2, October 1946.

7

SIR MANILAL B. NANAVATI

Tenancy and Agrarian Reforms
The Interplay of Factors

The year 1946 will go down in the history of India as an epoch-making year. As a consequence of the last war, momentous changes are coming over the world. As a cumulative result of the rapid march of international events, the victory of the Labour Party in England and India's persistent efforts for achieving freedom extending over three quarters of a century, the British Government has called upon the representatives of the people of India to frame their own constitution. The destiny of India is now in the hands of her people and if all elements therein come together, there is nothing in the world that can stop her onward march towards freedom and prosperity. Stupendous are the problems facing this country—social, economic and administrative—and if the leaders of the nation cooperate, our hopes of building a great India will be materially strengthened. The events of the last few months have no doubt cast a small shadow on the political situation, but considering the complicated character of our social structure such obstacles are perhaps unavoidable. Communal hatred, however, can solve no problem and only when harmony and peace are established, the way will be open for us to reap the fruits of this freedom. No farsighted Indian today can overlook the fact that the several communities and creeds in this country will always have to live together and work together in harmony. Whatever progress or prosperity India hopes to achieve as a free country will be impossible if we endanger national solidarity and peace.

On the economic front, things are not more cheerful. With the occasion of the war, it was hoped, the strain on the economy of the country would be abated. But the expected relief has not yet come.

The inflationary tendencies that were operating during the war still continue. The cost of living indices are steadily mounting up, giving rise to a crop of labour troubles all over the country. The most significant aspect of this unrest is its spread to the salaried middle class which demands an adjustment of its income to the abnormal price level. We are not yet out of the woods and if the present inflationary tendency persists, these troubles, which appear for a time to be over, may reappear perhaps with a greater persistence and disrupt the economy of the country. There is considerable shortage of consumers' goods, especially of the prime necessities of life, food and clothing and building materials. As a result, the common man is silently suffering, whether he be on the farm or in the factory. With our restricted power of production no relief appears to be in sight. The coal production is still restricted and this in turn, affects adversely industrial activity everywhere, particularly the production of cloth. The capital goods necessary to build new industries or to step up old production are nowhere in sight. Unless, therefore, government take some very energetic measures the situation may not improve to any appreciable extent. Opening of pre-war trade channels promises possibilities of large imports of luxury goods in the near future, but such imports, while frittering away our external resources, will not solve our main difficulty which relates to the prime necessities of life.

Of all the problems facing the country today, the most urgent is that of meeting the food requirements of the people. The year started with a deficit of nearly 7 million tonnes as a result of the failure of rains in South India and elsewhere. This difficult situation arose at a time when the dislocation in the food supply caused by the war had hardly readjusted itself. Official procurement, compulsory levy and rationing were working satisfactorily in the deficit areas, both urban and rural. Further, as a precautionary measure, rations of cereals were reduced from 1 lb. to 12 oz. and in some places even to 8. In the earlier part of this year, India's appeal for foodgrains to the Combined Food Board in Washington was drowned in the general clamour for food from all parts of the hungry world. It was only in the latter part of the year, after considerable canvassing and strenuous propaganda, that India could get about 1.7 million tonnes from various countries at the cost of nearly 90 crores of rupees. Some more foodgrains were procured within the country from the surplus provinces for the benefit of deficit areas. Somehow the country has pulled through for the time being from what, otherwise, would have proved a great calamity. But this at best is a negative satisfaction. Malnutrition persists both among the lower and the middle classes who could not buy enough food at such inflationary prices. It appears that this food deficit has today developed into a chronic state. If during the last 70 years, the per capita consumption fell from 1½ lbs. to 1lb, it has now further gone down to 12 oz. With the increase in the population of the country by about 5 million a year, this shortage would become more and more intensified every decade. The Grow More Food campaign launched during the war and now being continued has brought in an increase of 2 million tonnes of foodgrains and has cost the country nearly 8 crores of rupees. It has brought about 9 million acres of lands under food cultivation, i.e., transferred 5.5 million acres from jute and cotton to food crops and increased the double cropped area by about 3 million acres of land and the rest from uncultivated waste. But compared with the 27 per cent increase in production in the United States of America (US) and 66 per cent in the United Kingdom (UK) during the war, the increase of 3.30 per cent in the food supply of India is insignificant, all the more so in a country where nearly three-fourths of the population is engaged in cultivation.

It is no small compliment to the government now in power that but for their zealous efforts, the country would have been in the grip of a famine at this time. But the problem of chronic food shortage has not yet been permanently solved. The Grow More Food campaign of the future has not only to raise the required quantity of food but also the required type and variety so as to make a "balanced diet" possible for all. As much attention, thus, will have to be paid to the quantity as to the nutritive quality of the food crops to be raised from the soil. The problem does not end here. The purchasing power of the people will have to be raised also so that they may be able to command food which both in quality and quantity would come up to the standards of nutrition. This problem, therefore, has to be tackled from several sides. Things cannot be allowed to continue as before when government have assumed the direct responsibility to see the country adequately fed as a matter of their fundamental duty. The entire agricultural policy will have to be remodelled in such a way that land produces to its highest capacity and the consumers have the necessary purchasing power to buy all the necessaries of life. The development must be both vertical, giving the maximum production, and horizontal so that each one may get what is necessary for his primary needs. It is a tremendous task, especially when we have to deal with 28 million landless cultivators and 24 million farm labourers who have to be helped not merely to tide over the present difficulty, but who also need to be given a secure means of livelihoods.

Already, one hears of vast schemes for agricultural development receiving consideration almost in every province. Large sums are to be spent not only to make two blades of grass grow where one grew before, but also to give the people amenities of civilised life. The Bombay Post War Reconstruction Plan, for example, intends "to raise the economic level of the countryside, increase the well-being of its inhabitants and generally to improve the Province in every aspect of social, economic and cultural life, both urban and rural." The execution of the plan is to be divided into 3 five-year periods. Each aspect of the country's economy, agriculture, veterinary, forests, industries, cooperation, education, medical relief, public health, labour welfare, communications, irrigation, electricity, etc., has a number of well-defined schemes for its development. Thus, 'agriculture' has no less than 29 schemes requiring a capital expenditure of about ₹ 8 crore and a recurring cost of ₹ 3 crore. Similarly, there are 22 schemes under 'irrigation' demanding a total expenditure of ₹ 14-19 crore. Since the advent of Popular Ministry, these plans are being further extended in scope and the funds allotted previously have been considerably increased.

Development plans in other provinces envisage action in similar directions and would involve expenditure on an equally liberal scale. The vast schemes of irrigation in South India costing nearly 160 crores for irrigating 6 million acres so as to produce 6 million tonnes of rice will make the province self-supporting in the matter of food. In Punjab, the Nangal and Bhakra hydro-electric projects are expected to bring more lands into cultivation. Smaller projects are being conceived in Bombay and other provinces. The Central government is projecting the Damodar and Koshi flood control schemes. These projects when completed will add materially to the prosperity of Indian agriculture besides generating large power which may be helpful both to industries and agriculture. All this is very welcome news indeed.

At the same time, schemes for the reform of land tenure are being prepared in most of the provinces. In Bengal, Bihar, the United Provinces (U.P.), the Central Provinces and Madras where the *zamindari* system prevails—whether permanently assessed or temporarily—they have decided to do away with the *zamindari* system. In the U.P., they have appointed a special committee to work out the details for such a scheme. In Bombay the old Tenancy Act has been revised. Though it does not go the full length, it restricts the share of the landlord to one-third of the produce. A new bill for the consolidation of holdings and the prevention of subdivision is planned and published for inviting public opinion. This bill, when passed, will establish a new landmark in agrarian legislation. In the Central Department of Agriculture, the three sub-committees appointed by the Policy Committee on Agriculture have submitted their reports on the stabilisation of prices, agricultural finance and cooperative planning.

But all these plans, in my opinion, go along the same old beaten tracks of agricultural policy of the post-depression years. Better seed, better manure, increased water supply, consolidation of holdings, better implements and such other technical improvements along with the schemes for debts adjustment and land reforms are the major planks in these plans of development and reconstruction. The schemes at last are complete— if not perfect—on paper. But may one ask why some extremely simple measures like distribution of better seed and better manure have for all these years not taken root ? If I may venture to suggest, the flaw is in taking purely an objective view of schemes of reform. A plan may be perfect, but many other factors have to be coordinated, if it is to take root in the soil. Therefore, the other and perhaps the more intractable aspect of the problem is the subjective element. A few years back, Sir John Russel complained about the gap between the farm and the experiment station. Today, there is an equal danger of such a gap between the farm and the secretariat.

The point is, before any scheme of reform is launched, ground has to be prepared for its reception by the people. This preparatory work would necessitate action in three directions: firstly, we would have to improve the human factor in agriculture in order to raise the agricultural efficiency which in this country is very low. This would necessitate, on the one hand, absorption of the surplus rural population into urban vocations such as industry, trade, professions and services and, on the other, adoption of measures for weeding out the unwanted population in agriculture. Secondly, before such constructive measures are planned, ways and means would have to be devised to ensure intensive study of the rural problems by Central and Provincial governments, by universities and private institutions so that every aspect of rural life is properly investigated into and diagnosed and appropriate remedial measures are worked out. Thirdly, we would have to develop such agencies as would ensure the cooperation of the people themselves in implementing these various schemes. Only thus can we remove the inertia of the individual which has stood in the way of his taking advantage of the little that has been offered to him for his own benefit. This, in my view, is a fundamental step in the programme for rural reconstruction.

It is generally recognised today that one of the surest means to make the tiller get the utmost from the soil is to make him the owner of the land he cultivates. I am glad to note that some provinces are already taking steps in this direction—it is only a beginning. The next and equally important step is to remove from land the inefficient classes of agriculturists who have neither the necessary equipment not the aptitude to pursue this occupation profitably. The magnitude of the loss to our agriculture owing to the presence of these classes is seldom realised, but I must say that it is greater than that caused by absentee landlords. The latter assumes no direct responsibility for the use of land whereas the former assume this responsibility but fail to act up to it. To that extent, they not only depress the level of national income but not unoften, through misuse, cause positive damage to land which is the foundation of our nation's wealth. The Census Returns, for instance, reveal that, of the leather-workers, barbers, potters, weavers, tailors, scavengers and other rural workers, only 27 per cent are today in their traditional occupations. Of those who have given up their traditional occupation, no less than 64 per cent have taken to agriculture and allied pursuits. In this age of specialisation, it is difficult to plead for their retention in our scheme of agriculture. Indeed, agriculture requires as much traditional skill and aptitude as any other highly specialised calling. Apart from skill and aptitude the patience and industry which this occupation calls for can be found only on those classes which have for generations laboured to acquire these traits. However dire the compelling circumstances may be, we can never expect these artisans to be efficient farmers. This is a significant fact which we must comprehend fully before we plan for agricultural development. Without the removal of such classes, we cannot ensure efficient use of the soil nor effective spending of money on land and other improvements.

I may give here a rough indication of the issues involved in such a plan. According to the Census of 1931, India had over 75 million persons engaged in cultivation. This vast number comprised 3 million absentee landlords, a quarter million estate agents, clerks and private farm managers, 20 million cultivating owners, 28 million tenants and 24 million labourers. The removal of the parasitic interests numbering 3 million and a quarter would be relatively a simple task and measures to this end are already in the making. But in weeding out the inefficient workers from the remaining classes on land, we will have to proceed more cautiously. In this case, we have to bear in mind that the influx of outsiders into agriculture has been larger in the lower strata. Accordingly, we will have to weed out a larger number of tenants than that of cultivating owners, and a still larger number of labourers. In other words, the process of diverting the surplus population on land to non-agricultural pursuits cannot proceed in a haphazard manner. Those classes which are responsible for depressing the level of agricultural efficiency must be helped to quit the field first, so as to ensure that only those who have the traditional aptitude and the necessary equipment are retained on land.

With regard to the second problem of reliable data on which plans can be based, the state of affairs in India is pathetic. Economic planning, to be successful, has to be based on facts and figures regarding the country's national resources, manpower, finance and such other factors which condition the nature and scope of the plan. It also requires a complete knowledge of the conditions of the people and their requirements. In the absence of such basic data, any attempt at planning may be well compared to a blind man's bluff. But in India statistical investigation and economic research have been relegated to such an unimportant place that one wonders how much of the present talk of planning especially in rural life has any basis in reality at all. The Report of the Royal Commission on Agriculture drew pointed attention to various shortcomings in this respect and made valuable suggestions for removing them. Six years later, Dr Bowley and Prof Robertson still described the statistical compilation in many cases as "unnecessarily diffuse, gravely inexact, incomplete or misleading; while in many important fields, general information is almost completely absent." The agricultural statistics, according to them, were quite insufficient to determine whether or not food is increasing in proportion to population. Recently the Report of the Foodgrains Policy Committee, Dr Burn's Report on Technological Possibilities of Agricultural Development and Government of India's Manual on Food Statistics of India published as late as 1946 have further deplored the unsatisfactory position with regard to agricultural statistics.

It is interesting to know what the official itself is regarding these statistics. According to the Manual on Food Statistics 1946, data regarding areas and yield are neither sufficiently comprehensive not reliable and as such leave much to be desired. According to Dr Burns, they are extremely unsatisfactory. The trade statistics "arise out of compilation made in the course of administration of laws... and, from compilations made by Departments of Government for their own use" and, consequently "cover only a small fraction of the field of exchange of good." We cannot have a correct picture of the several agricultural strata because "the occupational statistics are too vague and imperfect to afford the detailed information required for public and administrative purposes." Our vital statistics, it is complained, "lead to quite unjustifiable conclusions." Above all, we have few statistics which would help us to assess the poverty of our rural people, still less to find out the cause of this poverty. The size of holdings, the size of the unit cultivated by farmers and their equipment, indebtedness, farm costs and incomes and such other subjects bearing on rural welfare are at present quite outside the purview of our official statistical compilations. While report after report has gone on lamenting the dearth of these valuable facts and figures, little has been done to improve the statistical organisation to meet the requirements of practical planning.

Government in every Province have now under their consideration various plans for raising the standard of living of the people and for reorganising agriculture on sound lines. This, however, would not be possible unless they have at command such data about our agriculture as would enable them to know in precise and quantitative terms the conditions to which the proposed measure is to be applied. There is no organisation at present, either at the Centre or in the Provinces which is regularly carrying on economic research except of course, in the Punjab and to a very small extent in Bengal where Boards of Economic Inquiry are doing investigational work. The activities of the Imperial Council of Agricultural Research are too narrowly conceived and confined to technical research. So also are the activities of the Departments of Agriculture in the Provinces. Two years ago, we were told by the Member for Planning and Development that a Central Research Department would be projected at the Centre but nothing has come out of this assurance as yet. A similar move was contemplated by the Central Department of Agriculture, but it now appears that it has been shelved. Further postponement of this question is bound to handicap the governments in giving effect to their various plans for economic development.

It is worthwhile pointing out in this connection how highly economic investigations and research are valued in the US. It is impossible to give here a detailed account of the innumerable subjects which agricultural statistics in that country cover. But it would not be an exaggeration to say that there is hardly any aspect of agriculture on which a research worker in America has to despair for want of data. This would be evident even if we confine our attention only to the Agricultural Census of 1940. This Census brought out 17 volumes of no less than 20,000 closely printed pages—a wealth of data throwing light on the minutest problems of agriculture. How intensive these investigations are would be obvious from a mere glance at the list of contents in some of these Reports. Volume I which runs into 6 parts gives products and crops. Volume II which runs into three parts classifies farms according to major sources of income and total value of products. Volume III gives detailed information relating to: (i) farms and farm property, (ii) size of farms, (iii) colour, and race of farm operator, (iv) farm mortgages and farm taxes, (v) work off farm, age and years on farms, (vi) cooperation, labour, expenditures, machinery, facilities and residence, (vii) livestock and livestock products, (viii) field crops and vegetables, (ix) fruits and nuts and horticultural specialties, and (x) value of farm products. In addition, a number of studies and monographs are published from time to time covering such important subjects as irrigation, drainage, dairy products, major crops and abandoned farms. The last study, for instance, gives, by States and by countries the number and acreage of idle farms. And yet the Hand-book of Agriculture expresses a feeling of dissatisfaction when it says that "with the urgent need for exact basic information on everything pertaining to food production even the 20,000 pages statistics... are not enough".

The value of these statistics can be seen from the numerous studies made by the department of agriculture and the universities and other research institutions in the State. The entire programme of the rehabilitation department which covers practically every side of the rural life is based upon the results of these statistical surveys. For example, they have been able to formulate the normal standard of life for rural population based on American experiences and have classified farms and farmers according to whether they are above or below the poverty line. For the latter groups, they have made out elaborate plans for rehabilitating them. Those plans comprise supply of food to meet the deficit in the family budget, better health and sanitation facilities, improvement of housing condition, provision of medical aid, schools community centres, organisation of the boys and girls, clubs for useful community services, improvements in domestic science through home management section. In each section, appropriate plans are made out for granting loans or subsidies or giving technical advice and guidance. Nothing is left to chance and the results of state assistance are watched and noted for further guidance.

Similar intensive surveys covering every holding were recently conducted in England and Wales. The survey in Scotland was comparatively limited in scope, as it focused attention on the land itself where as in England and Wales the emphasis was as much on the farmer as on the land. In the latter case, considerably more attention was paid to roads, power, fences, water-supply, convenience of lay-out, etc., which have no direct bearing on the inherence productivity of land and yet are of great help to the farmer. The survey has thus brought out a permanent record of conditions of practically every holding in the country. The data collected are tabulated and analysed under the following main headings: classification of occupiers by economic type and of agricultural holdings type of farming and by size; tenure, rent and length of occupation; convenience of layout and situation of holdings; condition of farm cottages; type and natural fertility of soil; managerial efficiency of occupier and the condition of cultivated land; farm water supply and the use of electricity.

This farm to farm survey was conducted by the technical staff of the Country Agricultural Executive Committees, while scrutiny and collation of the data were carried out by the Advisory Economists attached to the 11 Provincial Agricultural Advisory Centres in England and Wales, located at either universities or agricultural colleges. The survey was undertaken on realising that without detailed information regarding the potentialities and deficiencies of each farm it would not be possible to develop the potentialities and to correct the deficiencies of the agricultural system as a whole. It is on the basis of these data that systematic efforts would be made to correct deficiencies. The point to be noted is that in those countries every individual and every farm receives adequate attention with a view to reaching highest production and permanently raising the standard of life of the worker. In contrast to this, surveys in our country are based on estimates and vague generalisations accumulating authenticity as they go down from one blue book to another.

The need for organised economic statistical research in this country, therefore, cannot be too strongly stressed. The lead, must of course, come from the Centre. The problem is so vital that the Government should appoint a Committee of experts to go into this question. The terms of reference, however, will have to be far broader than those placed before the Bowley-Robertson Committee. They should include more important question such as ways and means for establishing Government agencies and semi-official bodies for economic and statistical investigation; broadening the scope of the present statistical compilations such as the Census and the food statistics of India; institution of a decennial agricultural census in India on comprehensive lines as that in America; endowment and other State assistance to universities and private research organisations which undertake to interpret available statistics and to conduct special studies bearing on the economic well-being of the people.

In the light of the recommendations of this Committee, Government should suitably expand the activities of the Imperial Council of Agricultural Research and the Central and the Provincial Departments of Agriculture. Agricultural research in this country is understood in a very limited sense and, consequently, the activities of these departments are confined to improvement of soil, manure, seed, farming technique, and implements. These activities have, no doubt, their own utility but they would not yield any appreciable results unless they are conceived as a part of a larger and more comprehensive programme of agricultural reconstruction. It is for this purpose that a close coordination between this technological research and economic research into the conditions of the agricultural classes becomes imperative.

These departments must therefore, have attached to them a section which would mainly concern itself with investigations into the economic conditions of the farming class. They should take up for their study the farmer not merely as an entrepreneur but also as a member of the society whose welfare is governed also by various factors other than the farm. Education, health, leisure, recreation, level of rents and wages, cost of marketing and trend of prices have as much bearing on his wealth and welfare as the technological improvements. In fact, unless the former are properly attended to, the latter can make little headway. The studies of these departments must be not only analytical but also offer constructive suggestions for the uplift of the rural class. As the Russel Report rightly points out, the problem of improving Indian agriculture is in fact the problem of improving Indian village life and this must be studied as a whole.

Similar work must be undertaken in the Provision by semi-official agencies such as Boards of Economic Inquiry and the universities. It is nearly two decades since the Royal Commission on Agriculture recommended the establishment of semi-official research organisation on the model of the Punjab Board of Economic Inquiry in every province but no province, except Bengal, has given effect to this suggestion as yet. Even in Bengal, no substantial work is reported to have been done by the Board. This matter demands immediate consideration of all provincial governments. Similarly, every university must maintain a research section in agricultural economics and initiate studies of rural problems. The contribution of these colleges to economic research has been very meagre in the past. If the universities come forward and cooperate in this scheme, they will have not only made valuable additions to useful knowledge but also helped the younger generation to equip themselves for more responsible duties in executing the various plans for national development. I am glad to note that the Bombay University has opened a section of agricultural economics in the school of Economics and Sociology with the help of the Bombay government. The Bombay University has thus set an example which other Universities should follow without delay. I do hope that this committee would consider this suggestion and formulate a workable plan for the study of agricultural economics, specially in their Post-Graduate courses.

It must also be noted that as important as this fact finding work is that of checking the accuracy of officially collected statistics, so is the interpretation of available data, and undertaking special studies bearing on the various aspects of the farmers' welfare. This, in my opinion, can be most efficiently done by non-official institutions which alone can view the problems unbiased and are free to criticise government policy in any of the measures undertaken by them. Such institutions are, at present, too few in India and their growth must be encouraged by the government through suitable grants, at least until they become self-supporting. From our own experience we know such institutions do not get the same official cooperation in India as similar bodies in other countries receive. Costly official reports have to be paid for and, even then, can be procured after much delay. Requests for unpublished information for our research are attended to, if at all, not before some months have elapsed. In fact, we have been able to get innumerable publications of the government and universities in the US without any trouble or cost and more promptly than from any Provincial government in this country. I do hope that this official attitude will change for the better under the influence of the popular ministries and that such private institutions will get from government every possible encouragement and cooperation in their activities.

Finally, to organise investigational work on sound lines, we would have to build up a Department at the Centre to guide, supervise and coordinate the work of Provincial governments. The trend of events in India, however, has been somewhat in the opposite direction. Soon after the cessation of the war, the Planning Department of the Central government was abolished and its work was distributed among other departments. This had the effect of putting into the background the organisation of a Central Economic Research Council for the whole country. India is a vast territory with diverse forms of administration in British India and in the Indian States, with people at different stages of civilisation and culture and widely varying economic and social conditions. Unless these research and development activities are suitably coordinated, it would not be possible to formulate policies for the uniform development of these diverse units. The organisation of the United States Department of Agriculture side by side with similar but smaller departments in the member-states lends support to this point of view. A strong central department for research and planning, therefore, is an imperative necessity for the country's uniform development.

While on this subject, I would like to sound a note of warning to the framers of our new Constitution. We shall be shirking our duty if institutions like ours do not emphasise the All-India character of the task of planning and research. If one had any doubt on this question, our experience of war-year should have dispelled it. Nature has conceived India as an economic unit and its disruption will be suicidal. The problem of our economic reconstruction and development is so vast and complicated that to leave it to provincial units and the tiny Indian States will be a deliberate act of sabotage. They will have neither the talent nor the resources to cope with the problem. At a time when the shortage of trained personnel is so acute, it will be suicidal to decentralise and duplicate research and planning. I earnestly hope that no political consideration will drive us into adopting a scheme of constitution which will be detrimental to our economic progress and thus delay the redemption of our masses from poverty and degradation.

The next step is to make an effective use of the statistical and economical research on which so much labour and money will be spent. To begin with, the agricultural census and socio-economic surveys will give us a reasonably accurate idea about the extent to which the cultivating classes include non-agricultarists. From this we will know how much of the population has to be pumped out to raise the efficiency of agriculture and increase the share of each in the total agricultural income of the country. They will also indicate the extent to which sub-division and fragmentation, indebtedness, transfer of land, etc., have been keeping farm incomes low and will thus facilitate drawing up a programme of land reforms. Studies relating to the economic condition of the farmers will point out how their resources may be strengthened so as to make them more efficient in their business.

In the absence of such complete statistics and comprehensive economic studies, the adoption of suitable measures to deal with the problem in proper sequence becomes impossible. For the same reason, we have no alternative but to make spasmodic efforts to deal with each problem separately as if it were an independent phenomenon by itself. This faulty approach explains the failure of improvement measures in the past. Take the case of agricultural indebtedness, for instance. It was hardly realised that this indebtedness is a symptom of low economic and social conditions, not its cause. No efforts were made to remove the root cause of these debts. Instead, attempts were made—not always with success—to reduce the rate of interest, to scale down debts and to save the debtor from being exploited by the creditor. In fact, there is abundant proof to show that these measures have restricted rural credit and made matters more difficult for the farmer. The economic and social forces which drive the farmer to the money-lender still continue to operate. The same may be said of the Tenancy Act. Rents have been restricted by law but so long as nothing is done to prevent the ranks of the tenants from growing, the landlord would have no difficulty in continuing to take advantage of the keen competition between the tenants. Consequently, the law will be circumvented in various ways. What these ways are I do not want to suggest here, but, I am sure, our landlords know them all very well. I can cite many such instances.

The defects in such a compartmental view towards national problems were realised by the western countries after the economic depression and during the last war. They have, therefore, evolved a new technique of approach to these problems which is based on two different lines of action. The first consists of determining the general standard which the people should attain with regard to food, clothing, shelter, and social and cultural life. Thereafter, measures are taken for helping them to attain these standards, both through developments on the farm and through the provision of certain facilities outside the farm; the latter include amenities of modern life such as good roads, electricity, medical and social services and centres of cultural life.

The second line of action consists of measures for dealing with the submerged class which cannot be approached en bloc, but has to be treated individually by an effective aid to make each self-supporting so that, in course of time, this class as a whole is placed on the road to prosperity. As we already saw, both the UK and the US have made such an approach to the problem of rehabilitating people in the low income groups.

Now I come to the third point regarding planning for reconstruction, *viz.*, the agency for implementing the plan and the enlisting of cooperation of the rural public in this effort. In our book, "The Indian Rural Problem", we examined the various types of organisations working in the country—both official and non-official—and how far they have been successful in influencing the village life and we had come to the conclusion that no drive for agricultural or rural development is possible except through the prime agencies of rural reform, *viz.*, the school, the *panchayat* and the multi-purpose cooperative society. These agencies have the capacity to revitalise the warp and woof of rural life—economic, political and social—and to pave the way for the smooth operation of other reforms of a far reaching character.

The fundamental issue in the programme of rural development is how to enlist the cooperation of the people themselves in implementing this plan. The Grow More Food Campaign, as has been already pointed out, has not yielded appreciable results in this country mainly for the reason that the question of enlisting peoples' cooperation did not receive due consideration from the government. For instance, the reduction of area under cash crops by compulsion to increase that under foodgrains was much disliked by the farmer over a larger part of the country. Mere placing of a benevolent law on the Statute Book, however imperative it may be in the interest of the public, produces no effect on the villager's mind unless he is fully convinced of its necessity and is himself made to participate in the working of the scheme. In all the schemes of technical improvements in agriculture so far tried by the departments of government for the introduction of improved seeds, manures and implements, the results have not been commensurate with the time and energy devoted to them. It would be, therefore, necessary to form all over the country organisations of village people which will, on the one hand, interpret to the people the provisions of the various improvement measures passed or suggested by the government, and, on the other, keep the government constantly informed of the villagers' requirements and grievances for necessary consideration and redress. Such an institution may represent either a single large village or a group of smaller villages. It is imperative, therefore, to create an effective agency in every village which will enlist the cooperation of all elements in all measures—social, economic or administrative—intended for their development.

We had an occasion recently to study the working of some organisations or institutions working in rural areas and we found that their activities had influenced only a very small number of people and that too to a very limited extent. There was no sign of these bodies having achieved any lasting results. From this we came to the conclusion that such an organisation to be successful in its efforts should fulfil the following requirements:

(a) There should as far as possible be a single organisation covering a village or a group of small villages very near to each other.
(b) The organisation must try to get in it most of the population.
(c) The organisation must be a simple one and must be managed by the people themselves.
(d) It should be multi-purpose in scope.
(e) It should be supported and supervised by another institution at a higher level working to the same end and assisted by the Departments of State concerned with the social and economic welfare of the people. These should be located in *tahsil* or *taluka* headquarters.
(f) It should be equipped to render every assistance to the common man—financial, technical, administrative—to each man according to his need.
(g) It should be manned by specially trained men who can win the confidence and cooperation of the people.

(h) It should work on a programme set according to the needs of the people and their capacity to absorb new ideas and adopt new methods.

(i) That these village and *taluka* organisations should be supported by specialised government departments and should work under their guidance.

(j) That these organisations should be adequately financed by the State for the first few years till, in course of time they become self-supporting.

The success of all government measures will depend on how soon these organisations are created and how best they are manned by very highly qualified personnel. The ordinary man in the village is not accustomed to taking advice from too many agencies. He needs a few trusted men to guide him; whom he know well and can follow with confidence.

Our main problem is to approach the largest number of men in the shortest possible time. It is then that our programme will be effective and create an impression on the economy of the country. In this effort, government should try to enlist the services of the largest number of men from all areas—rural and urban.

In short, the immediate problem before the country is that of increasing its food production and placing it within the easy reach of all strata of the people. This would be possible only when the efficiency of the agriculturists is raised by weeding out the inefficient classes which have crowded in agriculture under the stress of circumstances. The drawing up of the actual plan for this purpose would not have much basis in reality, unless we improve our statistical machinery and extend the scope of economic research so as to touch not merely the business of farming but to cover the whole field of rural life. Lastly, agencies must be set up in the village to prepare the ground for these reforms to take root and to bring the government and the public in closer contact for mutual understanding of the intention of the one and the difficulties of the other so as to develop in both a spirit of mutual faith and active cooperation.

Indian Journal of Agricultural Economics, Vol. 2, No. 2, August 1947.

8

V.L. MEHTA

Achieving Self-Sufficiency: The Next Step

AGRICULTURE is the basis of our national economic life; and in planning for its progress and prosperity let us plan keeping our gaze fixed, always, on the high ideals. Although I happen to be a member of a government which is in charge of the administration of a Province, I feel that my approach to the subject of agricultural economics is still that of a student.

Hardly any of the manifold aspects of the science of economics has such a fascinating interest for the student of economics as that branch which relates to agriculture and concerns itself with the conditions of work and life of the toiler in the field. In scarcely any other department of economic studies, has the student so much to deal with the forces of nature and with the aptitude and intelligence of human beings. Even in countries where agriculture is a highly mechanised enterprise, in the ultimate resort, results are largely dependent both on nature and on the skill and resourcefulness of the tiller of the soil. It is the interplay of these factors that invests agricultural economics with special importance, especially in a country such as ours with its preponderating agricultural population.

Unfolding his plans for national reconstruction after the Chinese Revolution, Dr Sun-Yat-Sen, the great Chinese patriot, enunciated the three principles which were to provide the foundation of a resurrected China—of nationalism, of democracy, of livelihood. In India, too, we have to built on a similar basis, the inculcation of the spirit of Indian nationhood, the acceptance of self-governance and equality in the field of politics and economics, and the raising of the standard of living of the people—enabling them to get the wherewithal to feed, clothe and house themselves adequately, decently and comfortably. In as much as it is agriculture that provides us not only with food but with various other primary products that are the raw materials for several of our essential industries, agriculture naturally occupies a predominant place in our national economy.

The foremost task before the country today—national leaders, administrators and economists tell us—is to increase agricultural and industrial production, to counter the evil effects of inflation and to rid the country of the shortages of all types which appear to have become a chronic feature of our economic life. The decision that has been taken recently to reduce the import of foodgrains from abroad emphasises the need for promoting self-sufficiency in the matter of the nation's food supply.

Coming as I do from a Province in which failure of the monsoon are a frequent occurrence, it is but natural for me to accord the place of price in any plan for increased production to the provision and extension of irrigation facilities. As a long term measure, the construction of canals provides the best insurance against scarcities and famines. But our need is urgent today; and Governments have, therefore, to decide whether for meeting the emergency, we should not encourage, forthwith, the construction of a chain of wells and the promotion of a number of minor irrigation schemes such as the excavation of tanks, the erection of *bunds*, *bandharas* and embankments, life irrigation from rivers, gully plugging and the like. The initial outlay of funds will, of course, be lower; but that, to my mind, is comparatively a minor consideration. The time factor is the primary consideration, and not less in importance is the fact that in all major irrigation projects the immediate result will be the throwing out of cultivation of large areas of lands required for construction of the works. Moreover, the present system of administering irrigation works leads to considerable wastage of water and occasionally to water logging. It is, besides, expensive, being highly centralised in its administration. To a certain extent, a premium is put on the capitalist farmer growing cash crops; while under a system of well irrigation or minor irrigation works, the agriculturist feels a sense of social responsibility and does not abandon the traditional forms of crop planning. Again, while land values rise and the large-scale farmer derives considerable benefit at the cost of the general tax-payer, the small-scale farmer tends to get weeded out. Lastly, oftener than not, food production gives place to a state of commercialised farming. Not that, on these grounds, I disfavour large irrigation works expenditure on which must, surely, receive the highest priority. I only urge that with the projected extension of major irrigation works, it is necessary to ensure that these evils that have ordinarily so far followed in their train are averted.

There are not a few in India who cannot conceive of a progressive agriculture except in terms of mechanised farming. They base this view mainly on the experience of foreign lands. Undoubtedly, the more elaborate the machines the fewer the plough cattle and men on the land. Where there is a temporary shortage of cattle, it may be expedient for the time being to resort to the machine, but the cattle have, so far as I can see, a value in rural economy in India second only to the labour and skill of the individual farmer. After a proper appreciation of agricultural economy we have still to be convinced that the power-machine is more efficient than the plough and the bullock. Let us not forget, too, that the power-machine consumes petrol which is not a product of our country and that no residues are returned to the land as is the case with plough cattle. Secondly, with our growing population there is not dearth, as is the case in other countries, of men and women to work on the land. Moreover, the primary concern with us, at the present moment, should be not to secure the maximum return for the minimum effort, but as Prof Tawney says of China "to distribute limited and unexpanding resources among the largest possible number of human beings." Only that which is best in the interest of the land and those serving it is in the best interest of the nation.

Another thesis which was propounded by some is that fertilisers are the salvation of Indian agriculture. Undoubtedly, again, the output per acre is lower in India than in any other country in the world. But those who build their arguments on this as the basis are apt to overlook the several factors that are responsible for the comparatively low yields we get in India. When we think in terms of output let us judge not merely in terms of quantity but also of quality. Recently, we have had imported into India foodgrains produced in countries where farming is both mechanised and scientific. How does the grain compare in quality with the grain we produce in our country? And what does it cost to produce in those countries the kind of foodgrains we produce in our country without the aid of artificial fertilisers? Moreover, agricultural experts have uttered the warning that with the introduction of artificials there has been a continuous increase in disease, both in crops and in livestock. Hence, in the years to come, observes Sir Albert Howard, "Chemical manures will be considered as one of the greatest follies of the industrial epoch." "When the finance of crop production is considered together with that of the various social services which are needed to repair the consequences of an unsound agriculture," he adds, "and when it is borne in mind that our greatest possession is a healthy virile population, the cheapness of artificial manures disappears altogether".

For ages, lands which have been brought under the plough in India and in China have preserved their fertility in a manner which elicited the envy of the West. We maintained, first, a balance between livestock and crops. The dung and urine of the livestock supplies even today, though to a reduced extent in India, the humus deemed essential for preserving the fertility of the soil; while a system of traditional rotation through the growing of leguminous crops enriches it. When the machine replaces the animal, the dung and urine disappear. Besides, the machine either when it works or is scrapped, contributes nothing to fertilise it. No wonder soil fertility declines. It can be maintained only by feeding it from various organic sources such as are locally available. Among those, farm yard manure, leaf mould and the waste of the urban population represent valuable ingredients. The preparation of composts and the proper collection or human and animal waste should, as with us in Bombay, form part of a programme for enriching the soil, a programme which acquires special importance for those who subscribe to the views set forth above about the limitations and dangers of resorting to the use of fertilisers, except on the advice of an agronomist.

The small-scale of the operations of the individual cultivator and the small size of his resources represent, Prof D.R. Gadgil remarks, the central feature of Indian rural economy. This should not, however, lead us to the superficial conclusion that no improvement in agriculture is possible unless we have their large-scale estate farming or state farms. The optimum size of a farm will differ from province to province, and from tract to tract. It may depend on the type of cultivation pursued, the kind of staple crop raised, the availability or otherwise of irrigation, the practicability or otherwise of mixed farming and similar other factors. But having determined the optimum size, how is the state to ensure that it becomes the unit of production? The average size is today much smaller all over India; besides, what a cultivator normally holds lands that are fragmented and scattered. Apart from other remedies, the first need is, therefore, to check further sub-division and fragmentation and then to promote consolidation voluntarily and on a cooperative basis if possible; if not under the provisions of a legislative enactment. In Bombay, we have recently enacted legislation for this purpose, in the enforcement of which we shall certainly try to benefit by the experiences of Provinces like Punjab and the Central Provinces and Berar and the advice of bodies like the Indian Society of Agricultural Economics.

That by itself will, however, not be enough, because, even the size of the consolidated individual holding will be smaller than that of an economic holding. Anxious as we are to secure not so much increased production in the mass as increased prosperity for the tiller of the soil, we have to devise some means of promoting cultivation through optimum units of production. The potentialities of the introduction of mixed farming, the extension of irrigation facilities, the development of local industries and secondary occupations allied to agriculture and animal husbandry should all be explored to provide for the absorption in remunerative employment of as many persons as possible out of those who may be displaced from the land. As Mr Tarlok Singh suggests, this has to be properly planned and arranged for in advance before any large-scale attempt is made in the direction of enforcing standards of economic efficiency in our agricultural life.

Presiding, recently over a Conference of Revenue Ministers convened by the Central Government, the Hon'ble Dr Rajendra Prasad observed:

> The rights of the intermediaries should be acquired by paying equitable compensation taking care that the incidence of the State acquisition on the intermediaries is followed by a programme of agrarian reforms which would be reasonably uniform in all Provinces. The agrarian reforms should aim at improving the lot of the tenant through various measures, including the fixing of the price of holdings and consequent redistribution of land on a very big scale and introduction of cooperative farming as well as increasing the efficiency and yield of agriculture.

All over the country today there is, obviously, the ardent desire to replace landlordism by a system under which the individual who labours commands the product of his labour. Will a system of peasant proprietorship as we have ordinary known it so far in the *ryotwari* provinces provide a good substitute? The legislation that has been enacted and is on the anvil in various Provinces will, undoubtedly, aid in the process; but it has to be seen to what extent it loads to a progressive agriculture. In his Minute of Dissent to the Report of the Famine Inquiry Commission, Sir Manilal Nanavati defines the following as the objectives of a policy of agrarian reform in India:

(a) All lands should be owned by government, i.e., the intermediate feudal interest should be removed.
(b) Occupancy rights should be given to cultivators with certain reservations such as restrictions as to sub-division of the plot, transfer of land, etc.
(c) These holdings should be economic units and should be as far as possible in one block.
(d) The land should be cultivated by the man who owns it.
(e) The cultivator should live on the farm or as near it as possible.
(f) The land should not be used as security for unproductive purposes.
(g) The holdings should be taxed on a graduated scale.

Each one of these objectives may not be acceptable to every agrarian reformer; taken together they provide the touchstone for judging the effectiveness of a sound agrarian policy. Any large-scale attempt in this direction, however, postulates not only action on the part of the state but the organisation of the farmer so that not only the individuality but the community in general stands to gain through increased production and well-diffused prosperity.

What form should such organisation take? Farming conducted by the state has to be ruled out, for it has hardly been successful in any part of the world. A centralised agriculture is, to my mind, a contradiction in terms. Husbandry calls for leving management and for that reason it may be urged that collectivisation of the type which has developed in Russia may not prove suitable for the purpose in view. Palestine with its cooperative farming presents a different picture. Describing this, a recent writer states:

> Their achievements are certainly impressive, and much more extensive than most people outside Palestine imagine. The Jewish collective and cooperative settlements are responsible for more than a quarter of the agricultural production of Palestine. They have turned many of its deserts into fertile farms by means of irrigation and have contributed greatly to agricultural research and development. They have shared their knowledge and facilities with their Arab neighbours, thus raising the level of Arab farming and providing a basis for Jewish Arab friendship which is in sharp contrast with the bitter conflict that arises elsewhere in Palestine. They have provided homes and a new life for many Jewish refugees, and have offered a practical contradiction to the ever-recurring myth that the Jew is incapable of 'honest manual work'. Lastly, they have actually succeeded in forming an internal organisation which contains no privileged class or individuals, and which really applies to 'each according to his needs' in a practical manner.

In a democratic order such as we seek to establish in India, the lines along which we seek economic advancement should be such as to promote the cultivation of a sense of individual responsibility as well as communal obligations. The limited experiment that has succeeded in Palestine has, from this point of view, a lesson for us in India. How that lesson should be applied will be for every Province to determine. Appreciating the need for examining and drawing attention to this agrarian development, the Central government such a deputation to Palestine under the leadership of Dr T.G. Shirname, the report of which deputation provides valuable material for the administrator and economist. In Bombay, government appointed a special officer, Captain S.P. Mohite, to study this question and several of his recommendations for State aid which cover wide ground have already found acceptance at the hands of government. This, however, is a matter where action by the state alone is not enough, particularly if a policy of compulsion is not favoured. Public opinion has to be educated and the minds of the agricultural classes, especially, have to be aroused to the necessity of some form of joint endeavour in the field of agricultural production, if with the end of landlordism we are to go on to a progressive rural economy.

The form of organisation will depend, again, on numerous factors not the least important among which is the cultural and social life of the farmers sought to be organised. The system of cultivation, the relationship between the state and the individual holder, the form of tenure previously in existence, the pressure of population on the land, the desirability or otherwise of recognising the right of individual ownership in the holding, the availability of various aids to agriculture, these will be among the considerations which will determine whether the organisation will be of the pattern of a collective farm, a cooperative farm, or merely a society for joint farming. The scheme of joint village management, as expounded by Mr Tarlok Singh, contains several of the features of the types of common endeavour in the field of agricultural production which have developed in other countries. Recognising as it does the right of ownership, it may be more acceptable to the rural community than collective farming of the type with which Russia has made us familiar. Except in so far as cooperative farming represents in theory voluntary effort, there is little to choose between it and joint village management. Either form can suitably be introduced under state initiative where agriculturists are to settle down on lands brought newly under cultivation or on land where, owing to the disappearance of the *zamindari* system, a new form of settlement takes place. Where displaced persons have to be settled on farms as is likely to be the case in East Punjab and in the western districts of the United Provinces, the new colonisation should, so far as it possible, be on a joint management or cooperative basis. This, as I understand it, is already part of the programme formulated by the Central government.

Elsewhere, should governments wait for public opinion in the country side to be ripe for this form of development or should they resort to a policy of compulsion? It is true that we need an all-round drive to rid our rural economy of the evils which hamper progress and to promote measures which, being based on the maximum utilisation of the skill and resources of the individual, ensure not only increased production but the moral and material advancement of those engaged in the common enterprise. Taking into account the general individualistic outlook of the rural community, it may be difficult to enforce action which encroaches on the traditional sentiments attached to rights of ownership and rights of inheritance. It may also be useful to respect the desire of an individual to control the operations of a particular farm and to diffuse the sense of ownership especially where the unit of cultivation is small. But after having duly recognised the value of individual responsibility and accorded respect to traditions, the state should be at liberty to ensure that cultivation is carried on a basis which is consistent with efficiency. If inefficiency results from lack of resources, such resources may have to be placed at the disposal of the farmer through the joint farming or multi-purpose cooperative unit which he should be more or less under an obligation to join. In order that he may get the fullest benefit from his labours on the land, the state may, in addition, not only provide facilities for regulated marketing, including the provision of warehousing, but may even direct that particular varieties of produce should be marketed through a cooperative organisation. Similarly, so long as rights of individual ownership inland are continued, the land should not be allowed to be charged and alienated at the discretion of the individual but should serve as security for credit along well-organised lines and through approved credit agencies.

Among the factors affecting agricultural progress the prevalence of a state of indebtedness is nowadays apt to be ignored. Persons with a perfunctory knowledge of the state of rural economy take the superficial view that the high price obtained by agriculturists have had the result of extinguishing the entire burden of their debts.

The only authoritative survey we have had of the effect of wartime prices on agricultural indebtedness is the Report submitted to the Government of Madras last year by Dr B.V. Narayanaswami Naidu. Despite the favourable circumstances which obtained during the period under investigation, Dr Naidu's conclusion is that the volume of debt has not shown any substantial fall and that the fact that it still remains 22.5 times the land revenue assessment gives cause for serious consideration. What is a matter of graver concern is the deterioration which, according to Dr Naidu, has taken place in the economic position of tenants and landless labourers. These along with the small landholders constitute the bulk of the rural community; and the latter, too, have derived little, if any, benefit from the high level of commodity prices. The big landholders have fared well, better than the medium landholders, the former having a reduction of 39.9 per cent in their debt and the latter of 24.6 per cent.

The results of a similar, but somewhat restricted, inquiry carried out under the guidance of Prof Gadgil among the members of agricultural cooperative credit societies in the Province of Bombay have not yet been published. But from the perusal of an advance copy of Prof Gadgil's review of the statistical data, the same conclusion is drawn as by Dr Naidu. A more elaborate and comprehensive investigation has recently been ordered by the Agricultural Credit Department of the Reserve Bank of India (RBI), with the aid of Provincial Governments, but these investigations will relate the conditions somewhat different from those of war-time economy. The facts that have been disclosed by the investigations that have been completed so far provide justification for the pursuit of measures for the relief of agricultural indebtedness such as were either embarked upon or contemplated in various Provinces before the outbreak of the war. No new measures have been forged in any Province; but in Madras, Orissa and Coorg, certain amendments have been carried out in the legislation adopted previously. In Bombay, the Bombay Agricultural Debtors Relief Act has been made applicable all over the Province and as a result of experience, the Act has undergone certain radical amendments to make it more workable and effective. Legislation for the regulation of moneylending which was introduced before the legislature ceased to function has now been put through and has recently been brought into operation.

Among the resources which the average agriculturist fails to get on fair terms and which he should obtain through some organised agency in a well-ordered society is credit. It is now nearly four-decades and a half since the foundations were laid in India of a system of cooperative credit. It is because this system does not fully meet the requirements of the agricultural producer in the mass that the Gadgil Committee has recommended that there should be in every Province and agricultural credit corporation for short-term, intermediate term and long-term credit, established by the State and operating under general official supervision and direction. It is only in Provinces which have well developed cooperative financial agencies that, according to the Gadgil Committee, the need might not arise for a parallel system of agricultural credit. The Committee insists, all the same, that means should be found for making finance available to all credit-worthy borrowers through cooperatives. The Saraiya Cooperative Planning Committee has expressed the view that if the same measure and type of aid recommended to be given by the Gadgil Committee to the Agricultural Credit cooperative financing institutions in a Province, they will be in a position to meet all legitimate requirements. In a note of dissent, two of the members of the Saraiya Committee have elaborated the view held by the main body. A perusal of both the dissenting note and the main report leaves, unfortunately, the impression that the responsibility of the State for developing a well organised system of agricultural credit has been somewhat inadequately recognised.

An entirely new situation has been created in most Provinces by the enactment of legislation for the protection of the tenant, the small holder and the agricultural debtor, and for the control of money-lending. The combined effect of all these measures must be to disturb the relations between the agricultural debtor and the lender and to impair the flow of credit from certain sources such as the landlord. For this reason, I, for one, hold the view that it is necessary for the State to assist actively in setting up an alternative credit agency—the services of which should be available to all creditworthy borrowers.

It is argued that the system of Tagai loans under the Agriculturists' Loans Act or the Land Improvements Act enables Government to satisfy this need wherever it exists outside the cooperative fold. Apart from the defects that have bee noticed in the past and are still prevalent such as inelasticity and dilatoriness, I accept the dictum of experts in the subjects that in the field of agricultural credit, the State may be a benevolent lender but it is always an inefficient lender. Modifications are possible in the Tagai system; and in Madras, the Central Provinces and Berar and in Bombay suitable adjustments have either been made or are contemplated. Nonetheless, the system of Tagai loans will and should remain as the last resort, to be availed of on occasions of emergency or in times of distress. It can never function as a normal source of credit.

The Gadgil Committee recognises that cooperative credit is ideally suited for the purpose in view. Its pace of progress is, however, dependent on voluntary effort; further, cooperative institutions of all types are autonomous bodies; and, lastly, compulsion cannot be exercised in respect of the financing of persons who choose to be or are kept outside the circle of members. In the final analysis, it is the voluntary principle recognised as the basis of cooperative organisation, as also the insistence on self-governance and freedom from outside control, which may prevent the cooperative credit structure being utilised fully by the State in building up a sound system of agricultural credit. In Bombay, government appointed early this year a Committee under the Chairmanship of the President of the Indian Society of Agricultural Economics, to examine and to recommend on what lines the cooperative financial structure should undergo adjustments and alterations to enable it to subserve the purpose in view. A compromise with the principle of voluntary action and autonomy is inevitable, if this task is to be accomplished quickly and satisfactorily.

It may not be out of place if I refer here to marketing. To indicate what may be necessary, I mention the organisation that government hope to see built up in the Province of Bombay. In the first instance, we have an Agricultural Produce Markets Act (APMA) which seeks to establish a form of control over the trading transactions taking place between the producer and the local buyer at all important centres of trade in the Province. Although the number of markets brought under regulation so far is not more than a score, it is government's intention to see that over a 100 leading trade centres are served by market committees established under the Act within the next quinquennium. This will ensure healthy conditions of bargaining, stop unauthorised deductions and exactions, help in maintaining proper weights and measures, provide simple machinery for settling minor disputes about quality and enable grading and classification to be introduced. The market committees that are constituted are expected, besides, to have market yards where agriculturists can unload their carts, can have shelter for themselves and their animals and can deposit their produce.

The last named service is of considerable importance. To supplement the meagre provision that is ordinarily available, government has recently enacted legislation for the regulation of licensed warehouses for the storage of agricultural produce. These licensed warehouses can issue warehouse warrants or receipts which are negotiable. They rank as documents of little to goods such as can be availed of for marketing credit from the RBI if under section 17 (4) (d) of the RBI Act. The warehouseman under this scheme has to be independent of the buyer or seller of the goods or of the borrower or lender against its security. An exception, however, has been made in this respect in favour of cooperative societies. It is proposed to link up these warehouses with regulated markets, on the one hand, and with cooperative marketing societies, both local and regional, on the other. At the top, there is a provincial society to facilitate and coordinate their operations and to establish touch between them and the provincial or all-India commodity markets and exchanges. In the village itself the local multi-purpose society will act as a collecting agency for the produce of its members, ensuring the recovery of loans from out of the proceeds of sale.

Marketing and credit are closely associated operations in any sound system of agricultural credit; although I believe that under the unregulated individualistic type of credit as it developed in India in recent years this association has lent itself to much abuse. A chattel mortgage is a common feature of the machinery or short term credit in several countries. In a system of controlled credit such as it envisaged by the Gadgil Committee, restrictions will have to be placed on the right of a borrower to sell his produce without reference to the lender. There is a provision to this effect in the Bombay Agricultural Debtors Relief Act which makes violation of such restrictions a panel offence. While it may be neither expedient nor feasible to enforce penalties, it is certainly desirable to coordinate the operations of the marketing and the credit agencies at all stages of the structure.

In view of the state of organisation achieved by industrial labour, on all occasions when questions relating to labour are under consideration scant attention is paid to agricultural labour. Even the recent Asian Regional Labour Conference held in New Delhi, the measure of attention the question received cannot be deemed adequate. There are parts of the country where a state akin to serfdom prevails, not only among the aboriginal and other backward class rural population but also among the landless agricultural communities. Now that in most Provinces there has been an attempt made to protect the tenant and to afford him security the protection and promotion of the interest of the labourer on the land must claim the attention of Provincial governments. On the subject of village *panchayat*, legislation of a comprehensive type has recently been enacted in the Central Provinces and Berar, the United Provinces and in Bombay. The Society has published an exhaustive and authoritative survey of the development of the system of *panchayats* in Madras Province. These should provide, apart from the papers presented to the Conference, ample scope for a fruitful discussion.

Lastly, the problem of population and its pressure on the land is ever present with us. The shortage of food, cloth and other essentials of the life remind us all the time—apart from lack of shelter for thousands who flock to towns and cities—that our population is on the increase and the increase has outstripped our productive capacity. What should the State do to ensure that the two proceed apace? The transfers of population that have occurred in North India and a few other parts of the country present a new set of problems which claim urgent attention. In our sheltered existence, we believed in the immobility of our population; and the settlement of large masses in the land was never attempted or achieved in the country. To the reform, I have made only a passing reference. The considered thought which this, Conference will bring to bear on the recent enactments will, I am certain, prove helpful to those in charge of the administrative machinery in implementing the various pieces of legislation or in introducing necessary amendments.

I am afraid that some of the points of view which I have brought to bear on the consideration of the aspects of agricultural economics that I have dealt with in my address may not appeal either to the economist or to the technologist or to the administrator. But these views I have urged because I believe that those responsible for moulding the economic destinies of rural India should keep ever in mind two primary considerations, namely, "that the whole employable population is at work and that its productivity is as high as possible." We want to usher in a social order in which no individual should be left with an income which is below the level of subsistence. This objective we wish to achieve by means of action along lines which help in building up in equalitarian, classless society, the twin watchwords of which are economic self-sufficiency and the dispersal, and not concentration, of political and economic authority. I am fully conscious of the fact that this is a harder task than that which, for instance, confronted Russia after the Revolution of 1917 because building an authoritarian state, as history amply bears out, is, for the time being, an easier task than building, for permanence, a democratic one. We shall, however, be false to those who have laid the foundations of a Free India if, deterred by its immensity, we flinch from the task or forsake the paths by which they won freedom for us.

Indian Journal of Agricultural Economics, Vol. 3, No. 1, April 1948.

9

S.V. RAMAMURTHY

The Five Year Plan and Rural Improvement

WE have been accustomed until recently to look at agriculture as an art and a way of life. India has a bountiful supply of food through her natural resources, a tropical climate, an intelligent and industrious peasantry, and a well-knit social organisation in villages. The rural civilisation of India was the stable and basic part of Indian civilisation. Indian villages were as prosperous and as cultured, as well calculated to provide good life for their people as anywhere in the world. The political instability of India was due to her kings rather than her teachers and peasants, her artists and craftsmen and merchants. Kings came and went. They made new cities and new empires. But the village life flowed evenly and peacefully with a spiritual rather than a material tempo.

The contact of India with Europe 200 years ago had upset the orientation and balance of Indian life. Science and the organisation it made possible placed the emphasis on a centralised direction of life and administration and urban life overshadowed rural life. Today, India is better off in the amenities of life in towns than it was in pre-European days. But conditions in the villages have steadily deteriorated and it is sadly true that the slums of India are not urban but rural. Intelligence has been drained from villages to towns to the clerical and other subordinate tasks required of Indians by their new masters. The use of science was in the main applied to the needs of towns and the imperial masters who made and maintained them. Food was less important than industrial raw materials. Tanks and canals were less important than roads and railways to take to the coasts products for export.

The last war was a turning point in many ways. It ended the political empire of England in India. It ended colonialism in Indian economic life. It ended the cultural subordination of India. It has released the social energy of India. India has regained her freedom of spirit under the leadership of Gandhiji. Today the problems of India are that of her recovered freedom. But these problems of India approaches with a new richness and strength of the spirit, thanks to her very tutelage under England. The three elements of her gains from England are law, science and democracy. The maintenance of law and order, under masters who among their subjects were impartial and from, has given to India a stabler floor of life than for instance our great neighbour China has been able to achieve under her own national guidance. India too has been able to imbibe more of the spirit and results of science through an orderly flow from a well-intentioned administration in England than has been the case in fellow countries in Asia. The sense of freedom which India developed in the spiritual domain has been strengthened by the method and content of political freedom, which England developed for herself and imparted to India. Although in material fields, there have been clashes between English and Indian interests, spiritually there has been kinship that has grown with close association with mutual benefit. It is good that India has shaken off English tutelage. It is equally good that they should remain friends, coordinating their life and joining in common service to the world.

I have said that agriculture was traditionally an art and a way of life in India. Today, it can no longer afford to continue as such. In the present context of the world, agriculture in India needs to develop as a science and as a business, if it is to sustain the growing population of the country and establish the position due to her among the nations. It is for this reason that the study of agricultural economics has an important part to play in present-day India. The tempo of the world has changed. Humanity is taking large strides both quantitatively and qualitatively. Science enables health to be maintained and death rates to be reduced. Science also yields a larger production of food so that population for a time can grow more rapidly than was possible in previous epochs. There is need to analyse the various factors affecting population, food production and nutrition so that in the orderly internal life that the country can provide, with the heightened social conscience that democracy produces, the maximum of progress and welfare may be achieved. Agriculture can no longer be left to be dealt with in crude and unstandardised forms. It has to be studied and built with the help of natural sciences and the science of economics.

The particular problem that I wish to deal with today is the problem of food for India. Before the War, its importance and urgency were hardly recognised. Famines were regarded as accidents due to seasonal failures of rains. Measures to deal with such emergencies were embodied in Famine Codes and kept ready. Hunger in normal times did not affect the vocal parts of the population. If there was no more food, it was regarded as due to unalterable conditions of nature. The war shook this complacency. War conditions produced deficits of food more severe than in famines. The maintenance of law and order needed measures to prevent starvation on large scale. The Bengal Famine showed what a dangerous level poverty had reached in India. With a 5 per cent deficit of food supply and a consequent large increase of prices in parts of Bengal through an inadequate administrative machinery, there was death on a terrible scale and the country was roused to the seriousness of its food problem. A statistical study of the position was made. It was found that India which could provide an average cereal ration of 1.5 pounds per person per diem at the beginning of the centre could now provide no more than 1 pound. Bearing in mind that cereals provide 80 per cent of the calories of an average diet, a pound of cereal indicates 2,000 calories a day. It was found that for some 25 years, the production of cereals was practically stationary while the population was growing and that the amount of cereals available per person has in fact gone down by 25 per cent in the last 25 years. The area of cultivated land has become as low as four-fifths of an acre per head. The average ration that could provide 2,500 calories at the beginning of the century, could provide only 2,000 calories before the War. Efforts to grow more food were made during the war and since, but their effect has been negligible. The population of India has been growing at the rate of one per cent a year. If this rate of growth keeps up, then every year we need additional food for 3 million people in India or half a million tonnes of cereals at the rate of 1 pound a day. The production of cereals in India is about 40 million tonnes a year. One imports of rice before the war were about 2 millions tonnes a year. During the War, we were not able to get even a million tonnes. In 1946, we were able to get an allotment of 1.5 million tonnes. This year, we have been able to obtain about 3 million tonnes. The position thus is that with an internal supply of about 40 million tonnes of cereals for 300 millions of people, we need to maintain our arrangements for rationing an import of about 3 million tonnes costing about ₹ 100 crores a year. Allowing for friction and leakage in procurement, the total supply available can provide a pound of cereals per adult a day, but this may be reduced if seasonal conditions are adverse. In 1946, we felt we had been specially hit by drought and cyclone. Unfortunately, since then the season has been failing over large parts of North or South India and the tension of 1946 continues unabated. There is thus growing anxiety in the country over both the low ration which a statistical analysis has brought vividly to attention and the need for imports by organised efforts of the government instead of traders, at a time when the former have to pay large sums to hard currency countries. Payments for food by the government has retarded the purchase of capital goods for the development of industry.

Let me put together the elements of our food problem.

(a) Fifty years ago, we had a daily average ration of 2,500 calories. Today, it is 2,000 calories. The quality of the food available is not nutritionally adequate.

(b) Before the War, deficits due to a bad season could be made up to the necessary extent by imports of rice from Burma and Siam. This year, supplies from West Punjab and Sind are also cut off. We now get only what we are allotted by an International Council, where big powers with their own colonies and protégées do not permit India to get what she needs. Before the partition of Burma and Pakistan, India was self-sufficient in rice and wheat. Burma's secession from India produced a deficit of 2 million tonnes of rice and Pakistan's separation nearly a million tonnes of wheat a year. It is these million tonnes of rice and wheat that India now needs to import each year.

(c) Agricultural production in India is low as it is carried on in very small units and on obsolete methods. The production of rice per acre is hardly half of that in other Asiatic countries. The results of scientific research are not readily adopted by cultivators. The application of water through large irrigation schemes needs 10-15 years to materialise and become effective. Chemical manure, which could appreciably increase production at once, is not available in any large quantities. Imports of such manure are rationed by an International Council. Machinery needed to improve methods of cultivation is similarly rationed. India cannot get either fertilisers or machinery which can increase production immediately. The land tenure of India is the result of social organisation for which short-range remedies are difficult to apply. Nor are methods for such change yet worked out. Efficiency of agricultural production involves the lowering of the percentage of the population employed in agriculture as against industry, and this cannot be achieved without social disturbance unless there is industrial development on a scale to absorb an appreciable part of the rural population.

Thus the food position of India is poor in regard to:

(a) The area of food cultivation per head of the population.
(b) The production per acre of such area.
(c) The availability of water, chemical manure and machinery for improvement of cultivation.
(d) The availability of imports.
(e) The high ratio of the population dependent on agriculture; and
(f) The low standard of living of the people due to paucity of industrial employment.

Let me examine whether and how these difficulties could be met. The deficit of cereal production is, at the present rates of consumption, not more than 10 per cent of the present production. Agricultural experts are confident that the deficit can be made up by reclamation of wasteland, increased supply of water through scheme of irrigation—small and large, use of good seed, use of chemical manures, use of machinery to bring new or difficult land under cultivation, efficient use of methods to deal with pests, improved method of storage, methods of maintenance of the health of peasants by anti-malaria measures and improvement in land tenures. These methods have been publicised for several years. Yet the position continues to be bad. Grow-More-Food schemes seem to produce no results. The organisation of measures recommended by agricultural scientists seems to yield no results as well. The efforts to combat malaria have not been sufficiently successful. We know our needs. We know the remedies. We have the material, land and human resources to apply the remedies to our needs. Yet we fail to do so. Scientists and economists, farmers and administrators are left with a sense of frustration and failure.

To suggest ways of meeting this situation, I fall back on my own experience. Before the War, I visited Italy and saw the magnificent work done by Mussolini in the battle of wheat and the reclamation of the Pontine marshes. The production of wheat was raised by 30 per cent and the marshes which baffled efforts of reclamation for 500 years were converted into smiling fields maintaining a proud and happy peasantry. I put it to Mussolini that in India, we had as good scientific research as any I had seen in Italy and yet we could not get the results translated in peasant's practice. He replied that in Italy they found no such difficulty and that they attacked agricultural problems in a comprehensive manner. His reference was to what was called the "integral principle". Italy has a proverb that it is the hundredth soldo that makes the lira or in Indian terms, it is the last pie that makes the rupee. Unless an effort is cent per cent complete, it cannot be successful. I applied this to the irrigation of land under the Mettur Reservoir on the Cauvery. The reservoir provided enough water for 3 lakh acres. There was commanded area of such extent. Yet for 10 years after the reservoir was ready, only 2 lakh acres were irrigated. The rest of the water was wasted and the commanded land remained a waste. In the War period when we had the need to increase food production, I turned to this absence of equation between needs and resources. I appointed a Committee of a Revenue Officer, an Engineer and an Agricultural Officer to examine each acre of the commanded but unirrigated land, find out what stood in the way of its being irrigated and remedy it. Was it because water did not reach the field? If so, channels should be dug. Was it difficult to find finance? If so, government loans should be provided. Was it the difficulty of finding labour? If so, arrangements to organise the import of labour should be made. Were there difficulties of tenure? If so, by conciliation and government assistance, the differences between landowner and cultivator should be resolved. I had set a target of 40,000 acres for the first year. The member of the Board of Revenue did not believe that the target could be reached but offered to do his best. He was told that unless he had faith he could not succeed and that if he did not succeed, he would be considered unfit to be the member of the Board of Revenue. At the end of the year, 60,000 acres were brought under cultivation. It was a great pleasure for me to see land which had been wasted or covered by scrub converted into smiling field of paddy. The land was not only cultivated but cultivated well with what the cultivators called the "London Medicine". They meant by the term, ammonium sulphate. An integral effort succeeded fully and in a short period as well, as in Italy. The Indian cultivator living at the margin of subsistence dare not make experiments the results of which are doubtful. But he has intelligence and resource enough to try new methods if he is confident of their success. The use of chemical manure in the newly irrigated area was the result.

There is another attempt I successfully made to increase food production. In the beginning of 1946, I had to go to UK and USA as a member of the Indian Food Mission. The central districts of Madras province had suffered from intense drought. I visited them before leaving India so that I could speak with force abroad of what I had seen. I found that in the desert of dried fields and tanks, there were green spots. Wherever there was a green spot, there was a well and where there was a well, there was a green spot round it. I visited many villages in the course of two days. I asked the villagers what help from the government would make them dig wells. I told them that the government had offered a quarter of the cost of digging a well and yet hardly any one made use of the offer. If the government gave them the full cost, they would surely dig the wells. What fraction of the cost would make them dig the well? The answer to which everywhere promptly was half the cost. This answer was as ready as if they had been agricultural economists themselves. Only one man asked for a three-fourth grant. Yet when I told him that it was the people's money that would be given to those who dug wells and when asked whether three-fourth grant would be fair to the general public, he too asked for a half grant. In the week before I left India, I sanctioned as Adviser to the Governor a crore of rupees from the Madras government finances; a crore was thereby secured from the Central government. In six months, the people dug 60,000 new wells and repaired 30,000 old wells at a total cost of about ₹ 5 crore. All this was done by the villagers with their own technical resources and materials. I attribute the success of the effort to the fact that I gave the people what they felt they needed.

These two efforts show that for successful economic development on a large scale, we need both good government and self-government. The country needs rulers and administrators of capacity and character to serve the people according to their instincts. British administrators did not pay full heed to the interest of the people, their needs and their welfare. Under present administration, we have yet to provide an adequate number of Indian rulers and administrators who could act with vision, faith and drive in realising the potentialities of India. It was an Italian diplomat who told me before the war that it was easy to get power, but asked, "Had we men to administer it when we got it?" I said then we had them. We have some men of as high a quality as anywhere in the world. But for the tasks of India, we do not have enough of them at present. Given time and experience, I have every hope to have enough of them. With the assistance of such men, the remedies that technicians have evolved for increasing production, namely the use of good seed, of available manure, of food cultivation practices and disease prevention and of good storage methods can more than make up for the present deficit of 10 per cent of our cereal production, which compels us to beg for rice and wheat abroad and yet pay a hundred crores of rupees a year for their import.

This however is not enough. We want more food, better food and an increased standard of living. The long-range method of achieving them is the execution of multipurpose river schemes for the production of water and power besides large schemes of land reclamation. India is a great country with a great future. To realise that future, we need to plan big and build big. We need to develop life in a multidimensional manner—economically, politically and culturally. Not only do we need strong bricks of personality but also large girders of economic and political life. Of all the great natural resources of India—mountains and rivers, forests and mines, seas and winds—it is the rivers that embody best the dynamism of India. It was in the river valleys of India that the Sun filled the river beds with water and sustained the adventure in spirit of thinkers and seers. The river of India will again witness a phase of that adventure, the adventure in matter of technicians and statesmen. It is this belief that has sustained my efforts to build the Ramapadasagar Reservoir of the Godavary. Such schemes not only make more food, but better food by stimulating horticulture and livestock production. Increased production of rice and wheat can release millets for livestock feeding.

India has a large area of culturable waste, estimated to be nearly a fourth of its total area. The density of population in India is only 246 per square mile. This is lower than that in most European countries. Only America and Australia have vaster unused spaces than India. Land reclamation in India has great scope. But this needs mechanical aids to agriculture such as the ordinary Indian cultivator cannot provide himself with. Agricultural economists as well as statesmen have not yet decided whether communal farms, cooperative grouping of farms or economic holdings offer the best alternative to the very small units in which land is now held by individual peasants in India. The unit holding is on the average 5 acres in India, 20 acres in Europe and 150 acres in US. A question is likely to arise in the near future: whether cereals in 5 acre farms with the aid of human and bullock labour can stand competition with that produced in 150 acre farms with mechanical cultivation. Are we heading towards an Agricultural Revolution which may compare in scope and consequences with the Industrial Revolution? In the latter, small-scale industry running on obsolete methods was swamped by machine industry. In America and Australia, the agricultural potential was largely increased during the last war to meet the food needs of Europe and Asia. But as peace restores agriculture in Europe and Asia, a surplus cereal production is arising in America and Australia for which markets have to be found. Can hand-made wheat and rice in India compete with imported machine-made wheat and rice? Can India protect herself against a ruinous reduction in price through the availability of imported cereals, particularly in the coastal areas? Or will different price levels be maintained for internal production and for imports? Today, there is only one cottage industry alive in India. It is not handloom weaving. It is agriculture. Is this also to be killed with other cottage industries? If it is, the social disturbance in India will truly be of the nature of a revolution.

To meet this prospect, how are we to rearrange our methods of production? It seems to me that methods of cultivation are not necessarily more productive per acre, but are less costly per unit of production. Perhaps the application of a cooperative method and measures to make holdings economic will be appropriate for new cultivation. This is an issue on which the country needs a finding urgently from agricultural economists.

With more land brought under cultivation and under irrigation by land reclamation and river reservoir schemes, we also need the power provided by the latter to yield a more variegated economic life than a mainly agricultural economy. In India, 75 per cent of the population are engaged in agriculture, while the US has 15 per cent so engaged. In a balanced world, it seems to me that agriculture should be the base and industry the apex, and not *vice versa*. 60 per cent of the people could I think be a safe population for agriculture. The rest have to be arranged in layers of industry—cottage, small scale and large scale, with the learned and artistic professions at the top. For reaching such an industrial orientation, electricity is bound to play a great part. Large-scale industry grew in Europe because coal yielded power in large units at centralised factories. Electricity can be applied in small units, although produced in the mass. The spread of electric power can be as wide as the spread of population. The large potentials of hydro-electric power in India are an augury for a strengthening of the country's economic life over the whole of its range. The needs of agriculture, industry, commerce and social life are all served conveniently by electricity. Fertilisers and agricultural machinery which we cannot get from the West, we can make ourselves. The multipurpose river schemes of India are thus of a multidimensional value in the growth of Indian life.

It seems to me that the challenges to India in regard to food production can be accepted and met. Our deficits can be wiped out and advances made both by short range and long range use of our human and natural resources. Commercial agreements with Burma, Siam and Pakistan can assure us of the present food deficit being met by imports from nations who are no more likely to be interested in a world war than us. The problems of economics are capable of politico-economic solution. The cultural activities of India are even better adapted to international than national ends.

Today, India stands on her own legs, free to develop herself according to her instincts, needs and resources. With the power that freedom brings, India once again dons her world responsibility. She stands between the two contending ideologies of US and USSR, and it is in the interests of the world as her own that she should be able to reconcile the differences that threaten to submerge the world in a war more terrible than any before. Great rivers of life in the world are in flood. But one river cannot check other rivers. A synthesis of rivers can be provided only by the sea, absorbing rivers and lapping many shores. India's need is to be a sea of life. Only then can she save the world. If in addition, India is a sea not of brine but of milk, she is then also the womb of a new creation. The world has need for the spirit of India.

The body of India is a function of the spirit of India. I believe that India occupies a place of destiny in the life of the world and that India will live because if the world is to live, it is necessary that India should live. Let us, economists and scientists, citizens and statesmen take up the task of securing food for India with care, energy and confidence.

Indian Journal of Agricultural Economics, Vol. 4, No. 1, March 1949.

10

RAJENDRA PRASAD

Technological Upgradation: The Next Step

INDIA is a vast country and although its two outstretched wings on the north-west and north-east have been cut away and fallen into Pakistan, it still covers 1,573,000 square miles and has a population of more than 320 millions. It has a variety of climate —varying from extreme cold of the snow-capped Himalayas where there is perpetual snow to the burning heat of the torrid zone— although the predominant weather in a greater part of the country may be said to be tropical. The land has all varieties of soil from the rich and fertile river valleys to ones that are more or less rocky or sandy waste. The people generally speaking are hardworking, resourceful and cultured, although not equipped with the art of reading and writing except for a small percentage. It is an agricultural country supporting and maintaining 70 per cent of its population by agriculture. The method of cultivation is as varied as its population and ranges between the primitive method of shifting cultivation still followed by the tribal people in hilly tracts known as the Jhum system and the modern method of large-scale farming with the help of machines and chemicals. The bulk of the cultivation, however, is of smallholdings carried on according to old traditional methods. Let no one run away with the idea that the traditional method is primitive or ignorant. If that were so, it would not be possible to grow all kinds of crops, herbs and fruits. The fact is that there is no variety of crop or fruit that cannot be and has not been raised according to the nature of the soil in some part of the country or the other. The Indian peasant knows the art of agriculture and has practical experience. He knows all the techniques required, such as rotation and combination of crops, the value of allowing land to remain fallow to regain its fertility, the value and method of irrigation and manuring, the art and method of ploughing and cultivating the land in diverse ways and directions so that the entire soil may be turned upside down and so on. He does not know, except in a small way the use of modern manures and modern machines. That is because they are beyond his resources. He has invented various devices for drawing water from rivers, tanks and wells for irrigation, and some of the old works of irrigation can compare with the most modern irrigation projects of today. They are all the more wonderful as they do not depend on electric or steam power and have been operated by man since time immemorial. The Indian peasant is not unreasonably conservative, although he is cautious. If he can be convinced that a new method can give him better return or save labour or is otherwise useful, he does not hesitate to adopt it, if it is within his resources. I may mention one or two instances to show this as it is very often said without justification, I believe, that he is conservative and does not care to adopt improved methods. In the province of Bihar from which I come, sugar cane has come to be extensively grown during the last 20 years or so on account of the establishment of large sugar factories. Not that sugar cane cultivation was not known before. The fact is that sugar cane has been grown in large tracts from time immemorial and the soil is also suited to it. It was because of these facilities that industrialists established their factories and this has led to an extension of sugar cane cultivation. When once the peasant saw that sugar cane cultivation brought in more money than other cultivation, he changed over to sugar cane from other crops without hesitation. He is calculating and shrewd enough to increase or decrease the area under sugar cane as per the rise or fall the price of sugar cane from season to season. He has shown a remarkable adaptability not only in adjusting area according to the market requirement but has also adopted the variety of cane that has been demonstrated to be most profitable in existing conditions. Just as the area varies from year to year, the quality of the cane also varies almost similarly. In the whole province, one can hardly find the variety of cane that used to be grown—say 20 or 25 years ago—which had softer rind and was more suited for crushing in mills worked by bullock power and now we have the modern varieties with harder rind, but having large sucrose content capable of being crushed easily in power-worked machines, but most difficult for bullock-operated mills to crush. Similarly whenever he has seen a new variety of cotton or jute or wheat or paddy which brings in better return, he has not hesitated to adopt it. What is needed therefore is to demonstrate to him and assure him that a new method or a new variety gives a larger yield and profit, and he will adopt it.

With all these advantages, the country is faced today with the problem of shortage of food which may well appear inexplicable. Such statistics as are available make out that there is a shortage of cereals and as a matter of fact we have been importing large quantities of foodgrains from foreign countries, especially during the period following the World War II (WWII). We used to import rice even before the war very largely from Burma in pretty large quantities—to the tune of 1.5 million tonnes on an average according to the nature of the season and weather and the consequent crop affected by it in the two countries. But it used to be done in the ordinary process of trade and the ordinary man never thought or cared about it. But during the last four or five years, we have been hearing such a great deal about the shortage of food, the necessity of importing it, the steps taken to import it and the hundreds of crores required to pay for it and subside its distribution. Our imports have gone on increasing. In 1946, they were 2.25 million tonnes, in 1947, 2.32 million tonnes, in 1948, 2.83 million tonnes and in 1949, 3.69 million tonnes, or 11.0 million tonnes in four years or nearly 2.8 million tonnes annually, which is very nearly double the average of previous imports. The price paid for it in 1946-47 was ₹89 crores, in 1947-48, ₹108 crores and 1948-49, ₹131 crores. The amount paid by the government to subsidise food was ₹20.59 crores in 1946-47, ₹17.12 crores in 1947-48 and ₹26.50 crores in 1948-49. We have a very large staff to deal with food, its procurement from within the country and from outside, control of its price, its storing, stocking and movement, distribution and rationing. With all this, prices have risen higher and higher. There is so much talk about hoarding and black market, control and corruption, scarcity and disappearance of stock from the market.

It is therefore worthwhile probing into the causes of this most remarkable and apparently inexplicable problem. We have many difficulties that render such effort almost fruitless and futile. There is generally speaking a lack of reliable statistics. This arises from the very nature of the material which has to be handled by statisticians. We have millions of small tiny holdings on which all kinds of crops are grown. Not an inconsiderable number of these tiny plots are left fallow. Similarly a large number of them have double crops on them. Only a small proportion of the cultivated land can depend on irrigation, but has great bulk of them depending entirely on the monsoon. Plot-to-plot survey and recording of crops actually raised are most difficult, if not impossible, on account of vastness of the task. Nor is the staff employed in collecting the basic information adequately well trained. Such information as is collected is not available for a great part of what used to be Indian states in the country where no attempt is made at collecting the information. Even in the provinces, there is no uniformity; nor is every province equipped with a regular department for this purpose. In many places, the work is divided between several departments. The method of collecting information by random sampling which statisticians regard as capable of giving fairly accurate information has recently been adopted in some provinces, and it is hoped that with its extension more reliable statistics will become available. I have stated this difficulty about correct and reliable statistics at length only to point out that planning becomes difficult, if not possible, both in respect of any satisfactory scheme for increasing the yield and for its proper distribution.

I do not desire to burden this paper with figures that can be gathered from government publications or from the Easter Economist which has brought together and published very valuable material in a handy and easily accessible form. It is generally assumed on the strength of such statistics that there is an overall deficit of 4 million tonnes of foodgrains. In the face of these figures and statistics, it may appear fool hardy for anyone to challenge the existence of a large deficit. But one thing is known to be the experience of anyone who comes to verify facts for himself. Go to any town or city and you can buy the grain required in the black market, if you are prepared to pay the price. Prices are high not only in cities but also in villages for those who have to purchase, but there does not appear to be absence of stock generally speaking. In any case there is no relation between the deficit that exists and the very high prices that are charged under control, not to speak of the black market prices. So many factors enter into the price of a commodity that it is difficult to analyse and distribute the increase under various heads. But this much can be said without fear of contradiction that shortage of supply is not responsible for the 300 per cent rise in the prices of foodgrains as compared to pre-war level. Even before the war, India imported annually nearly 1.5 million tonnes of rice on an average. That deficit did not affect the price structure to any appreciable extent; it is difficult to understand why the present deficit should upset it so completely. It is clear that the present production with the import of 4 million tonnes of foodgrain should not leave any food problem to be solved, and there need not be any occasion for control and rationing. But the fact remains that in spite of this huge import of over 2.8 million tonnes in 1948 and of 3.7 million tonnes in 1949, control and rationing on a very extensive scale are considered to be necessary. If our production together with the import nearly covers and meets our requirements this year, why should it be necessary to have controls at all and why should it not be sufficient to arrange for the imports and leave the process of distribution of the produced and imported grain to work itself out? The fact that control and rationing are considered necessary even when we have a sufficient supply (taking internal production and imports together) shows either that the deficit of 4 million tonnes is an underestimate or that in spite of procurement, control and rationing distribution is defective and inefficient, and whatever element of scarcity enters in the price is due not to an inconsiderable extent to the whole of the production not being made available for distribution—in other words, due to hoarding or large consumption by certain section of the community. The problem therefore reduces itself, as of old, to one of distribution, if the imports are assumed.

In the above discussion, I have taken the deficit to be 4 million tonnes and the imports in 1949 being nearly that, it is clear that the deficit has been wiped out. We cannot afford to continue these heavy imports which cost is no less than ₹108 crores in 1947-48 and ₹131 crores in 1948-49, and the subsidy on these imported foodgrains met by the Government of India (GoI) came to no less than ₹20.69 crores in 1946-47, ₹17.12 crores in 1947-48 and ₹26.59 crores in 1948-49. Steps, therefore, have to be taken so to increase our own production as to cover this deficit. Now the average production of all foodgrains for triennium 1946-1949 comes to 42.6 million tonnes, and the deficit is therefore less than 10 per cent. It means that even if no lands are brought under cultivation, and if only 11 maunds are produced where 10 maunds are produced today. The experience of anyone engaged in agriculture will tell us that provided he can be assured of water supply and manure, more than 1 maund can be added to 10 maunds. We can also take advantage of improved seed and better methods of cultivation with the help of better tools and better organisation even within our present resources. Apart from theses, there are vast tracts which can be reclaimed and cultivated and further vast tracts which can be cultivated if only irrigation can be provided. It is, therefore, worthwhile considering in detail how this improvement in yield can be affected.

Water supply for irrigation where we have to depend on the vargaries of the weather and also to supplement what is given by nature in the shape of rains hold the key to more yield per acre. We have a long array of big hydro electric projects for controlling our rivers costing hundreds of crores, some actually in hand and others being investigated and planned. They will take time to be completed and will require enormous costs and amounts of material and equipment which is difficult to procure. We may not therefore take them into account for meeting this deficit which is an urgent and pressing necessity. Our population is also increasing and they may very well be left over to cover future deficits, which are bound to arise on account of increase in the population. This present deficit should be met by small irrigation projects, such as sinking ordinary wells and tube wells, utilising the water available from small rivers, rivulets and channels, repairing and putting up bunds and embankments for storing rain water to be utilised when required, and draining off water from where it accumulates and damages crops or renders their raising impossible. I can say from the experience that I have of villages in my province and generally from what I have seen in other parts of the country that there is very extensive scope for this kind of small projects, and I have no doubts in my mind that if we could concentrate on this and give it top priority in every respect, we should be able, to a very considerable extent, to implement such projects in a short time. We have had plans and the government has been spending large amounts on such projects and on subsiding them. But we have not so far succeeded and we have to find out what the causes of failure have been. Speaking as a layman, I can say that our efforts have not been coordinated to give us the best results.

The next in importance is the question of manure. Here also we have a large project of producing chemical manure which the government has in hand. It is costing an enormous amount to put up the factory at Sindri and it is taking longer time than was expected. Even when it is completed, it will meet but a fraction of our requirement and can be more easily and locally produced without anything like the heavy investment required and all that is implied in concentrated production of any article on a large scale. I am glad that increasing attention is being paid to the utilisation of what is not only wasted, but is also a source of danger to the health of the people by converting it into manure. People have to be familiarised with improved methods of utilising this waste and provided with the wherewithal to do it. The excreta of human being and animals, together with vegetable matter that is allowed to rot and go waste, give us very valuable material for compost. Human excreta as well as cattle excreta are not fully used for this purpose. A great part of the latter is used as fuel or not used at all for any purpose, or if used at all, not in the best way. We must find and supply to the people cheap alternative fuel and demonstrate to them better methods of utilising them. We must teach them how to utilise human excreta and try to get over the prejudice attached to their handling by many persons, by way of teaching them better methods and providing them with means which may make their handling less irksome and dirty. I have no doubt that a drive on the right lines approved by scientific research in this direction can give us immediate results as nothing else can. Compost, green manure supplemented by chemicals and various kinds of oilseed cakes can work wonders in increasing production.

Similarly improved seeds and methods of cultivation with improvised implements can also help much.

But in all these, we should be able to utilise the results of scientific research, which should be directed in this direction also. As I have said earlier, our peasant will adopt all or any of these, if he can be assured that these will be better returns for his labour in adopting these. We have our research institutions but they have on the whole devoted more attention to and therefore succeeded in what may be called money crops, even though sugar cane is incidentally also a good crop. Our research institutes have given us numerous varieties of sugar canes which have become more and more money crops on account of being better suited for large factories. It is a curious phenomenon that in a province like Bihar and in large parts of the UP, where sugar factories abound, the ordinary peasant who grows the sugar cane eats less sugar today than he used to do when he grew much less sugar cane. When he used to crush his own sugar cane with the help of his own bullock *Kolhu* to convert it into *gur,* he and his children used to take sugar in all possible forms starting with the cane juice produced by the *Kolhu* or by human teeth and ending with the *gur*, and also to a small extent with refined sugar. He does not have that opportunity today, as the rind of the improved variety is very often too hard for his teeth and the bullock mill and the entire crop of cane passes on to the factory. His cattle also do not derive all the benefits they used to get in pre-factory days, as the cane has to be cut for supply to the factory in such quantities that his bullock cannot consume all the green leaves of it when they are available and the bagasse which also has some sugar in it is not available to them at all. Nor can he afford to buy refined sugar for his cattle which came a share out of the *gur* that was produced. We have by concentrating research on improving the cane for the factory thus succeeded in depriving the producer—both human and cattle—of the sugar that they used to get and in marketing for others sugar without that vitamin content and organic salts that used to form part of the *gur*. If a part of the attention had been given to improve the variety of sugar cane, I have no doubt that equally good results would have been obtained.

We have also succeeded in introducing improved varieties of cotton, jute and tobacco which are all money crops. Nothing like similar results have been obtained regarding food crops like rice, wheat, maize, *bajri*, *jowari*, gram and pulses. To the extent improvement has been secured as a result of research, our people have not hesitated to adopt the result of such research, if they are made aware and brought within their resources. I know of ordinary cultivators changing the variety of wheat and paddy which they grow if it appears to them that the newer variety will give better results and if seeds are made available. Other grains like maize, *bajri*, *jowari*, gram and pulses have received practically no attention at all worth mentioning. Our research institutions have therefore to devote more and more attention to needs and requirements of the small cultivator. It is for them to give him better varieties which will give greater yield. It is for them to find out the varieties best suited to the particular kind of soil; in every locality. It is for them to devise the best methods of converting waste into manure, to make irrigation from wells, tanks and channels and rivulets easily and within the reach of the peasant's resources. With small improvements, the instruments now used can be made to provide much better results and it is for research workers to invent and popularise them. Research in this direction should not be conducted with focus on the profit of the manufacturer of such implements, but for the benefit of the peasant. I know that researches in connection with compost, green manuring, mixing organic and inorganic manures, improving implements of cultivation and generally the method of cultivation have brought about good results. But there are some inherent defects from which many such efforts suffer. Many of them are beyond the resources of the ordinary cultivator and are too costly. Even where they are within his means, there is no systematic propaganda and practical demonstration to make him familiar with them. Our demonstration farms are not as successful and useful as they should and can be made. It is no use having a demonstration farm which only demonstrates its own inability to be run economically, that is, which costs more than it earns in spite of all the advantages it has in the form of technical skill, mechanical equipment, large size and finance. I consider that to be the best demonstration farm which can show how a farm of say 10 acres, if not even of less, with a pair of bullock, one or two milch cows and buffaloes, well and improved village implements of agriculture can with the labour of two or three adults of a family give the family sufficient grain and milk, edible oil, vegetables and local fruits, condiment like chilies, ginger, tamarind, etc., *gur* or sugar to feed the family and sufficient cotton to cloth them when spun and woven. It should supply further sufficient fodder with such additional grain, *gur* and oil cake as may be required to supplement the fodder for the cattle. It should also be able to give him sufficient cash to pay the rent or revenue and meet other requirements such as, purchase of salt, kerosene oil, etc., and something to lay by to meet emergencies and to meet religious, social and other expenses, in connection with birth, marriages, deaths, etc.

We know that 80 per cent of holdings of cultivators are less than 10 acres and 60 per cent less than 5 acres in many parts of the country. The average area under foodgrains is less than 0.7 acre per head in the country as a whole, and very much less in the most densely populated parts where the land is generally more fertile. No research can be of much avail, unless it can serve these smallholdings and its results can be brought within the means of the cultivator who owns them. It is a most difficult task to abolish smallholdings and amalgamate them so as to make large-scale farming with all the advantages of modern machines, chemical fertilisers and mechanical irrigation possible. Nor is it demonstrated beyond question that such large-scale farming gives higher yield of grain per acre than intensive cultivation by a peasant proprietor of a small holding, although large farms may in some instances be shown to have brought good dividends in money and to employ less labour, if machine are used. Mechanisation of agriculture on large scale in India is at any rate in present conditions impossible. The size of holdings and large-scale unemployment consequent on its adoption rule it out. Collective farming in Soviet

Russia has struck an average of 1,600 acre for a farm, the size varying in different regions from below 600 acres to three times that size. "For a 40 H.P. tractor to remain usefully employed for 2,000 working hours in a year a minimum size of 200 acres for a cultivable unit is irreducible" says Mr K.K. Banerji. "As against this a cultivator holding in the Gangetic plains of Bengal, Bihar and U.P. come to less than 3 acres on the average" and fragmentation is going on. If mechanisation is introduced, not more than one-fourth of the population now employed in agriculture will be employed. Further it is nor possible to get the machines in anything like the quantities and numbers required within a reasonable time even if we can find the exchange to pay for it, which is another almost impossible condition. But deep ploughing machines and use of chemical manures on an extensive scale are not regarded by all scientists as best suited to long-term agricultural industry. In adopting them, India has to circumspect and to see that what has been described in hectic language as the rape of the earth does not take place. Our agricultural economy has been raised on a well-worked-out natural balance of restoring to the earth what is taken from it and this is largely responsible for our land-bearing intensive cultivation for thousands of years without any considerable appreciable deterioration in its fertility if ordinary care is taken. This cannot be said about modern countries which have come under cultivation within the last 100 years or so and where it remains to be seen whether they are able to stand intensive cultivation by machines for any great length of time. We will do well therefore not to go in headlong for mechanisation and to the extent we consider introduction of modern power-driven ploughs and cultivators necessary, we must see to it that deep ploughing is avoided. To meet the present emergency, we cannot depend upon mechanisation and have to improve the yield of the smallholdings as best we may. The few large farms and new land freshly brought under cultivation may be used experimentally for mechanisation, if considered necessary, but it is certainly not worthwhile creating a crisis by attempting forcible amalgamation of holdings for the sake of mechanisation.

To my mind, cooperative servicing of farming by supply to the members of a village society improved variety seeds, improved implements which an individual may not be able to afford, movable water pumps and wells for irrigation purposes that are not within the resources of most peasants, technical, advice regarding utilisation of waste, preparation of compost and improved methods of cultivation offer likely to be accepted by the villager than large-scale farming necessitating amalgamation of holdings and consequent loss of personal interest by each peasant in the improvement of his holding making that which will be everybody's business to nobody's business. They will also offer better facilities for the utilisation of every material and labour now available in the family of the cultivator and reduce to a minimum the evils of division of holdings which is constantly going on by reason of the operation of the laws of inheritance. They may also, in course of time, lead to the emergence of larger holdings by amalgamation or otherwise without creating a crisis in the even tenor of village life.

All this will be possible only if the peasant is given security and continuity of his tenure, freed from illegal and unauthorised exaction and extortionate interest realised by the landlord or the moneylender. Congress governments have been trying since 1937, when they first came into office to give relief in all these directions. The tenancy legislation in the permanently settled parts of the country like Bengal, Bihar and Odisha had for a very long time given occupancy rights to what are called settled *ryots*, which made their holdings heritable, with rent enhancible within certain strict limitation and from which they could be evicted except when it was shown that they had rendered them unfit for cultivation. Arrears of rent could be realised only by intervention of courts and by sale of the holdings in execution of a court decree and could not be made a ground for eviction. But there was no right to transfer the holding without the landlord's consent, although courts had held that portions thereof could be transferred. Legislation undertaken by Congress governments in 1937-38 in Bihar reduced the rent by 25 per cent on an average in the province as a whole, gave free right of transfer of occupancy holding on payment of a small fee to the landlord, and made only such portion of a holding liable to sale for arrears of rent as would be considered sufficient by court to cover the amount due. In the United Provinces, occupancy rights were granted and eviction for arrears of rent abolished and other reliefs given. In Madras, the government appointed a Committee which after a prolonged enquiry suggested abolition of permanent settlement. In Bombay Presidency where the *ryotwari* system prevails, reforms were introduced that improved the position of sub-tenants and sought to prevent the creation of conditions similar to those in temporarily settled *zamindari* areas. After coming into power in 1946, Congress governments had undertaken and completed legislation for abolition of *Zamindari* in Bihar and Madras, and bills for the same purpose are before legislatures in the United Provinces, Odisha and the Central Provinces. Unions of States have not lagged behind and have under active consideration the abolition of Jagirdari. One weakness was discovered in the proposed legislation for abolition of *zamindari* in Bihar and Madras and that was that the bills made no provision for any land reforms after the elimination of *zamindari*. The effect of the absence of such provisions was to replace the *zamindar's* agency for collecting rent by a government agency without in any other way improving the position of the cultivator. As a result of a conference of Revenue Ministers of the various provinces, the President of the Congress appointed a Committee to inquire into and report about the condition of the cultivator and make recommendations that might be considered necessary. The Committee has after a prolonged investigation submitted a report, which is at present being considered by the Working Committee.

Our yield per acre of rice land is 716 pounds as against 2,112 pounds per acre in the US, 2,668 pounds in USSR and 2,767 pounds in Japan and of wheat 567 pounds as against 973 pounds in US, 1720 pounds in USSR and 1,508 pounds in Japan.

Government plans for growing more food have been conceived on right lines and aim at giving more facilities for irrigation by means of small projects, supplying better seed, encouraging the making of compost and supplying chemical manures, improves implements, etc., and giving financial aid to the cultivator. But the high hopes from these plans have not been realised and it is necessary to find out the causes of this failure and remove them. The causes are manifold, but can in one word be summarised to the failure of the human element to rise to the occasion. Government agency has been found insufficient to deal with this tremendous task. It is not its fault. It has never been required before to handle such a problem and is unable to get out of routine and red tape, and to establish that intimate relationship with the cultivator which is necessary to win his confidence and rouse his enthusiasm. It continues to be as wooden as it was more than 30 years back when Mr Montague, the then Secretary of State for India, described it as such. There is also lack of coordination among different departments, which renders quick decision and peremptory implementation of decisions difficult, if not impossible. Only one illustration may suffice to show this. It is admitted on all hands that irrigation facilities, agricultural implements whether improved or not, etc., are necessary for the success of any "Grow More Food Campaign". Irrigation can be improved by constructing new wells, embankments and sluice gates, opening new channels, power pumps, rahats, etc. These require bricks, which require coal for burning them, steel and cement. Those have to be transported from one place to another. For making bricks, coal has to be obtained and transported from the coalfields, and this requires transport facilities that can be given by the Transport Ministry. Steel and cement have to be obtained from the Supply Ministry and have to be transported with the help of the Transport Ministry. Both the Supply Ministry and the Transport Ministry have many demands to meet. Shortage of steel and cement requires rationing after adjustments of claims of industry, house building—public and private—defence department, agriculture, etc. Transport Ministry has similar adjustments to make and also to make provision for moving millions of tonnes of food from ports and from one part of the country to another. Delays and even conflict in viewpoints regarding priority have inevitably intervened with the result that either essential material and transport have not at all been available, or made available only too late for proper or any use at all. The human element has had its own share in creating or at least in not solving these difficulties. Apart from the method and procedure followed in Government Secretariats which lend itself to procrastination and delay where quick decision and action are absolutely necessary, it is a fact which cannot be denied that not in a few cases other considerations than the mere merits of a case do influence decision. This is what is called in ordinary language corruption, which is not confined to the government agency but today pervades practically the entire structure of our society. No one seems to be content with his legitimate remuneration for the work he does or service he renders, whether he is a government servant, a merchant, a trader, a producer, or a supplier. Even a consumer resorts to it to get preference over other consumers. It does not mean that every government servant, every merchant, trader, producer, supplier or consumer is corrupt; it only means that the honest ones too, are unable to prevent or even to check corruption. To make the "Grow More Food Campaign" successful, the following steps are absolutely necessary.

On the government side: (i) at higher levels elaborate redtapism should be avoided; strict coordination eliminating delay in decision and implementation of decisions should be ensured if it is found impossible to unify command in this respect; (ii) those in charge of execution should be required to show results which should be checked by an independent agency; (iii) complaints should be quickly attended to and redressed; (iv) corruption should be strictly and drastically dealt with where proved and not ignored or passed over when detected but not proved; all concerned should be made to understand that it is not enough to be above suspicion, but that everyone should in fact be above suspicion and should be acceptable to those with whom he has dealings as being above suspicion; (v) mere paper reports, however satisfactory and encouraging, should not be accepted in the face of facts which contradict them; (vi) research or at any rate some of it should descend from its high pedestal to the level of discovering what will suit the ordinary peasant with his smallholding and very limited resources; (vii) results of such research should be demonstrated and brought home to the small peasant in a practical manner in small government farms, run by people who can mix with the peasant, understand and sympathise with his difficulties and render him such help, technical or otherwise, as is required and in a way which he can understand, appreciate and adopt, and this should be done without any unauthorised quid pro quo; (viii) implements and steel from which it can be made available through channels which have no bottlenecks and through agencies which can be kept free from corruption and corrupt influences; and (ix) monetary aid whether by way of loan or grant where required, should be made available and that quickly and without leakage. On the side of the people, there should be a realisation that agricultural and food production are a national concern and should not be looked upon as merely personal affairs of the individual engaged in it who is free to produce or not produce, or to produce what and in whatever quantity he likes. It must be realised by every peasant that he has to produce food not only for himself but for the country at large and after having produced it, he has to make available what he does not need himself for the use of the nation, taking for himself only a reasonable remuneration for what he has done. He should be told not to yield to threat or temptation and while insisting upon his right not to be mulcted. He should be prepared to do his duty to the country by putting forth his best in its service. Much of what is required of government agency and of the people at large can glow only from a proper appreciation of moral values and discarding greed and low selfishness which is clearly distinguishable from enlightened self-interest—in other words, we require character and a tolerable, if not a high, moral standard. But organisational and institutional safeguards against lapses of individuals should also be provided by the government and by society generally. It is possible to provide them if we can get at least some people who can serve as leaven and bring about a general uplift. The future of the country in this as in all other respects, depends upon its capacity to produce such men of character. We have so many complaints about control and black marketing. Control and rationing have proved to be a blessing in other countries where there was a shortage of supplies. But if enforced when they are unnecessary or enforced inefficiently and dishonestly, they can also prove to be curse. Efficient, effective and honest enforcement of controls depends mostly on the quality of the agency employed and on how the various parties play their part. Apart from the governmental agency, the other parties are the producer, the distributor and the consumer, and the failure of controls to give satisfaction has to be attributed to all these parties in varying degrees, that is, to the human element. Let us hope and pray that a free India will throw up that human element which will be able to rise above the prejudices and legacies of the past and carry the country at large on the crest of a wave of renaissance—a new birth. Mahatma Gandhi's *tapasya* did create heroes out of clay and threw up elements which carried the country out of the slough of despond to the heaven of freedom. Let us hope that the moral and spiritual resources of Bharat will be able to meet the new challenge and render that freedom not only secure but also fruitful by providing means for comfortable contented life among which food is the foremost.

No discussion on the agricultural problem can have any claim to completeness unless it deals also with the question of animal husbandary both as a source of additional food and as supplying bullocks on which our agricultural economy depends to such a large extent. But that opens up a big question which may not be considered here beyond stating that the policy of developing breeds of cows which give more milk regardless of the bullock which you may get from then has to be reversed and we have to concentrate on double purpose breeds.

Indian Journal of Agricultural Economics, Vol. 5, No. 1, March 1950.

11

R.K. PATIL

Planning for Indian Agriculture

EVER since the separation of Burma, India has been a net importing country in foodgrains. This deficit has been accentuated by the partition which also lost us important jute and cotton growing areas. The persistent shortage of food and raw materials symbolises the failure of our agricultural economy. A substantial increase in agricultural production, as also in productivity in the shortest possible time must, therefore, be a chief objective in our agricultural planning.

Many circumstances have combined to produce a favourable climate for the development of Indian agriculture at present. Foodgrain prices are relatively high, prices of raw materials are much higher, and if increased prices alone could provide incentives to increased production we should by now have succeeded in improving our agricultural output substantially. State and Central governments on their part have been doing their utmost in this matter during the last few years. Loans have been liberally advanced to cultivators, agricultural services have been expanded, a large number of minor irrigation, land improvement and land reclamation works have been undertaken, and various seed and manure schemes have been launched. In fact, considering the resources available, it must be admitted that Governments in this country have been according high priority to agriculture. Even in the Five-year Plan, the main stress is on the development of irrigation and agricultural production, power generation being, in a sense, incidental to irrigation.

It must be confessed that in spite of our best efforts, the picture today is not reassuring. It is true that we had unfortunate calamities and bad seasons, which have blurred the true picture and a good season will be safely said that we would be far from attaining cereal self-sufficiency not only in March 1952, but for some years to come. The targets in the Five-Year Plan aim at securing substantial increases in the output of foodgrains, cotton, jute, sugarcane and oil seeds. With the fulfilment of the plan, there will be a substantial reduction in our dependence on supplies from abroad, and the balance of payments position will have been significantly strengthened. But, all this would still not assure a level of agricultural output which may be regarded as a satisfactory basis for a welfare state.

What can we do in such a state of affairs? Must we rest content with the self-complacent feeling that we have done all we could and must rely on the inevitability of gradualness for the results that we need? Or is there anything that we can and must do, to increase the tempo of production? There can be no question about the immense scope for development. Our agricultural yields are among the lowest in the world, and it is well known that a few items of agricultural development, if they could be universally applied all over would solve our immediate problem in no time. On the other hand, we must also keep an eye on the distant future and ensure that any steps which we take to solve our immediate difficulties act also, as far as possible, as foundations for future development. This is the question which the Planning Commission asked itself, and I will try to explain briefly the conclusions we reached.

I propose to deal with this subject under the following heads:

(a) Administrative reforms.
(b) Land reforms.
(c) Other miscellaneous items.

Administrative Reforms

With our present resources, and even with the increased taxation that we have envisaged in the five-year period I do not think there is much scope for any substantial increase in the personnel of our administrative machinery. It is, therefore, of the utmost importance that we utilise our personnel only for such duties as cannot be discharged by either non-official or semi-commercial bodies. From this consideration there is a very good case for taking away all supply schemes, for credit, manure, implements, etc., from the purview of the Agricultural Department and entrusting these functions to non-official cooperative organisations or semi-commercial bodies. Pure seed schemes are in a different category. Till we are able to evolve registered seed farmers, the Department will have to undertake responsibility for the production and distribution of pure seed. Otherwise too much time of the staff which should be utilised for genuine extension work is today lost in doing this supply and clerical work. The same remarks apply to other spheres of administration.

Apart from this rationalisation of functions, there is, in my view, considerable scope for rationalisation of personnel, especially in the lower cadres. All technical knowledge of the development departments must ultimately flow to the villages in a comparatively simple form capable of demonstration by a person of average intelligence and capacity. If such a person could be properly trained he should be expected to represent many development departments. Apart from the rationalisation of functions, such a step would produce more efficiency in administration. The only objection to such a course would be that in this process we are sacrificing, to some extent, efficiency. This may be so, in part, and for a time. In fact, the very basis of this arrangement is that not much specialisation is required at the village level. Such a multi-purpose village service would be able to give more satisfaction and render better service. Just imagine anyone in this gathering being required to discharge such duties for a village. It is ultimately a question of raising the level of the village worker and giving him good training. If any of you can be a satisfactory multi-purpose servant, others also can be so. In my view, this idea requires to be extended to other spheres also. Even in this non-development departments we have now a variety of officers—*patwaris*, secretaries of *panchayats* and irrigation subordinates can be cited as examples—whose duties could be combined and this will lead to efficiency and economy in administration.

The recommendation of the Commission is that the activities of all the development departments should reach the village through a common agent. At higher levels which should consist of a group of villages, say 50 to 60, corresponding to a Revenue Inspector's circle, there should also be a multi-purpose officer who would supervise the work of seven to eight multi-purpose village servants. At the district level, under the Collector of the district the District Development Officers should function as a team coordinating the whole development work and reaching the villages through the common agent. At the state level, there should be a development Commissioner with a Board presided over by the Chief Minister on which will be represented the heads of development departments. We expect that through such a set-up there will be greater co-ordination of the development activities and Government will be able to discharge its functions more efficiently and effectively. Experience will alone indicate if this expectation will be justified.

At the village level, we have proposed a Village Production Council (VPC). At present, all the development departments try to further their activities by trying to reach the villagers individually. There is no organisation in the village which concerns itself with the work of agricultural and rural development in its entirety. The *panchayat* where it exists interests itself in village administration and cooperative societies have often too restricted a membership to entitle them to act as representatives of the whole village.

We want an institution which will focus as through a lens all the activities of the development departments and give to the picture that point which a focused lens alone can give. The Council will thus help to mobilise village effort for purposes of increasing production and for promoting rural development. In consultation with the villagers, the VPC will fix development programmes, ascertain and provide for the removal of the difficulties of the cultivators, see that standards of cultivation are maintained, that all necessary land improvement is made and that local labour and initiative is moblised for the purpose of village development. It will also hold crop competitions and take other measures to kindle enthusiasm in the villagers. In order to enable a VPC to discharge these duties effectively, an organisation will have to be developed which will assist it in meeting all local needs in respect of improved seeds, manures, technical assistance, credit, etc. Some executive authority will also have to be conferred on it to enable it to discharge the duties which are expected of it.

The details of such an organisation have to be carefully worked out. At present, the needs of the cultivator in respect of credit are met by Government departments and cooperative societies and in respect of seed manures and implement partly by Government and partly by private agencies. It is necessary to amalgamate all these agencies into a single supply organisation. The provincial centre of this organisation will work under the Provincial Cooperative Bank and may have different branches for credit, supplies etc. From there, at each level, right up to the VPC, it will function through cooperative organisations affiliated to the Central Supply Organisation. We will then have a central organisation dealing with credit, seed, manure, implements and other requirements of the cultivators operating through a series of subsidiary cooperative bodies right up to the level of the village. All the needs of a village will be collectively estimated and the VPC will obtain all the village requirements from the nearest cooperative society.

The Reserve Bank of India (RBI) has recently liberalised its policy regarding the provision of short term credit for agricultural production and marketing. The Reserve Bank Act has been amended extending the period of accommodation provided to cooperative central banks from 9 to 15 months. Certain other matters regarding the provision of agricultural credit on a more adequate scale are under examination by the Bank. If a sound cooperative structure could be developed in a state with the VPC or any similar body as the base for providing short term finance for credit and supplies, and linked with a cooperative apex bank through intermediate bodies, the RBI could play a much more extensive role in meeting the requirements of rural finance. With a major portion of our cultivators living on subsistence farming, the distinction which is made in some countries between the development of agriculture through an extension service and the rehabilitation of low income farmers through a specialised agency loses much of its significance. The great majority of our farmers are really low in a service capable of rehabilitating these low income farmers. The extension service must be such as can meet the genuine needs of the people. It must be capable of delivering to the people what they need actually in the context of their local conditions and not an agency which tries to give them something which they do not and cannot appreciate.

This brings me to the land problem and the future pattern of our agricultural organisation.

Land Reforms

What is the nature of the land problem in India? It has been estimated that our net sown area works out at 0.7 acre per head. Assuming a family of five persons, this gives us about 3.5 acres net sown area per family. If in a property balanced economy, about 50 to60 per cent of our population is going to be on land, then each family can have a net sown area from 6.0 to 7.0 acres only. Allowing for land reclamation and fallow and grazing areas, it could be said that at the maximum it could be possible for each agricultural family in India to have from about 8.0 to 10.0 acres of land. So much about availability of land. But it may be emphasised, this is only a hypothetical illustration based on two assumptions: that the present population depending on land declines from the present level of 70 to 50 per cent and secondly that other non-agricultural avenues of occupation are found for the normal increase in population.

In view of this low availability of land, what should be the future pattern of our agricultural organisation? The pattern must be such as would:

(a) Lead to maximum agricultural production and maximum employment for the rural production.
(b) Promote intensive use of land and effective investment in agriculture; and
(c) Be capable of being put into operation through a democratic administration.

Should the land be cultivated in small family units, each cultivator cultivating his own plot or should it be cultivated in large units? What are the technological, economic, commercial and social advantages of family farms and large-scale farms? Cannot all the advantages of large unit of cultivation be obtained through an intensive system of cooperation on family farms?

It is almost impossible to answer these questions categorically as they are framed. There are advantages and disadvantages in large and small units of cultivation, and any conclusion in favour of a particular system will have to be related to the particular circumstances of each country, to the man-land ratio, technique of cultivation capacity for corporate organisation and other social and agricultural conditions, besides, the expressions 'small units' and 'large units' must have reference to a particular norm, which would provide an optimum size for maximum efficiency under a particular technique of cultivation. They could also be used with reference not to an optimum size but a smaller one which yet admits of sufficient surplus to enable cultivation to be carried on efficiently. This is generally known as an economic holding. From the technological point of view, it may be stated that there is no reason why the rates of physical output, measured in terms of crop yields depend not on the acreage of crop grown, put on the quality of the land, intensity of cultivation, seed, water, manure etc. And as both crops and animals respond better to individual care and attention, the small farmer has an advantage on account of the personal incentive factor. This is the reason why higher gross production per acre may frequently be obtained from small farms. And the commercial and financial advantages of a large farm can be generally obtained with an intensively developed cooperative activity. Further, the economies of large farms are to an extent due to mechanisation, and on account of the very large human and cattle population of India, it becomes necessary to adopt a selective approach to the subject of mechanisation. On the other hand, where increased agricultural production is dependent on the application of capital, which has to be found from agricultural surpluses, a large unit of operation has definite advantages.

If economic holdings were available, and if population were not outstripping the growth of resources and normal economic development—a pattern of agriculture, in which such economic holdings are cultivated in family farms, might be adequate. Though smaller than the optimum size, the various advantages of a family farm could be favourably balanced against a pattern which involves collective action, and a diminution of freedom of the individual farmer. This statement will however hold good for a particular technique of cultivation. As techniques improve and mechanisation progresses, an economic size tends to become uneconomic with reference to the improved techniques. And unless there is scope for an increase in the size of cultivating unit, all the disadvantages of an uneconomic holding, higher costs, lower production and insufficient employment would result. This fact has to be taken into account in formulating long-term agrarian policy.

Historically, in a free economy, where there is full scope for individual enterprise in agriculture, there is a growing tendency, with increased industrialisation for the size of holdings to increase, so that it can adjust itself to improvements in agricultural techniques and increase in farm costs.

The view is widely held that a pattern of agricultural organisation in which land is divided into small units, independently cultivated by individual families, would suit our conditions best, and at the same time give maximum of production and employment for our rural production. This view proceeds on the assumption that there is very considerable scope for increasing yields in India if the land is given to the actual tiller and circumstances are created which would enable him to obtain a large share from his labour that he does today. This view is correct to a small extent only as a good proportion of the cultivated land of India is already farmed by owner cultivators. This was so even before the abolition of the *zamindari* as in all the states where the *zamindari* tenure prevailed, the actual occupancy cultivator has been given a permanent and heritable right in the holding. The question of providing more incentives to the actual tiller arises only in the following categories of land:

(a) Land cultivated by tenants at will.

(b) Land cultivated by large owner-cultivators through hired annual labour.

In the first, case incentives to greater production can be provided if tenants-at-will are protected and given security of tenure, and gradually enabled to purchase the lands of which they have become protective tenants. In the second case, if the terms of remuneration are suitably linked with the gross production from land, the necessary incentives would have been provided to this class also. We shall revert to this later, but the advantages of family farms are supposed to be the following:

(a) Being owner-operated, such cultivators have the largest incentive to produce more.
(b) Small farms frequently yield a high per acre production and are capable of absorbing the largest number of people on the land.
(c) The disadvantages of a small holding could be got over by an intensive system of cooperation for farm and non-farm activities, and
(d) Any other system would be difficult to adopt in India, where there is a tradition of centuries of peasant cultivation.

The examples of China and Japan are frequently cited to show that small farms and high yields can go together.

It is possible to argue that there is no reason why the rates of physical output, measured in terms of crop yields per acre, should have any relation to the sizes of farm as such. This statement is advanced for large farms as well as small farms. It is true that intensive cultivation is generally associated with small farms. But it is not absolutely essential for intensive cultivation to be limited to small farms. Intensive cultivation can be carried on on big farms too and larger yields per acre—larger than small farms—can be obtained on them. As one agricultural economist has put it:

> It does not follow that larger farms could not shift the character of the crops grown, increase the stock or produce a greater output per unit, if it were necessary and the means were available to do so. Small farms do not appear to be better placed for producing a high output because they are small farms, and it seems that the necessity of having to earn a livelihood is the main reason for the phenomenon of their greater output per acre.

It is also not correct to say that small farms can absorb a comparatively large number of people in the land. This really depends on the man-land ratio that we want to have and not on the unit of farming at all. Also, this, so-called 'absorption' means in reality larger disguised unemployment, on going back to a comparatively self-sufficient economy.

It is true that at least some of the advantages of a large unit can be obtained by an intensive system of cooperation for farm and non-farm activities, and Denmark and Holland are countries where such cooperative activity has developed very well. But in its most extreme form when one has cooperation for all possible activities, the distinction between the individual operator and the cooperative activity of the group narrows down so much that for all practical purposes it becomes a large unit. Experience has shown that such cooperative activity is not easy to develop and requires a lot of painstaking spade work.

So far as the adoption of a new system is concerned, it must be a gradual process. It cannot be thrust upon the people by a fiat of the government. The issue here is what is the best pattern of land organisation suited to our country when we are unable to provide economic units for all the people who have to be absorbed on the land.

We must now see what conclusions follow from the arguments which we have discussed. We have already seen that even on the liberal estimate of our cultivation supporting 50 to 60 per cent of our total population, the land available to each agricultural family will not be more than 8 to 10 acres. Such a holding cannot be economically cultivated except in the rice tract. For the millet region, a pair of bullock would frequently require as much as 20 acres for economic operation. The advantages of small units which we have referred to above apply to economic family units and not to small fragmented units of 8 acres or so. Even this size is based on the assumption that a very large section of our rural population would be absorbed in non-agricultural occupations. This process must be spread over a number of years and would itself require a substantial increase in agricultural production to facilitate it. This may not occur with our present system of uneconomic land holding and in a sense this assumption presupposes conditions, the creation of which might involve giving up the present system of family farming in very small units. It is well known that a very large number of our agricultural holdings are so small that they are well below 2-3 acres. The most appropriate size of cultivating units must, while providing the maximum scope for economies under the plough and bullock technique, enable sufficient surplus to be produced for capital formation and its investment on land improvement and yet be not too large to deprive it of all the advantages and benefits of small unit. As the availability of land in India is too small to provide units of such size for each cultivating family, the unit will have to be cooperatively cultivated by a small group of families coming together. The size of the unit will have to be determined for each tract, but it may frequently be found to be in the neighbourhood of the area of land roughly equal to land which can be cultivated by 6 pairs of bullocks, under our present conditions.

By definition, "an economic unit" should over a period, provide a surplus after allowing for cost of cultivation, for land improvement, replenishment of live-stock and other repairs. If the size of the unit does not provide enough to cover the costs, then the farm will decline and the condition of the owner will become progressively worse. There is a long period of resistance and stagnation before this may happen. In the meantime, the economy continues to function in a suppressed and stifling atmosphere, unable to provide the surplus of food and raw materials, which are the prerequisites of a developing economy.

Besides, such an organisation is also satisfactory from the national point of view. A country where agriculture is organised in uneconomic family farms cannot create, ex-hypothesis that substantial improvement of the land which is required for increased agricultural production. The agricultural economy then becomes stagnant, leading, as has already been suggested, to disguised rural unemployment and low agricultural output. This, in fact, is the present situation in India.

When, therefore, we consider reorganisation of Indian agricultural, we must adopt a pattern which will lead to the creation of holdings of an economic size and if, for that purpose, we have to resort to group farming, then we can easily go a step further and create holdings of an optimum size. It is in the light of these considerations that the Planning Commission has framed its recommendations for the future pattern of Indian agriculture. It visualises existing small holders coming into cooperatives and forming units of a size, which may approach the optimum in relation to the present techniques. Farms above this size, which are already reasonably surplus, are to be registered and thus, become subject to social control. It is through organisation on these lines that in the immediate future, we can hope to create that surplus in agriculture which can be re-invested in land improvement and which is the very basic condition for further increasing our agricultural production.

Before we begin to consider how such transformation to which we have referred could be brought about from our present pattern of landholding in India, another question has also to be considered. Are we to plan for maximum production from the land and providing full employment for the cultivating population only? When we assumed a population pressure of about 50 to 60 per cent of the total population on the land, we were visualising a stage when our surplus population would have been absorbed in other pursuits. This will take some decades to bring about and in the meantime, productive occupations will have to be found for all the village population. This responsibility must be accepted by the state and indeed on our ability to implement this responsibility our future economic development largely depends. Our main problem is finding the occupations for our unemployed and under-employed rural population, not only from a humanitarian point of view, but also from an economic point, as it is mainly through such utilisation that we can find tha capital for our economic development.

If this is correct, then the land resources of the village must be managed for the benefit of the whole village population, so that the maximum amount of employment of a developmental character is provided in the village. This objective will not be conflicting with that of the optimum utilisation of the village land for securing maximum production, as it is primarily the surplus from agriculture which will finance the additional unemployment that we are contemplating. It follows that for such a utilisation, we must have an authority in the village which will have full control over the land resources of the village, and will be in a position to utilise them to the maximum advantage. That is, the tiller of the soil will have to cultivate it, not as its owner, which must be the village authority, but on certain definite terms laid down by the village authority.

This village authority will have to concern itself with making the best and maximum use of all the village resources, including land and human labour. It will select such number of persons for cultivating the land as is required for the purpose. The rest could be employed on such works of rural development as land improvement works, minor irrigation works, construction of roads, afforestation, industries subsidiary to agriculture like poultry, vegetable and fruit cultivation etc. Any labour which cannot be absorbed in these occupations will have to be first absorbed in such cottage and small scale industries as help to increase national production. Production of consumption goods like *khadi*, oil, paper etc., which are already produced sufficiently cheaply and efficiently, should be resorted to as avenues of employment, only when the possibilities under the first two categories are fully utilised. While the village is the unit, in the matter of employment, it is of course not enough to limit ourselves to it. We have to expand cottage industries to the full limit of demand and not only for residual employment.

In order that an organisation on the lines suggested above may work successfully, the village authority should not only have full control over the village land and human resources, but it will also be necessary to see that the levels of earnings in different occupations of roughly equivalent skill, are about equal. Only under these conditions can surplus labour in the village be mobilised for purposes of rural development. This is really the key to the problem of utilising fully the vast surplus human power in the rural area.

The proposal of the Commission for cooperative village management ultimately envisage this state of affairs. The whole village including the land owners, cultivators, landless labourers, artisans will set up a body whose task will be to manage the land and human resources of the village in such a way as to find the maximum employment for the village population. Such a development would of course take some time. In the first stage, we have the VPC, registered farms and cooperative farms. The first stage of cooperative village management is the management of the village land by all the cultivators giving an ownership dividend to the owners. In this stage, the landless people as a part of the village population, would have a voice in the constitution of the village body but it is the cultivators and the land owners who may dominate the counsels of the body managing the village. In this stage, cultivation of the village would be carried on largely in the interests of the tillers and the owners, but the problem of finding employment for all the people would also receive attention from the start. Gradually, however, the situation will change and as soon as there comes a realisation of the wider duties and responsibilities of the governing body of the village, the full implications of cooperative village management would be realised. This is the ultimate ideal of cooperative village management, in which the whole village land is cultivated, not only in the interests of the cultivators themselves, but the whole village population.

The question now is how and by what stages the present pattern of our agrarian economy is to be transformed into the final pattern of cooperative village management. Obviously, if it is to be done in a democratic way, such a transformation must be by stages and be contingent on the generation of enough consciousness in the village and capacity to shoulder the responsibility of cooperative village management. It is this body which will ultimately be responsible for creating this consciousness and capacity.

The prevalence of a large number of tenants-at-will in any agrarian system tends to depress the level of agricultural production. Insecurity of tenure prevents proper land utilisation, manuring of the land and other land improvement. The first step in any programme of agrarian reform is, therefore to confer security of tenure on the tenants at will and regulate their rents. It is difficult to state how much portion of the land in India today is cultivated by tenants-at-will. It may vary from state to state, but on an average, about 30 per cent area is cultivated by tenants at will and giving them more secure rights is bound to increase agricultural production to some extent. The proposals of the commission envisage that this should be done in all the states and a ceiling has been recommended beyond which such protected tenancy rights should not accrue and gradually protected tenants should be enabled to purchase the ownership rights in the land.

Another important problem which is very much before the public is the question of ceiling on holdings. But before considering this question, two preliminary objections have to be dealt with.

When we consider the question of land distribution, the question of the optimum size and technique of production immediately arises. We have already indicated above, that the maximum economies from the plough and bullock techniques are available from a unit which can be cultivated by 5-6 pairs of bullocks and this should be the optimum unit for this type of technique. As for mechanisation, except for special purposes, like land reclamation or special operations which cannot be performed efficiently by bullock power, or in cases where speed of operation is an important factor due to the shortness of the season, it confers no special advantages on the community in our particular circumstances. And for these special purposes, advantages of mechanisation can be had by owing cooperatively the instruments of mechanisation. On the other hand, mechanisation does not necessarily increase agricultural yields. It reduces costs, by economising on man and bullock power, both of which are in abundant supply in India. At present, it is the large farmers who are using tractors, largely to save costs and increase their profits. Mechanisation, therefore does not confer any special advantages, and the retention as a long term policy of big capitalist farms, to enable them to adopt this system of cultivation is not justified, either from the point of view of increasing agricultural production or creating more employment.

The next question is would this imposition of a ceiling on cultivation reduce agricultural output? This question is very difficult to answer. In all probability, if a person cultivating the land takes personal interest in his cultivation, he would produce more even from the land above the ceiling since he would be able to invest in it a larger amount of capital and seed and manure; but this also depends on the ability and capacity of management. On the other hand, if this land were to be made over in small parcels to people who have not enough wherewithal to cultivate it properly, production may fall.

If however, the owner is an absentee and does not take much care of his land, particularly the comparatively inferior or marginal land, that land or a portion of it in the hands of an enterprising farmer would certainly produce more than the original owner was getting. But this contingency has been provided for in the conception of 'registered' farms. These registered farms would have to conform to certain regulations which would secure that the owner takes proper care of his land and utilise it to the maximum national advantage. On his failure to do so, the land can be taken over for management. Assuming that this arrangement worked effectively and agencies are available to government to take charge of such lands which are not being properly looked after by their owners, a system of registered farms would be the easiest and quickest way to secure increased production. The operation of a ceiling on holdings and the redistribution of surplus land might not give comparable results in the near future.

The advantages of a ceiling on holdings and consequent redistribution would thus accrue, if at all, over a long period. The degree of social justice likely to be secured would depend on the limit at which the ceiling is imposed. We have already indicated that as a rule most of the economies in the plough and bullock techniques can be secured if the land is sufficient to utilise the labour of, say, about six pairs of bullocks. We have no statistics to determine how much land would be available for distribution, if a ceiling were to operate on this basis, but it would perhaps be about 10 per cent of the existing cultivated land. The surplus land would, therefore, not be sufficient to make an impression on the vast social problem created by the existence of extremely small and uneconomic holdings and the large landless population. It would be difficult to distribute the surplus land amongst the landless, ignoring the extremely small holders, anxious to cultivate their plots, and the surplus land available would be so small that even the small holdings would have only small accretions to their existing sizes. The problem of landless labour would thus remain untouched.

Were such a distribution to be attempted to all, it will have to be done through the agency of the VPC. The land above the ceiling, the surplus land, will have to be distributed in the first instance to those small cultivators, who having the agencies of production with them, e.g. land, bullock, implements etc., yet have sub-economic level. For this purpose, a norm will have to be determined, for each agricultural tract and possibly for each village or group of villages up to which the very smallest cultivators would have to be brought up before land could be allotted to landless people. In practice, this will raise many difficult issues and the procedure may create a great deal of unrest in the rural areas. In view, therefore, of the more fundamental reorganisation of agriculture proposed by the Planning Commission and also on the practical considerations mentioned above, an approach to social justice through land redistribution does not appear to be an attractive proposition.

It will thus be seen that the idea of imposing a ceiling does not take us any further on the way of securing an agricultural economy, the foundations of which must be laid by any system of agricultural reorganisation that we propose. It will not lead to more employment, nor will any substantial measure of social justice result from it. It is really proposed by those people who believe in small individual farms and who see no other way of achieving social justice. As already pointed out, such a course involves not only a risk of a fall in production and the marketable surplus, but also confers no continuing advantage to the economy.

All these drawbacks are not present in the system of cooperative village management. This secures to us all the advantages of large farms, and also the individual incentive of a small cultivator, who has a definite work unit assigned to him. It introduces in the present system a new mode, which can galvanise the people and revolutionise rural economy. It provides an organisation in the village which can not only plan for the village agriculture but also find occupations for the village people. That seems to be the only way to do away with the stagnation that has come about in our rural economy through the present system of peasant cultivation of very small farm and what is more, it can be brought about in a democratic way and in stages. It provides sufficient time for the different interests affected to adjust themselves to the changed conditions created by the introduction of a system of cooperative management in the village.

How can we in a short time bring this about all over India—what are the preliminaries to be gone through to achieve this? The first step is to introduce cooperative farming on a large scale. The real bottleneck here is the absence of trained personnel with experience of cooperative farming and capable of taking charge as managers and supervisors to guide such bodies of farmers, who come together for cooperative cultivation. The first step, therefore, should be the creating of cooperative farms in each state where theoretical and practical training in the art of running cooperative farms is given. Another essential step would be the passing of enabling legislation authoring a village to enter upon a system of cooperative village management, after a majority of the villagers agree upon such a course. The legislation will prescribe details about the manner in which cooperative village management can be organised and run. The third step will be a national effort on a large scale designed to bring about cooperative village management over the whole country in defined stages. A plan should be drawn up for each state, indicating the stages by which the villages in the state could be brought under this system. With a system of cooperative village management, other bodies in the village, such as *panchayats*, cooperative societies and marketing societies will be integrated into one unit, and the village will become a truly cooperative community. What is required is determination on the part of the people, the political parties in power and the administration to build up this system in a substantial number of villages in the states. Indeed, on a long view, this is the only solution. If we do not bring it about, there will be no alternative to the imposition of a ceiling and the running of Indian agriculture in the same age-old fashion of small peasant farms. It is only because people have doubts about our ability and capacity to run cooperative village management that the alternative of having a ceiling on farms is being suggested; but once we make up our mind on this question, the way would be clear for the adoption of the reform on a large scale. New ideas take time to sink in but it is new and bold ideas alone which ultimately produced large results.

Whatever system of land reform we adopt, its full advantages can only be realised, if it is followed up by an intense administrative effort designed to canalise the enthusiasm thus generated into creative channels. We are apt sometimes to lay too much emphasis on generating mass enthusiasm and too little on the equally important duty of utilising it through well ordered official and non-official effort. Judging by what prevails today, it must be said that such a determination to achieve results appear sadly lacking amongst political parties and official and non-official elements in our society. Orderly progress requires a state of society in which the masses will have confidence that social organisation and effort are geared to a progressive improvement of their condition. There is something today in our social and administrative conditions which prevents this feeling being generated and this the real bottle-neck to progress. Administrative efficiency has to be raised to such a level that what is intended by the legislators and ministers does actually happen at the village level. Mere rousing of mass enthusiasm is not enough. It will be temporary and die down, and we will be left where we were, none the better for the reform or it will lead to a violent revolution leading to disaster.

The question can be raised if it is at all necessary at our present stage of social, economic and administrative development, to embark either on a programme of land distribution or one of cooperative village management, and how far such a programme is necessary on economic considerations for increasing production and how far it is necessary for satisfying the demands of social justice. We have already briefly dealt with the first aspect of the question. Cautious people would certainly doubt if a Village Council or a VPC (both can and should be the same) can undertake a programme, either of land distribution or of cooperative village management, and under such conditions, such a deliberate disturbance in our agrarian economy when we are woefully short of food and raw materials may be questioned. They argue that the land that could be distributed would form about 10 to 15 per cent of the total and in respect of the rest all the incentives that an owner who cultivates his own plot of land can have are already there. Therefore, a land distribution programme would not, by itself, contribute greatly to increased production, particularly as it is based on certain assumptions of doubtful validity, i.e., the effective functioning of VPCs. Would not the next step, they argue, in agricultural cooperation be the application of cooperation to all items of farm business outside the farm boundaries? Cooperative procurement of all supplies by the cultivators, cooperative marketing, crop insurances, etc., would be such item. As and when the villagers get used to these forms of cooperative activity and get enough confidence and experience of running these cooperatives then either cooperative village management or cooperative cultivation of surplus land or land distribution through a village body could be tried.

We have already discussed the possible effect of land distribution or cooperative management on agricultural production. There is nothing inherent in either of these reforms which will depress production. If anything, both are likely to increase production. It all depends on how the village organisations develop and function.

But it is time we took some positive steps in the direction of creating those conditions, which point in the direction of further development. It is ultimately a question of political judgement, if the demand for imposition of a ceiling can be withheld for any length of time, or whether it can be stayed by the introduction of a system of cooperative village management in a sufficiently large number of villages.

Agricultural Planning

It will be seen that whatever system of land reform is ultimately adopted, the VPCs are going to play a very dominant role in initiating that system and in working it up. The success of any land reform will depend on the extent to which the VPCs have a stable base and function effectively. It is they who have to arrange for cooperative cultivation. It is they who will ultimately develop into Cooperative Village Management. They are thus the very basis of the whole system of administration and future and reform. They must also be the base for crop planning. At the present day our targets in the Five Year Plan are purely notional. The basic condition for their realisation, indicated by the Commission, was an integrated price structure. This does not exist today. But even if it did, by itself it would not have provided an adequate instrument for reaching a particular crop pattern after five years. Ultimately planned pattern of agricultural production can only be reached as a result of systematic crop planning and the maintenance of a price structure capable of securing it. Both the factors are essential for securing a particular patterns of agricultural production. The agricultural plan for the whole country has to be broken up into regions and finally into each village. In a system of numerous plot proprietors covering the whole village area, the degree and intensity of crop planning will vary with the amount of control the VPC succeeds in securing over the whole village land and the body of cultivators. In an extreme case, it could replan the whole agriculture of the village not only to suit a particular land pattern, but also to secure the best land use. It could rearrange the whole village land into cultivable land, pasture land, vegetable and garden land and land fit for grazing and afforestation, and see that the cultivated land has the crop pattern indicated for the village. Otherwise, it may simply lay down the proportions of areas under different crops for each cultivator.

No statement of future agricultural planning can be complete without a reference to the problems of agricultural finance and marketing and agricultural research. These can only be briefly referred to here. The main bottlenecks in the development of agricultural credit and marketing on sound lines are lack of organisation, finance and warehouse facilities.

Besides production finance, the cultivator needs assistance for marketing his produce. Regulated markets have been established in certain states. There is a considerable scope for increasing their number and it is also necessary to ensure that these markets are properly supervised. The regulated markets should also develop storage facilities for the grower as at present he often has to sell his produce even in a declining market. Warehousing Acts have been enacted in the states of Bombay, Madhya Pradesh, Madras and Travancore-Cochin, but their implementation has been rather slow so far. Regulated markets, warehousing and development of cooperative marketing are very much inter-related and need to be developed very considerably in the next five years.

The cultivator would also be able to get better prices for his produce if important agricultural commodities are subjected to compulsory grading and quality control. This would include quality control of such edible articles, as milk, *ghee*, oil and such exported raw materials as lac, bristles and wool, oilseeds and vegetable oils, cashew nuts, spices, hides and skins, etc. A beginning is to be made in the next five years in this direction.

Much has to be done in the field of agricultural research to enable the cultivator to adapt the results of research to his everyday problems of cultivation. The results of our research today are not available in that detail, which should admit of precise and detailed instructions to the cultivator for adoption in the particular circumstances of his field conditions. Improved varieties have been evolved, but evolution of these varieties has not gone so far as to enable the scientists to recommend confidently a particular variety for a particular soil. The composition of manurial doses for different solid conditions and different conditions of cultivation is not available. Agronomic trials on different soils which would admit of detailed instructions to cultivators regarding the time of sowing, spacing varieties, improvement in systems of cultivation, manure etc., have to be conducted in different localities to obtain the knowledge which can be passed on to the cultivators. Our agricultural research has given us broad items of knowledge which need to be brought to the notice of the cultivator and practiced by him, and it is necessary to develop an organisation properly trained and equipped to pass on those results to the cultivators. But future progress must lie not so much in new discoveries as in obtaining by practical experimentation that detailed knowledge which is required to adapt known scientific results to the cultivator's field conditions. From this angle, the problems of extension and research are similar. Both call upon the scientists and the administrator respectively to step out from his laboratory and his office to the fields of the cultivator and demonstrate to him there what they have to offer him. A series of experimental stations throughout India for obtaining this knowledge and a research organisation to lay down the lines of their work and the coordination of their results is the immediate task, which faces our agricultural scientists.

Before I conclude, I would like to emphasise what may appear to be an obvious point. A major step in planning is improvement in the quality of administrative and managerial personnel whether in private business or in Government. The paucity of such personnel is, I believe, a more serious limitation on our ability to plan than even the paucity of finance. With the formulation of the Five Year Plan, we now have certain definite goals to attain. It is proposed that adequate measures will be taken for raising the necessary finance. Steps have to be taken simultaneously to strengthen and expand our existing resources by way of administrative and managerial personnel. I am not blaming the present organisation of our administrative services, but I am convinced that considerable reorganisation is necessary in order to enable responsibility to be attached at different levels for failure to obtain results. This is not a matter of framing new rules and regulations but of creating a new outlook and of imparting new skills. The problem really links up with the wider one of improving the tone of public life in general and stressing those human qualities on the strength of which alone all national endeavour must be based. In politics, in business, in administration—in all walks of life—the problem before India today, in the last analysis, is the raising of the quality of personnel; for, what is planning but organised human endeavour in the interests of defined ideals?

Indian Journal of Agricultural Economics, Vol. 7, No. 1, March 1952.

12

C.N. VAKIL

Role of Agricultural Economics in Indian Economy

Paucity of Economic Data

As is well know, though efforts at planning in India were first made by private organisations, the first, official step in this connection was taken by the Government of India (GoI) in 1944 when the Department of Planning and Development was constituted with Sir Ardeshir Dalal as the Member in charge. Sir Ardeshir was one of the authors of the Bombay Plan which, along with the National Planning Committee, was responsible for creating planning consciousness in the country. Sir Ardeshir expected that the GoI would have at its disposal adequate material by way of statistical and other information for the purpose of planning, and that to the extent to which it was deficient, adequate machinery would be created to make up for it without delay. He naturally looked for guidance and help in this connection to Sir Theodore Gregory, then Economic Advisor to the GoI. Sir Theodore pointed out the paucity of data for the purpose of planning and advised that under the circumstances, planning in India was not possible. It was a sad commentary on the state of affairs in the country then, that a person of the eminence of Sir Theodore, who was specially brought out to train suitable Indian staff for giving economic advise to the Government and for creating suitable machinery for the collection of scientific data as the basis for such advice, should have damped the enthusiasm of the Planning Member in this manner. A Departmental Committee consisting of officers and Economists in Government Departments with Sir Theodore Gregory as Chairman was appointed to devise ways and means to improve the statistical machinery of the Government.

One does not know the fate of this Committee. In the meanwhile, several departments dealing with economic affairs were created in the different Ministries of the GoI. At the same time, several Provincial Governments created Departments of Economics and Statistics to help in the collection and interpretation of economic information and the Reserve Bank of India (RBI) enlarged the scope and activities of its Departments of Research and Statistics. This move, in a way, had some adverse effect on the Universities inasmuch as some of the trained staff in the University Departments of Economics was employed by government at more attractive scales of remuneration, and as the government departments began to increase their activities, younger qualified recruits in this subject found employment in these Departments more attractive than in the Universities. Research activities in the universities in economics in general and in agricultural economics in particular, suffered in consequence.

Tendency to Ignore Economic Facts and Analysis

Before things could take a better shape, we had the uncertain political situation in the country soon after the cessation of hostilities in 1945. This was followed by Independence in August 1947, accompanied by the partition of the country. The new economic situation which was thus created had to be handled with great care and foresight. Those responsible for the partition agreed to it on political grounds without realising its economic consequences and significance. Those who came to power were, however, anxious to bring about rapid reforms of the type that they were advocating while in opposition. Some of these reforms affected the basic economic life of the country such as changes in land tenure, tenancy legislation, relief from agricultural indebtedness and efforts to introduce schemes of improvement for the development of agriculture such as large irrigation projects, reclamation of land and development of rural banking. The stagnant condition of agricultural life in the country and the systematic neglect of the farmer and his work became the concern of the well-meaning politician now in power. It was believed that by passing quick legislation for this or that reform affecting land and farm finance will give the desired result at least in the next millennium. In his anxiety to achieve rapid results, the politician was impatient to the extent of ignoring facts or even of making efforts at obtaining facts. Wherever facts were available, he was equally impatient regarding its correct interpretation; he was willing to trust his own practical sense instead of the trained mind of the expert. As a result of this, we got legislation for the abolition of *zamindari* without the means to implement the same or without the realisation of its reactions on production under the existing circumstances. The effects of the various Debt Relief and Money-lenders' Acts were neither foreseen nor examined. We also noticed wild schemes of simultaneous self-sufficiency in food, in cotton, and in jute at a time when we were seriously in deficit in each of them, showing total ignorance of the elementary facts of the situation or want of capacity to appreciate the fact that with a limited area under cultivation it was not wise to make such an assumption. The inter-relationship between different crops, both with reference to production and prices, was ignored both in public discussions and executive decisions. Nowhere was this tendency more apparent than in the field of prices, where the changing mood of the political leader more than economic facts of economic interpretation decided the issue, either of fixations of prices or of changes in policy such as control or decontrol.

Confusion of Thought

In the complex economic situation created by the partition and later by the merger or union of former Indian states, the above tendency succeeded in adding confusion. The confusion was so great that anyone trying to understand in simple clear terms, the policy of the GoI as a whole or of any given State government in matters affecting agricultural progress in the country, found himself at sea because he was faced with contradictions and conflicting measures. It was under this background that the Planning Commission was appointed. It was charged with the function of evolving a coordinate plan of economic development for the country. The task of the Commission was not at all enviable. It did not have adequate data on many aspects of its work, particularly in the principal economic activity, *viz.*, agriculture. It was not allowed time to make up for this deficiency because it was asked to produce a report at short notice based on the existing in adequate material. Both in the Draft Outline of the Five Year Plan issued last year and in the Final Plan issued recently, the commission has proceeded on assumptions, perhaps with full awareness of the uncertainty involved in the same. For example, it has assumed that prices in general in the country will not rise; if at all, they will slowly reach a downward level. For this basic assumption they are naturally in favour of effective price control and have pointed out that as food happens to be responsible for 60 per cent of the budget of most people in the country, control of prices is essential for the success of the Plan. It was rather paradoxical for those interested to find that at the time when the Planning Commission was working out its plan on this assumption, a psychology of decontrol was being created in the country by responsible persons. If any evidence were needed of the chaotic conditions in which our basic facts are and of the queer lack of unanimity as to their interpretation it was provided by the recent debate on food controls in the Parliament. In view of this, the remarks of the Planning Commission on controls in their Final Report do not inspire confidence. If in matters so vital to the well-being of the people, we are so ignorant and so uncertain, one can imagine the extent of disregard of facts, in several other questions, which though less urgent are none-the-less significant for sound and successful planning.

Controls

In this connection, I would like to make a few observations on controls in general and price control in particular. In a country suffering from inflationary pressures and desirous of undertaking a large scale development programme, controls—particularly those governing prices—have a significant function. The Planning Commission has advocated the continuance of certain price controls. Sections of the business community, however, believe relaxation of controls would serve to increase industrial production. It is difficult to find much substance in their argument. For example, when the government experimented with partial control of the sugar industry, manufacturers had to sell a certain percentage of their output to the government at a fixed price; the remainder could be sold in the free market. A noticeable increase in the sugar production resulted, but at the cost of reduced acreage in foodgrains. In the same way, the increased foreign demand for oilseeds led to a reduction in the acreage devoted to cereals. Unless much land is reclaimed for cultivation, or unless new irrigation works permit an increase in the cultivated area and in the yield from existing acreage, the prospect for improved agricultural output seems to be limited in the short run; in other words, an increase in prices of agricultural commodities would not lead to a rise in aggregate output. Higher production depends upon large-scale capital investments of a type far beyond the reach of the ordinary farmer.

As long as aggregate supply of most commodities falls woefully short of total demand, and as long as the possibilities of a speedy expansion in their over-all production are sharply restricted, the nation must depend upon price controls to allocate resources in accordance with the pattern of social priorities established by the government. The complaints of private business on this score are justified in so far as the controls have operated unsatisfactorily because of administrative defects and a lack of integration among the various controls. An improvement in administrative efficiency would go far to abate prejudice against controls, which are necessary for the smooth operation of the economy.

Neglect of Agricultural Economics

In a country like ours which is predominantly agricultural and in which we are faced with the problem of providing the elementary needs of a growing population, one would have expected that systematic attention would be paid to the collection of correct data on all agricultural problems, to their analysis and interpretation on scientific lines and to a continuous examination of the inter-relationship of different agricultural problems inter se, as well as with the other sectors of the economy. One would also expect that those in authority would do everything in their power to encourage such work by all possible means so that the country might be fully aware of the exact situation, and whatever steps were taken either by government or by private parties were taken with knowledge, and therefore, with reasonable prospects of success. Some of the unfortunate tendencies referred to above would not even be expected under normal conditions. But they have to be mentioned to show that such unexpected things do happen in our country. It is a tragedy that a country which is spending crores of rupees for the development of scientific and technological studies and research, and can boast of some of the most imposing structures with the most well-equipped laboratories in the world, neglects the basic facts of our life in the field of agricultural economics and allows such vital problems to be the plaything of the politician, who thereby treats the vast mass of the population in the country as a living laboratory on which to experiment with his fads, which though exalted by the term practical wisdom, are sometimes based even on intuition. Already there is clear evidence of unemployment among technically trained persons; but without paying any attention to this economic fact more institutions for technical education are being created costing crores of rupees. Economic progress in the modern world depends considerably on technological studies and research, but it has to be remembered that unless there is a parallel progress in the field of economic information, economic analysis and interpretation, we shall have a lop-sided development resulting in shocks and inevitable failures.

In this connection, I may refer to the view of a foreign expert of the status of Prof Ashby, Director of the Oxford Society of Agricultural Economics, who was invited by the Indian Society of Agricultural Economics in 1949 to India. He referred to the amazing concern of the government in this country and observed:

> Having regard to (a) the area of agricultural land, (b) the size of the agricultural population, (c) the importance of agriculture in the national economy—it's actual and potential contributions to the national wealth—I am appalled at the small provision made for investigation and research in Agricultural Economics..... Recognising that India is a relatively poor country, it is still true that in comparison with other applied sciences of agriculture, Agricultural Economics has been starved.

The weakness of the existing organisation for the collection of basic information is well known. The persons in charge of the work at the village level are ill-qualified and low-paid. The work of collection and presentation is done mechanically. There is complete absence of the research angle. In fact, there is a tendency among government departments to confuse collection of economic statistics with economic analysis and interpretation. Though some efforts have been made in recent times to improve the collection of economic statistics, there is no change in the attitude of government, so far as improvement in arrangements for economic analysis and interpretation are concerned. It is not realised that the handling of economic statistics by the lay politician may do more harm than good, if the data are not properly analysed and interpreted by the economist.

In a way, the British administration took more pains in such matters. For example, the Settlements Reports—apart from their political implications and anxiety to increase revenue receipts—were very through and comprehensive and contained a detailed assessment of the prevailing economic conditions. The Deccan Agriculturists' Relief Act was subjected to systematic periodic scrutiny and results were published which revealed all its drawbacks. The volume of similar legislation has now increased manifold. But the corresponding efforts at collecting the basic facts relevant for the purpose have not increased. There is also a striking unwillingness to have a scrutiny and evaluation of the working and the effectiveness of the new legislation and other administrative acts. And when independent efforts are made at such an evaluation they are frowned upon. For example, the Bombay School of Economics in collaboration with the Reserve Bank conducted an investigation with a view to discovering obstacles to the Grow More Food Campaign. When the results were published, their validity was stoutly contested from official quarters. Subsequently, the Government of Bombay was compelled to appoint an official Committee to look into the question, and I learn that all its major findings amply supported our conclusions. People may lose faith in the capacity of legislation to effect change, if unimplemented or worse still, if unexpected results emerge from it. The need for continuous examination of the effects of rapidly growing legislation relating to agriculture is obvious.

By now some of the Departments of Economics and Statistics in the Central and state governments have been considerably expanded. The Department of Economics and Statistics in the Central Ministry of Agriculture and the Bureaus of Economics and Statistics in State governments have large staff. While considerable attention is paid to the compilation and publication of statistical data in a routine way, one wonders whether these departments are encouraged to do correct analysis and interpretation of the data and help effectively in the formation of policy. One feels that those in charge have to watch the changing mood of the politician and the administrator and adjust their work to suit the same rather than in the interests of truth. It would be appropriate to have a searchlight on the working of these departments including the methods of recruitment of their research staff, and opportunities available to them to do independent analysis of the problems before them, and the extent to which their work influences policy. These remarks also apply to the Department of Research and Statistics in the Reserve Bank of India (RBI), which has done some good work and which is perhaps the most expensive of similar departments in the country with an estimated expenditure of ₹10 lakh a year and is still going strong. I may be permitted to mention here that the country owes a debt of gratitude to Sir Chinaman Deshmukh who as Governor of the RBI took special interest in its Department of Research and developed it. Not being subject to the exigencies of the fluctuating finances of a government, the RBI could set apart liberal funds for the purpose. While some of the work done by this department is creditable and some of its members have proved worthy of their task, one may say that there is considerable room for improvement in the working of this department, including the methods of recruitment of its staff. The method of recruitment and promotion of the research staff both in the RBI and in government departments is a matter of concern to those of us who are engaged in the training of research workers in economics in general and in agricultural economics in particular. Instances are not rare in which persons with ordinary foreign degrees are preferred to those who have done approved research work in the subject concerned in Indian universities.

Economic Analysis and the Plan

In spite of the fact that more than 40 per cent of the proposed expenditure of the Five-Year Plan is to be devoted for agricultural development, few efforts have been made to carry out competent research both as a preliminary to the Plan as well as for its subsequent evaluation. A good deal of the work is based on vague assumptions and beliefs. For example, regarding rural electrification, it is assumed that once electricity is produced rural areas would be able to absorb the electric energy. The question whether the farmers will have the necessary equipment and resources to utilise electricity even if it is available at low rates is not asked. Similarly, the lack of coordination between electricity planning and industrial planning is well-known. This has resulted in the paradoxical situation in which large electricity schemes were planned in areas where industrial possibilities were not even considered, and expansion of electricity supply was not thought of in existing industrial areas where the demand was rapidly growing. It was left to non-official research workers to point out this anomaly before some adjustment could be made. Similarly, regarding the agricultural targets in the Draft Outline, their correlation with the available financial resources was not properly examined. The same is true of minor irrigation projects for which sufficient provision was not made in the Draft Outline. In this case also, it was left to private research workers to point out the defect. Similarly, the fact that the average cost for bringing an additional acre of land under cultivation varied widely in the case of tanks and tube wells, and also from one area to another was not studied. Though this was pointed out, it is not known whether these correlations have been properly examined. Though in the Revised Plan, the Commission shows awareness of the problem and realises the importance of detailed cost and benefit estimates, it does not seem to have considered how the work is going to be done.

The most glaring example of the absence of economic analysis and interpretation is in connection with the recent food policy to which I must again refer. The National Sample Survey (NSS) is reported to have calculated on a sample basis that agricultural production is underestimated by 25 per cent. One wonders whether we should rely on the accuracy of a small sample in a country with such diverse agricultural conditions for this conclusion. On the same analogy, production in 1948 must also have been under-estimated by 25 per cent and therefore, decontrol ought not to have led to higher prices. Apart from this, it may be pointed out that what is relevant is not merely the collection of statistics of production, but how the economy would behave and how the distribution of grain would be affected, if controls were removed. In other words, the question to be asked is, what is the amount of agricultural surplus that would be marketed in the community in the absence of procurement and comprehensive controls? The discussion of such a vital matter as food control by lay politicians in Parliament on the basis of such statistical calculations without proper economic analysis and interpretation of the relevant issues is a sad commentary in the prevailing confusion regarding the treatment of economic problems in the country at the highest level. In this connection, we may also draw attention to the fact that in the wake of the decontrol policy in Madras, economists predicted that prices would rise though the lay public were beguiled into the belief that they would not rise, so much so that the Madras step was hailed as a bold one to be imitated blindly by the rest of the country. It is well-known now that prices have substantially increased in spite of the fact that stocks were released on the market from imported grains.

Need for Independent Economic Analysis

It may be pointed out that economic problems are often mixed up with political issues and that it is not unusual to find that political considerations often weigh in deciding economic policy. The harm done to the country at large in consequences of this confusion is not realised. The lay politician often relies on economists and statisticians employed in government departments not only for his material but also for his brief. He does not always expect them to tell him the truth; he wants them to find arguments to support his conclusions arrived at on non-economic considerations. In view of this, in a democratic constitution, it is essential in the interests of truth that economic analysis and interpretations should be beyond the influence of those in power and should be conducted by non-official academic institutions on scientific lines. The research workers in such institutions should be placed in such a position that they would be free to tell the truth without fear or favour, so that all parties in the country might at least have a common basis on which to work. In view of the fact that the state in India is trying to develop into a welfare state, and that democratic planning is being rapidly organised, it is hoped that the importance of this objective will be realised by all concerned. If this is realised the importance of encouragement of research in agricultural economics in the universities and independent research institutions would be easily appreciated.

Arrangements should also be made to remove the existing barriers between government departments and the universities by organising more active cooperation and exchange of personnel. As in other countries, academic persons should have opportunities for consultation by government departments, and research workers in government departments should have opportunities to work in the universities to have that independent and scientific outlook on their work which they lack. One of the reasons for the existing barriers between these two types of workers is the wide gap in their respective emoluments leading to a superiority complex among those in government departments and elsewhere, which leaves no impetus in them to do quality work, and which is bound to be resented by the university research worker, who prefers his independence to money.

The Role of the Universities

It will be easily conceded that the most appropriate parties to undertake the study of and research in agricultural economics are the universities. It is their function to arrange for imparting instruction and spreading knowledge about this subject. It is also their function to train workers who can help in the dissemination of correct information and problems, relating to agricultural economics, and, above all, to carry out research in the subject. Besides the Departments of Economics of the universities, there are a few research institutions which can also do similar work of one type or another. So far as collection of primary data is concerned, the government must improve its own administrative machinery for the purpose. Some of the primary data are collected as a by-product of administration. In other cases, special machinery has to be attached to the government departments for the purpose. Whereas this is true of routine primary data, in those cases in which governments require special studies or surveys or enquiries for specific purposes, the universities are obviously the most appropriate parties for carrying out the work. So far as the interpretation primary data or of material specially collected in a special inquiry is concerned or their correlation, research workers in the universities should be entrusted with the task. Among the advantage of doing so are: (i) that the work in the universities would be without bias and, therefore, more accurate and reliable; (ii) that the universities would thus get an opportunity to train young workers in this type of work who would be useful to the country in similar work in future; (iii) that besides helping the universities to build up in this manner their research activities, on the whole the work would be cheaper, as any effort to create ad hoc research organisations by government proves not only costly but also involves administrative bias and, above all, unnecessary delays. In a democratic constitution it would be easy for all parties to have confidence in the results of the work of the universities in matters of this kind than in a government agency, which may be under the control of a given political party at any moment of time.

This suggestion naturally leads to the question: are the existing university departments of economics and similar organisations in the country equipped for the task? The only answer that can be given to this query is that the continuation of the present attitude of ignoring the universities in work of this kind, adds to the existing neglect of the universities in their efforts to develop research in agricultural economics. Apart from other special funds, which should be earmarked for this purpose, even the diversion of funds which are now being spent by the government for ad hoc enquiries to strengthen university departments would result in permanent advantage to the country as a whole. Whereas continuity and tradition would be gradually developed in this type of research work done by the universities, in the case of ad hoc enquiries by government such experience is often lost. I would, therefore, suggest that immediate steps be taken for strengthening the existing departments of Economics in the universities, by the addition of the separate sections in agricultural economics to them. The courses and syllabus should be suitably revised to enable those interested to pursue this study. Scholarships and fellowships should be liberally endowed to enable qualified persons to undertake research in agricultural economics. Small sample research projects should be undertaken by these sections on their own in their respective areas to train research workers. For this, they should have adequate funds at their disposal. Both the Central and state governments should first think of utilising the resources of the universities, for any given enquiry in which they might be interested, because by doing so they would not only get their work done at smaller cost and more efficiently, but would also provide training to young workers in future. Once this arrangement is set on foot in a proper manner, it would be possible to expect private donations to flow into the same channel.

Finance for Research in Agricultural Economics

I may be reminded that this is all a question of funds and that the funds at the disposal of government are limited. But if the basic facts and tendencies in the economic life of the country are not correctly known and interpreted for those in authority, the expenditure which they want to incur for development would be wasted. Compared with the total expenditure earmarked for agricultural development in the Plan, the percentage that would be required for the purposes of developing study and research in agricultural economics would be insignificant, if we take into account the immense advantage likely to result from that effort. What is lacking is not so much funds for the purpose, but the desire and the will to understand and appreciate the problems as well as the approach suggested here. Instead of keeping the universities as water-tight compartments and condemning them as academic or unpractical, if efforts are made to see that the life of the people is reflected in the work of the universities and that the results of such work and such training are utilised with advantage in government policy and measure, we shall have that harmonious and well-balanced development for the progress of the country for which we all aspire.

In making this suggestion, I may refer to the experience of the United Kingdom (UK) and the United States of America (US) where universities and research institutes are liberally assisted by government and generously endowed by private donors for work of this kind. It is also true that the work of the universities is kept free from political bias so that all parties are able to rely on their conclusions. Besides, there is close cooperation between the research departments of the universities and those of the government which is conspicuous by its absence in this country.

I have said that government does spend sometimes lavishly on some pet schemes of research in an uncoordinated manner. It sets up ad hoc bodies to do the work because it does not have the correct approach on the best methods of carrying out its work. In order to make my point clear, I shall try to give some illustrations. Several illustrations could be cited of this tendency; I shall, however, satisfy myself by referring briefly to only four of them.

Inquiry Regarding Agricultural Workers

The Ministry of Labour, Government of India, undertook an All India Enquiry into the conditions of agricultural workers in June 1949. Among other things, the enquiry was intended to collect data regarding unemployment, earnings, cost and standard of living and indebtedness of agricultural workers. After a preliminary enquiry which covered 27 villages and 3,000 families, the method of enquiry was revised and the rest of the enquiry was conducted with the help of state governments. I understand the total expenditure spent on this enquiry exceeded ₹10 lakh. Instead of entrusting the work to different university departments of Economics, it was laid down that the organisation created by the Ministry of Labour for the purpose should ask university professors and students to give voluntary assistance in the collection of data. I remember, while this was in progress, that the person in charge of the Bombay office of this organisation wrote to me letters asking me in the name of the GoI that a university professor and some students should be at his disposal for collecting the necessary information. The person who wrote did not seem to be competent to handle the subject-matter of the enquiry, much less to deal with scientific workers in seeking their cooperation. He was also not aware of the conditions under which university departments functioned and the way in which they could cooperate in such an enquiry.

National Sample Survey (NSS)

In connection with the work of the National Income Committee, I understand the data are to be supplied by a separate organisation called the NSS. Among the fairly wide objects of this survey is the collection of dependable data on different socio-economic aspects of rural life in the country. The number of schedules to be filled is large, and some of them are very detailed and quite impressive. What I am concerned at the moment is the way in which this is organised from the point of view of the encouragement of studies in agricultural economics in the universities. I understand that a large sum of ₹35 lakh had been set apart by the GoI for the purpose and has been placed at the disposal of the Indian Statistical Institute, Calcutta. As I do not have first-hand knowledge of the way in which things are done, I have tried to find out the position as best as I could, and subject to correction, I would like to make the following observations. In the first place, this enquiry should have been conceived not merely as a statistical enquiry. It is essentially an economic enquiry, though its form may be statistical. I am not aware whether competent economists have had any effective share in the planning and execution of the work. I am not aware whether the work is being carried out on some uniform pattern of organisation in different parts of the country. I can, however, say with confidence that the University Departments of Economics have in most cases been kept in the dark. One and a half years have already gone since the survey started its work. We do not know to what extent it has helped the work of the National Income Committee. We are not aware whether this survey is likely to become a permanent organisation, because we find that under this assumption, it is trying to take up other work for which it is obviously not suited. One would like to know how the large funds at the disposal of this organisation are spent, whether there is any financial control over it and whether still more funds are likely to be given to it.

Rural Credit Survey

The third enquiry to which I would like to make a reference is the Rural Credit Survey now being carried out by the RBI. Arising out of the recommendations of the recent Committee on the Development of Rural Banking, the RBI is now anxious to play its part so far as rural credit institutions are concerned. It appears that the RBI was anxious to complete the enquiry in record time, and came to the conclusion that the only way to get its research staff and other parties associated with it to do the work in time, i.e., in 9 months from the start, was to provide the necessary drive. I understand that for this purpose, a small committee of five persons has been set up by the RBI two of whom are administrators and one of whom is the Chairman. I am aware that in a few centres the cooperation of the universities or research institutes has been sought. But these appear to be exceptions. It is not easy for me to understand why the RBI which has a very costly Department of Research and has done some useful work, should not have thought of parallel departments of research in the universities as the appropriate parties to cooperate in this work. So far as the expenditure incurred for this purpose is concerned, I understand that it is likely to be about ₹15 lakh.

Evaluation of Community Projects

The Planning Commission has recently set up an organisation for the Evaluation of the Community Development Projects in the country with the help of the Ford Foundation. I am glad that we have at the head of this organisation one of our valued colleagues, Prof D.G. Karve. This organisation is supposed to be independent, though it is a branch of the Planning Commission. If the Commission really wanted it to be independent, I submit in all humility, that the work ought to have been entrusted to the Universities, with a small staff at the Centre to coordinate the work, Prof Karve would have been really independent in his work if he were functioning in cooperation with the universities, with whom he has worked all his life. I am also surprised that the American authorities in general and the Ford Foundation in particular, did not think of this, in spite of the fact that the universities would be first considered for such work in the US. I venture to suggest that even now it is not too late to make the necessary change, because the work has hardly begun. It must be conceded that the different universities in the country would have greater and more lively interest in the projects in their respective areas, that they would have greater local knowledge and facilities and that they would be able to recruit and train in this way would have been utilised with greater effect, better results, wider training, and above all, for encouragement to the universities to develop research in agricultural economics.

Indian Society of Agricultural Economics

In the somewhat dark picture regarding the difficulties experienced by universities to develop the study of agricultural economics that I had to paint, there is a silver lining. This is provided by the work of the Indian Society of Agricultural Economics. For the last 12 years this Society has made an attempt to convert itself into a research organisation. It has published several valuable research publications and its quarterly journal has now an established reputation both in India and outside. It has also created interest in the subject by its Annual Conferences, and by keeping contact with similar bodies in other countries including the International Conference of Agricultural Economists. In spite of indifference or want of response from the authorities, the Society has continued to persevere in its efforts to draw pointed attention to the need for suitable encouragement of the study of and research in agricultural economics. Not being disappointed at the apathy of government, The Society at the instance of its President and Mr Elmhirst, President of the International Conference of Agricultural Economics recently made concrete suggestions to government calculated to rouse interest in agricultural economics, not only in the country generally but also among those in authority. The first suggestion was that the GoI should invite a well-known agricultural economist to advise them regarding the role of agricultural economics in the Ministry of Agriculture as well as in the rest of the country. The Society suggested the name of Prof Theodore Schultz in this connection, and we are happy to find that as a result of this Prof Schultz is likely to be in India soon. Let us also hope that the advice that he may be able to render will be followed up by the GoI without delay.

The second suggestion made by the Society was that the GoI should sponsor an invitation to the International agricultural economics Conference to hold its next session in India. The object of this suggestion was that the meeting of International Experts on this subject in India to discuss world problems in agricultural economics would rouse the necessary interest in the country in general and in those in authority in particular, so that it might be possible to remove the prevailing apathy towards the subject and bring about the desired progress. I am glad to be able to say that the GoI did send its representative to the recent International Conference at Michigan and an invitation was extended. Unfortunately there was some delay in doing so, and it was not possible for the International body to accept this invitation for the next session which they have decided to hold in Finland. I have no doubt however, that the invitation of the GoI will stand and be repeated at Finland for the meeting thereafter.

The third suggestion of the Society was for making a special financial provision for research in agricultural economics. So far as the source from which the necessary finance should come, Sir Manilal Nanavati has repeatedly made a useful suggestion *viz.*, that a certain percentage of the profits of the RBI should be constituted into a reserve fund for creating the necessary organisation at the Centre for the development of agriculture, including research in agricultural economics. This principle is in operation in Australia. In fact in an agricultural country like ours, what is novel is that such a provision has not yet been made. By doing so the work for the development of this subject will be independent of the changing financial situation of the government, and there would be continuity of resources from which to finance such a desirable object. It is expected that there would be unanimity of opinion on such a proposal, and that all concerned would wholeheartedly support the early implementation of this suggestion from so experienced a person both in agricultural economics and in central banking as Sir Manilal Nanavati.

In the first place, the government should see that its own work in this connection is put on a proper footing. There is long-standing need for having an Economics Wing of the Indian Council of Agricultural Research, so that the economics of the various scientific schemes suggested for the improvement of agriculture are properly studied as well as certain problems of an all-India nature are effectively examined. The same applies to the Department of Economics and Statistics in the Ministry of Agriculture and similar organisations in the states and the RBI.

Besides, the government should encourage independent research. The GoI would be failing in its duty, if they did not rise to the occasion and did not decide without delay to provide adequate funds to existing university centres and organisations like the Indian Society of Agricultural Economics for the development of the study of and research in agricultural economics. Besides providing funds in this manner to strengthen the existing University Departments, I do hope that instead of indulging in the costly luxury of *ad hoc* economic enquiries in rural problems by administrative personnel, who have little interest of capacity to do the work, the governments at the Centre and in the states will appreciate that to get such work done with the help of the University Departments would be not only cheaper, not only more efficient and more accurate, but would also be in the national interest inasmuch as it would provide training and experience to a large number of young workers.

Conclusion

In conclusion, let me say that in this rapid survey of the role of agricultural economics in our country, if I have been frank, it is in the interests of the subject for the development of which we have met today. I have no malice against any institution or person whom I may have criticised directly or indirectly, and if I have emphasised the part that the universities should play and can play in this national work, it is based not only on the experience of other countries, but also on my own experience and observation both as a research worker in the University of Bombay and as one who has had opportunities to know first-hand the working of government departments and institutions. I am fully aware of the existing deficiencies in the universities. But it is more a reflection of the studied neglect of the past, than a cause for ignoring the universities in this national work.

While emphasising the role of the universities and academic workers, I would like to address a few words to brother-economists. We should be aware of the increased responsibilities that we have to shoulder under the new environment in the country. We are in a position partly by influencing the work of the younger generation trained under us and partly by contributing to the knowledge on the subject by our own researchers, to render substantial help in the formation of correct economic policy in the country. This requires devotion to the subject and hard sustained work. The time has come when the universities and other employing authorities in academic institutions should give to the economists working under them adequate opportunities for research work and make it worth their while to do independent work of this nature. Though there is considerable room for improvement in the outlook of the employing authorities on the one hand, and in the work of the academic economist on the other, I venture to suggest that some existing institutions have reached a high level of efficiency and standard of work to be able to undertake research projects of the type referred to above. In fact, they have been entrusted by government with some research projects and their work has been appreciated. I need not give illustrations, let I may overlook deserving cases, but I may say that there is no reason why many more such institutions cannot be developed in the country in the different university centres.

In all such development, agricultural economics should have an important place. In the existing Departments of Economics in the universities, we find that due attention is not given to this subject by those in charge. Whatever may have been the position under the British regime when some felt that there was no room for original research work in Agricultural Economics due to the static condition of our rural economy, we must recognise that the entire scene is now rapidly changing. The various reforms for improvement of agriculture which have been already introduced and the various schemes of development of agriculture included in the Plan indicate the pace at which changes are occurring and are likely to occur in the life of the people. Add to this the consciousness among the rural folk due to adult suffrage and the work of the political parties, thus increasing their desire and urge for a better existence which were hitherto absent. The land policy as outlined in the Plan, the various schemes of irrigation that are contemplated and particularly the river valley projects, the community development projects which aim at all-round development of total economy including the outlook and methods of work of the farmer, are but few illustrations of fundamental and rapid changes which are about to take place. This opens up a vast and almost unlimited scope for research and investigation beyond the capacity of government departments. Those interested in agricultural economics are now provided with huge experiments the like of which they have not seen before. If we can utilise these immense possibilities for work each in our respective areas, we shall be able to contribute to a better understanding of the new forces which are set in motion. Because of the nature of the problem, there is also a room for joint efforts in this work for social scientists not only in Economics but also in Sociology, Psychology and Politics. We must be ready to view these dynamic forces in their proper perspective. We have thus opportunities to contribute in our humble way to the thought on the subject so that timely changes in policy and measures may be adopted. I have every hope that in spite of difficulties, we shall rise to the occasion and prove worthy of our new responsibilities.

The time is also overdue when workers in this field should organise themselves more effectively so that their voice can be heard in the appropriate quarters. Our existing arrangements, by which we meet once a year in Conferences of this type though useful, are not adequate for organising scientific work and scientific opinion in

Economics in particular. Some more effective machinery is called for the purpose, and I trust both the Indian Economic Association and the Indian Society of Agricultural Economics will give due attention to the problem of coordinating research activities in the country and of bringing about greater cohesion and contacts among research workers.

Let the academic worker in economics have his due place in the new scheme of things in the country; let him play his legitimate part as an independent worker above party politics, and let him contribute to a correct understanding of the complex forces which operate in the economic life of the country in a detached but independent manner. All thinking persons will agree that for the building up of new India, unless we strengthen the universities to do this work properly, we shall not have that independent critical examination of economic problems, which we need so badly. We boast of political independence but tolerate intellectual paralysis; we confuse independence of thought as hostility to the government of the day and therefore unpatriotic; we want to abolish caste from our social system and yet create new castes in the intellectual sphere; we treat the academic research worker in economics as a person whose shadow shall not fall on the politician or the administrator, lest the practical wisdom of the latter be polluted by the theories of the former; we do not realise that the so-called wisdom of practical man is the unsystematic and often unconscious theorising on his limited individual experience, and that the so-called unpractical theories of the scientific economist are the result of systematic thought based on the widest possible experience in the subject concerned. I trust this unfortunate tendency is only a passing phase and that the academic workers in economics in general and agricultural economics in particular will have their rightful place in the growth of new India.

Indian Journal of Agricultural Economics, Vol. 8, No. 1, March 1953.

13

V.T. KRISHNAMACHARI

The Five Year Plan and Rural Improvement

THE basic problems of rural India are the enormous volume of unemployment and underemployment that exists in agriculture and its extremely low productivity. The factors that contribute to this are well known—they have been brought together in the recent Report of the Census for 1951—but will bear restating. The population has been growing steadily from 1921 to 1951. In consequence of this, there has been a progressive decline in these years in the cultivated area per head of the population (111 cents per head in 1921 to 84 cents in 1951); in double crop area per person (13 cents in 1921 to 10 cents in 1951) and in irrigated area per person (18 cents in 1921 to 14 cents in 1951). These figures are for districts, specially studied in land where records are accurately maintained and may be accepted as representing conditions throughout the country. In the words of the Census Report "the study decline of all the factors during the three decades since 1921 is an all pervasive fact".

Taking the agricultural population as a whole (249 millions), constituting 69.8 per cent of the total population, 71 millions are self-supporting persons, 31 millions are 'earning dependants' and 147 millions are non-earning dependants. These 71 millions come under the following categories:

Owner cultivators	45.7 million
Tenant cultivators	8.8 million
Cultivating labourers	14.9 million
Agricultural rentiers	1.6 million

The figures show not only the disproportionately large part of the population which agriculture has to support but also the relatively low proportion of workers. There is also the fact to be remembered that the extents cultivated by a family, small as they are, do not lie in compact holdings, but are often dispersed as isolated plot in different parts of the village. The broad picture thus emerges of over 70 million families cultivating about 250 million acres of land, only about 20 per cent of which is double crop or irrigated area, distributed in fragmented holdings and supporting large numbers of non-earning dependants, many of whom continue on the land as they have nowhere else to go and make only a nominal contribution to agriculture.

Nor is this all. Over about three-fourths of the area of the country's agricultural operations depend on the monsoon which, even when it is regular, is confined to 3 to 4 months in the year. Rainfall is precarious over 30 per cent of the land area of the country—ranging from about 30 to 10 inches. These are the 'scarcity areas'. Even in areas of higher rainfall there are years in which rains fail in the crucial period before the ripening of the crop. Moreover, from about the 30 September to 1 June (8 months) there is practically no rainfall except in the small parts of the country which benefits from both monsoons. The rainfall in these months varies from 1.5 inches (out of a total of 32.9 inches in Gujarat) to 4.4 inches (out of 33.7 inches in Hyderabad) to 7.9 inches (out of 56 inches) in East Madhya Pradesh. This long rainless stretch of 8 months—the dreary period of waiting between crop and crop—has a most depressing effect on the rural economy. Under the best of conditions, agriculture is thus possible over 75 per cent of the cultivated area only for 4 to 5 months of the year. In the rest of the area, which is irrigated, it is possible for double the period. During the rest of the year, employment opportunities are limited and intermittent for the millions in the countryside. This is confirmed by a detailed census conducted by the Labour Ministry.

To complete the picture, it must be mentioned that every year three millions are added to the population whom land has to support. Further, practically over the whole area 'subsistence agriculture' is practiced. It has been estimated that, after several years of intensive effort under the Grow More Food campaign, only 2 to 4 per cent of the cultivated area came under the influence of improved scientific methods.

The facts set out above disclose the colossal volume of the unemployment and under-employment among large proportions of the population in rural India, resulting in extremely low standards of living. This is not merely an economic waste. It is more wasteful in terms of human values. The Plan seeks a solution for this the most deep-seated evil in our economy, in two directions:

Firstly: Recommendations are made for the widest possible application of research in agriculture in a broad sense including animal husbandry, fisheries, etc. The intention is to promote a system of intensive farming and the evolution of a diversified economy by means of cooperative effort. This is assisted by the provision of irrigation and power facilities on a scale hitherto unprecedented, as these afford the readiest means of lengthening the period of employment in agriculture and increasing production.

Secondly: The Plan seeks to provide occupation for the surplus labour force on the land by encouragement of—industry occupations, subsidiary to agriculture, cottage and small scale industries, large scale industries and tertiary occupations. There have been varying estimates of the size of the surplus population in agriculture. I attempted one in the chapter I prepared for the *Report of The Fiscal Commission*. Obviously the size of this will depend on the extent of rationalisation in agriculture. As rationalisation progresses, the surplus will become larger and larger. The two processes are closely interlinked. By and large, it is correct to say that "for every great region of the world, living standards tend to be the higher, the smaller the relative importance of agriculture as field of employment".

The Plan also stresses three fundamental conceptions in all programmes:

Firstly: The cooperative principle should be applied, in its infinitely varying forms, for solving all problems of rural life.

Secondly: The motive force for improvement should come from the people themselves. Self-help is at the root of all reform, the state assisting with supplies and services and credit. The vast unutilised energy which has been referred above should be harnessed for constructive work. Every family devoting its time not only for carrying out its own programmes but also for the benefit of the community.

Thirdly: The state should ensure for the people the largest measure of social justice.

From what has been said above, the pattern of rural society that is envisaged by the Plan becomes clear. It is that of millions of families of farming owners organising themselves in cooperatives of different kinds and taking decisions as free agents, practising scientific agriculture on the largest scale possible and finding supplementary occupations in a variety of decentralised cottage and small-scale industries; while the State assists by organising research, supplies and services and credit. The basic considerations are not merely economic: there are non-material values and social gains which are equally important. The Plan emphasises the need for improving the techniques and organisation of cottage and small scale industries by continuous research and in other ways. It is also intended that large-scale industries and tertiary occupations will find employment for others displaced from agriculture. It is not without interest that the Fiscal Commission, attempting an answer to the question—what is the kind of society which the Constitution of India contemplates in the Directives embodied in it?—outlined the same pattern as that envisaged by the Planning Commission.

When the initial steps to establish responsible government in India were under discussion, there was much talk of 'the placid, pathetic contentment of the masses'. This is no longer true. The depression of the thirties, the Second World War, the coming of Independence and the general elections under adult suffrage have brought about a 'dynamic situation, ever-changing'. Everywhere there are new hopes and aspirations; rural India is seeking to build up a new life for itself. I need not say how revolutionary is the transformation implicit in the pattern of rural society placed before the nation by the Planning Commission. What it calls for is a change in the mental outlook of the people, the instilling in them of an ambition for higher standards of life and the will and the determination to work for such standards. As I have said elsewhere, this is "essentially a human problem—how to change to outlook of the seventy million families living in the countryside, arouse enthusiasm in them for new knowledge and new ways of life and fill them with the ambition and the will to live a better life". This is indeed one of the most difficult problems that a nation ever faced in the history of the world.

In the Commission's view the far reaching reorganisation of rural life such as it proposes can be achieved only if India transforms itself speedily into a Welfare State as enjoined in the Constitution. This means a change in spirit and outlook as well as in structure. The Commission has put forward proposals under which the administration, at all levels, including the district, will become the agency for rural improvement—an 'extension' or rural welfare agency. The functions of an extension service are well known. They are summarised in the following extract:

> Extension or advisory services and other allied services have the following indispensable functions—instructing farm people in agricultural and domestic science and in agricultural affairs; bringing to them the latest results of research in these fields, on quality and cost as well as on the quantity of production, and teaching them improved techniques of farming and rural living; calling the attention of research institutions to the agricultural and home problems that require study; and providing opportunities for farm people to meet together for the purpose of learning from each other and developing leadership in agricultural affairs.

Alongside the administrative organisation, there are proposals for the closest cooperation with the best non-official leadership at every stage. In essence, the National Extension Organisation is both official and non-official: both have to work together. Programmes for improvements should be made after the widest discussion with the agriculturists who should feel a sense of proprietorship in them. An important part of the set-up is the village level worker who is the link between the villager and the development departments. A friend and guide to whom he looks for advice and for the supplies and services needed for implementing programmes. It should be again emphasised that the whole movement is based on self-help: the active interest and support of the people should be enlisted in the task of bettering their own condition.

To facilitate the introduction of the new order of things, the Central government has undertaken to assist state governments with grants of half the additional cost of setting up extension staffs and also to make every effort to find by the end of the Plan period short term credits for ₹100 crores a year, medium term credit of ₹25 crores a year and long term credit of ₹5 crores a year through the cooperative movement and other agencies. The National Extension Organisatioon is expected to cover 120,000 villages—nearly one-third of the settled villages in the country—by the end of 1955-56. The training of the staffs has been commenced and in the Plan over ₹100 crores have been allotted for giving effect to this programme and an additional development programme in selected areas for a three year period. These are over and above the amounts included for promoting the Grow More Food campaign.

Thus within the Plan period, intensive development covering all aspects of rural life is to be organised over an area, eight to ten times that covered in all previous years and it is hoped that, before the end of the second Five Year Plan, the whole country will have come under the movement.

An equally substantial measure of advance is provided for under irrigation and power. In the programme multipurpose projects are included for irrigation and hydro-electric power: large and medium irrigation projects and minor irrigation works. On the first two classes of projects, about ₹560 crores are to be spent. I may mention, for purposes of comparison, that the book value of all major works in India at the time of partition was ₹110 crores. Quite recently about ₹40 crores have been added to the Plan for medium irrigation projects to give permanent protection to scarcity affected areas, thus bringing the total expenditure to ₹600 crores. For minor irrigation works over ₹110 crores are provided. Then there is a programme of tube-wells. To the 2,500 tube-wells that existed when the Plan commenced, 5,000 are to be added in the five-year period which will irrigate 1.5 million acres. The irrigation schemes in the Plan will add about 17 to 18 million acres to the irrigated area by the end of 1955-56 and another 9 to 10 millions after they are completed. It should be added that, under Irrigation and Power, the Planning Commission has recommended a 15 year programme which would double the area under irrigation and increase electrical energy about five times and that a Technical

Committee has been set up to examine projects for inclusion in the second Five Year Plan.

The National Extension Organisation as it develops will create increasing demands for research. The Plan includes additional provision for promoting research and the assistance of experts from outside is obtained wherever needed. There is special stress on dry farming research and soil conservation. It is recognised that however much irrigation may extend, large areas in the country will have to be under dry-farming and that the techniques of such farming need to be improved. Soil erosion is, again, a serious problem in many parts of India and the intention is to set up an expert organisation which will train workers, and assist in surveys and in preparation and execution of schemes.

I have referred to the importance attached to cottage and small scale industries. Here, the Plan has suggested changes in policy. The principle of a cess on the corresponding large scale industries to support cottage industries has been accepted. It has also been agreed that these industries should receive protection by reservation of spheres and in other ways, until by improvement in techniques and organisation they can face competition. Special autonomous boards have been established for promoting them and these have been endowed with adequate financial resources for efficient working. Recently, a team of foreign experts selected for their knowledge and interest in small scale industries have been invited for conducting intensive studies of such industries in four regions and suggesting programmes for developing them. Indians possessing competence in this field are working jointly with this team.

The Planning Commission has made far-reaching recommendations in regards to the land reform for ensuring social justice. These include laws for abolition of intermediary tenures and for reform of tenancies for guaranteeing security and fair rents and enabling tenants to become owners on payment of moderate compensation spread over a period of years. Recommendations have also been made on the question of imposing ceilings. Space does not permit a detailed discussion of these proposals. I hope those who are interested in the subject will study the report and the relevant laws in States. The land reforms legislation in India is impressive, judged by the far-reaching character of the changes, the areas and populations involved and the time and manner in which it has been carried. For a variety of reasons, however, actual implementation has not been as speedy in all areas as might be desired but everywhere beginnings have been made. The Planning Commission is now having the social and economic effects, caused by these measures, studied objectively by universities and other research organisations.

Connected with this is the question of reasonable prices for agricultural produce. It is obvious that scientific agriculture on the scale contemplated—every family having its own programme of improvement-will be impossible if prices are allowed to fall below remunerative levels. This is not to say, that cost in agriculture cannot be reduced or that productivity cannot be increased. Cost reduction and improvement in productivity are, of course, essential. At the same time, in a country like India where agriculture is the mainstay not only of the agriculturist but also the whole economy, the question of maintaining incomes in the agricultural sector at a reasonable level has to be constantly kept in mind. What the reasonable or remunerative level would be is a matter for detailed examination. One has to consider not only what the farmer should receive but also what he has to pay for the tools and implements and the consumer goods that he needs. Stabilisation of agricultural prices cannot, obviously, be divorced from the wider question of the stabilisation of the entire price level. Moreover, the price for any particular agricultural product must be related to the price obtainable for alternative crops, so that parities have to be worked out not only between agricultural products on one hand and the products of manufacturing industry on the other, but also as between agricultural products inter se. In this particular field we need a great deal more information than we have at present. Studies in farm costs and farm management are necessary as a first step in this process, and it is proposed, under the auspices of the Research Programmes Committee to which I shall come presently, to inaugurate such studies.

I now come to the most disquieting and in many ways the most difficult of the problems facing us. The Planning Commission has drawn attention to the increase in population in the last 30 years and the pressure exercised on the limited resources available and the urgency of measures for family planning and population control. It has also made recommendations in regard to such measures. In the Census Report of 1951, there is a study of the problem from the point of view of trends in food production from which the following conclusions among others are drawn:

Fourthly: It is possible to overcome food shortage permanently, by taking suitable measures over an extended period of about 15 years.

For this purpose, development of agricultural productivity on an even larger scale than that undertaken in the Fist Five Year Plan will be necessary, but it will not be sufficient. Effective steps should be taken concurrently to limit births to approximate parity with deaths and thereby achieve a substantially stationary population.

Fifthly: Births will get limited to approximate parity with deaths, if what has been described as 'improvident maternity' is avoided by all or most married couples. ('Improvident maternity' consists of all births occurring to mothers who have already had three or more children, when at least one of them is alive).

Sixthly: A combined approach to the solution of our problem calls for the following targets:

(a) Increase of agricultural productivity from the present level of approximately 700 lakhs of annual tonnes to about 940 lakhs of annual tonnes.

(b) Reduction of 'improvident maternity' from over 40 per cent (its incidence at present) to under 5 per cent.

If these two targets are accorded equal priority and ranked higher than all other development targets, they can be attained before population increases beyond 45 crores—our total number will then become substantially stationary round about 45 crores.

I think all of us will agree, without committing ourselves strictly to the statistical data and to the definition of 'improvident maternity', that no increase can be achieved in the present standards of living, which are so low as to constitute a serious menace, unless the growth of population is controlled. This is eminently a matter in which non-official bodies like yours can assist by studies and propaganda and creating an informed public opinion and I hope that, in the coming year, the Association will devote special attention to this subject.

An essential condition for success of the Plan is that the administrative machinery should be improved in efficiency and adapted to its changed functions. There is need for creating a new outlook—a new conception of a public service—and also for strengthening the administration for its new tasks. There are also the questions connected with the harmonising of the activities of the Central and state governments in regard to development and ensuring that they are based on common, clearly understood principles. On these aspects, recommendations are contained in Chapters V to VII of the Report.

I have attempted to place before you in broad outline the proposals embodied in the Plan for rural India. The Plan should not be regarded as a series of projects and programmes. As the Prime Minister has said, "It is much vaster...the mighty scene of a nation building itself. All of us working together to make a new India." It is a democratic Plan. It is based on the dignity and worth of the individual. Its basic idea is to bring about a change in the outlooks of the millions of families in the countryside and to assist them in their efforts to build up a new life. The foundations are now being laid for a National Extension Organisation—the Welfare State in action—which will cover about one-third of India by 1955-56 and the whole country in as short a time as possible after that. A pattern of social and economic life has been placed before the country which will be the aim of the Welfare State to bring into existence. There are instances of communities organising themselves on this pattern but nowhere has the experiment been tried for a vast rural population like ours and in an under-developed economy with so many intractable problems. And we cannot afford to fail; for failure will mean the destruction of values which we all cherish. The results so far obtained have been on the whole encouraging, and as administrations and leaders and the village and other associations gain more and more experience there are grounds for the hope that the movement will gather momentum. What is essential that, the movement, while basing itself on a sound system of cooperative finance, should always retain its vigour and dynamism as a genuine people's movement, ever adapting itself speedily to the new demands and new situations that will arise.

Before concluding, I would like to refer to the position of agricultural economics in India. I think it is correct to say that, speaking broadly, agricultural economics in India is still in its infancy. It is only recently that some of our universities have established chairs or readerships for teaching and research in agricultural economics. The problems of agricultural economics are varied, and they are difficult because one has to take into account not only general casual relationships in the field but also the entire institutional framework in which these relationships operate. On many aspects of the problem data are lacking. I do not refer merely to statistical data relating to such matters as cultivated area, production and prices but also to mental attitudes, incentives and disincentives, etc. In saying this, I do not ignore the intensive studies conducted in the past by individual economists or research centres. But these were isolated efforts confined to a narrow range of problems. In recent years, however, systematic studies have been and are being undertaken under government auspices or with government assistance, the conclusions derived from which will advance our knowledge of problems connected with agriculture in India. I refer to the enquiries conducted by the National Income Unit, the National Sample Survey, the survey by the Labour Ministry of the condition of agricultural labour and wage levels, and the enquiry into all aspects of agricultural finance which has just been completed by the Reserve Bank of India. The material collected in these enquiries is being sifted and when reports analysing them are published, there will become available data of much value to agricultural economics on which policies can be formed and actions taken. Your own Society, I must not fail to mention, has been doing excellent work in spite of its limited resources in finance and in research personnel, and the record of work you have done shows how much the efforts of a single energetic person like your president Shri Manilal Nanavati can achieve. But there is still a vast field remaining to be explored. Throughout its work, the Planning Commission found that on vital matters it had no reliable information on which to base valid conclusions. It therefore, decided to set up a Research

Programmes Committee of well-known economists, practically all of whom are connected with the Association, for advising on a programme of research which will be of assistance in development policies, to arrange for their conduct by universities and research organisations on a carefully thought out plan and to collect and study the results. The Committee met some time ago and, as a result of its deliberations, researches will now commence all over India on three general categories of subjects:

(a) Land reform and cooperation.
(b) Employment, Savings and Investment; and
(c) Regional Development

The Planning Commission has every confidence that these studies will maintain a high level of thoroughness and objectivity and that, as time goes on, they will cover a wider range of subjects so that a reliable volume of verified knowledge will become available to students of Indian agricultural economics. I may also mention that the Evaluation Organisation recently set up under Professor Karve for studying programmes and methods in community projects and national extension areas is also undertaking fundamental studies of rural problems which will yield results of great value. In all these directions we are making, I believe, a reasonably good beginning. We need a great many research workers and we must strengthen our universities and our research institutions to enable them to expand their activities in this sphere. An underdeveloped country is inevitably backward also in the field of research, and agricultural economics, like our agriculture itself, has suffered from long neglect. Now that we as a nation are launching upon a career of rapid and many-sided development, all this has to change and change quickly; for, without knowledge, you cannot build firmly. A distinguished economist has characterised economics as a fruit-bearing science, and agricultural economics must be particularly so. I hope that your Society will grow from strength to strength and play its part in building up a new body of knowledge relating to problems of development. What we are seeking is not merely new techniques for raising more crops or for increasing productivity in the narrow sense; what we desire is a better economic and social order and a fair deal for the millions who inhabit our countryside.

Indian Journal of Agricultural Economics, Vol. 9, No. 1, March 1954.

14

D.R. GADGIL

Land Reform or Agrarian Reform

THERE are many aspects of agrarian reform and this paper deals only with what I consider to be the core of agrarian reform in India, namely organisation of agricultural production with particular preference to the size and structure of the unit of agricultural production. I need not emphasise that I shall confine myself to the problem as we face it in India today, but may add that I shall deal with the Indian problem chiefly in the context of or as part of, a plan of rapid economic development.

The main constituents of the programme of agrarian or land reform currently undertaken by the states in India are classified as follows in the progress report for 1953-54 of the Five Year Plan :

(a) The abolition of intermediaries;

(b) Tenancy reforms designed:

   (1) to scale down rents to 1/4th of 1/5th of the produce;

   (2) to give tenants permanent rights subject to the landlord's right to resume a minimum holding for his personal cultivation within a limited time;

   (3) to enable tenants (subject to the landlord's right of resumption for personal cultivation) to acquire ownership of their lands, on payment of a moderate compensation to the landlord spread over a period of years;

(c) Fixing of ceilings on holdings;

(d) Re-organisation of agriculture including the consolidation of holdings, the prevention of fragmentation, and the development of cooperative village management and cooperative farming.

Many of the items included above have been subjects of attention and activity on the part of governments for many decades past. However, a number of important elements are of recent introduction, and the programme has begun to look like an integrated attempt only within the last few years. An important feature which has received emphasis only since the attainment of independence is of *zamindari*. Theoretically, it is possible to argue that it is not necessary to abolish the *zamindar* or other intermediary or intermediaries between the state and the actual tiller of the soil. The state could effectively regulate activities of the superior holders so that all the conditions required for an efficient conduct of productive operations by the tenant-cultivator are obtained. Indeed, it has been suggested that a programme of tenancy legislation such as that adopted in a *ryotwari* state like Bombay could have done equally well in *zamindari* states like Uttar Pradesh and Bihar. However, landlordism, where it is a recent product and holds only a part of the field, in an essentially owner-cultivator community, and operates vastly differently from that in an environment where it is of long duration and well- high universal. Experience of legislation in the recent past has shown how it is difficult to stop even irregular exactions by landlords in the *zamindari* states. As in all other spheres of Indian economic activity, in this also, it remains true that it is easier, and in the long run more profitable, to abolish a powerful private interest whose activities are apt to run counter to desired goals of national policy, than to attempt to regulate it. The abolition of intermediaries, especially in areas in which landlords has been all-pervading in extent ad influence is thus no more than a primary step in making possible the creation of a politico-economic democratic society.

While emphasising the primary importance of the abolition of *zamindari* it is necessary to remember that, in India, it affects, in the main, the distribution of agricultural production. Barring a small number of exceptional areas, to which reference is made later, the typical agricultural producer in India both in the *ryotwari* and in the *zamindari* States is a peasant family operating with a small and fragmented holding. The abolition of *zamindari* may put heart into the erstwhile tenant, may give him hope and will most probably reduce the share of the agricultural produce with which he has to part in terms of revenue and/or rent; but his total working capital resources. This is because, in the first instance, the abolition of intermediaries does not mean the break up of large farms or farming estates or the redistribution of land and secondly because, even if this had been a part of the programme, there are, in fact, with the exception of a small number of regions, no large farming estates in India. The very description of the reform as abolition of intermediaries emphasises this aspect of the redistribution of the product and reduction of the burden on the actual cultivator. It is conceived of essentially as establishing, as far as possible, a direct relation between the actual tiller of the soil and the state.

Tenancy reform or tenancy legislation has a much wider sphere of operation than legislation for the abolition of intermediaries. It has significance both in the *zamindari* and the *ryotwari* states. Indeed, tenancy legislation is found necessary wherever private ownership and operation of land exist, though its direction and content may vary from one system of land tenure to another. The abolition of *zamindari* may result in landlordism ceasing to exist, as a general feature of the situation in an area; it cannot however, entirely and for all times prevent the emergence to some extent, of temporary or permanent tenancies which must be regulated. In fact, the very effort to keep in check the emergence of tenancies postulates existence of appropriate tenancy legislation. Tenancy reform is essentially directed towards obtaining for the tenant, who, in the present Indian context is generally supposed to be the actual operating cultivator, security in the cultivating possession of his holding and reasonable, if not generous, terms relating to share of produce with which he has to part in favour of his landlord of other superior holders. Where it provides for the acquisition of rights of ownership by the tenant it seeks to establish ultimately a direct relation between the cultivator and the State. It may, however, be doubted whether in areas where landlordism has not been deeply rooted and tenancy legislation can, therefore, be effective in protecting the tenant, it is necessary to divert, for acquisition of ownership rights, such financial resources as the tenant cultivator may possess.

Tenancy reform also, it will be obvious, does not affect the size and shape of the agricultural holding. It brings about, in the main, a redistribution of the total produce in favour of the tenant and also gives him a sense of security regarding the future which should react favourably on the economic and technical operation of the tenant-cultivator. However, there is one important difference between tenancy reform and the abolition of intermediaries. The latter is, for the most part, a once-for-all operation; the former, on the other hand, has not only continuous effect but has further to adapt itself constantly to a changing situation. In consequence, though tenancy legislation may not operate directly on the unit of agricultural production, the framing of tenancy legislation is always influenced by total land policy including policy relating to the size and structure of the unit of agricultural production.

The fixing of ceilings on holdings is likely to affect the size of the unit of agricultural production much more directly than either the abolition of intermediaries or tenancy reform. Before proceeding further I may note that the omission to distinguish clearly between the ownership holdings and the cultivation holdings leads often to a confusion in thinking and exposition of the subject of land reform in India. I shall concern myself in this address, except where explicitly stated otherwise, with the cultivating holding. The adoption of a ceiling may be with reference only to acquisition of additional land in the future. This would act as a restraint on the activities, subsequent to the passing of the legislation, of persons who in the absence of such legislation might have increased the size of their holdings above the ceiling limit. If in any area a trend towards formation of very large holdings has been established or was likely to be established a ceiling on future acquisition would prevent its operation. The immediate effect of the adoption of a ceiling for the future, on the size of the production unit would, on the other hand, be negligible, except to the extent that it would encourage actual or notional division of existing large holdings among family members so as not to be affected early by the operation of the legislation. The long term effects are problematical and would depend on the extent to which the existing of future situation otherwise favoured the formation of very large holdings.

The idea of ceiling has, however, been developed not in the context of checking future possibilities but in that of granting immediate relief. It had origin in lands where very large landlord estates existed side by side with very diminutive peasant holdings and in which there remained some vestiges of serfdom or in which the memories of a serf society were yet alive. The situation created a land hunger and also suggested a not unwelcome manner of satisfying it. The process was best exemplified in the countries of South and East Europe, both within and without the Communist belt, during interwar years; and it was finally worked out in these countries in the years immediately following the end of the Second World War. In India there are few parallels to this situation. The closest resemblance in all features, social, economic and political was in Kashmir. It is here alone also that the ceiling has so far been imposed so as to bring about some transfer of cultivated land included in large estates to actual cultivators. However, it would appear that even in Kashmir the effect of the legislative reform was mainly that of bringing about the abolition of intermediaries. This was because even here the element of direct cultivation in big estates was limited, the bulk of the land in them being already cultivated by tenants. The only other state in India which has adopted the ceiling as an immediately operative concept intended to bring about redistribution of cultivated land is Hyderabad. However it is impossible to judge at present what, if any, redistribution would come about as a result of the somewhat complicated piece of legislation that has finally been adopted in Hyderabad. Even the comparatively large estate of Kashmir and Telengana do not or did not contain elements of direct cultivation large enough for redistribution to affect sensibly the problem of the small peasant holding, in even restricted areas. Elsewhere the contrasts are much less glaring. The concept of the ceiling, it is to be used in India must, therefore, be different in content and operation from that in countries with large landlord estates, plantations or *latifundia*.

The concept of the economic holding or a minimum holding called by any name, like the concept of the ceiling, can form the basis of an immediate operation of reorganisation or can be confined to setting a limit to future transactions. In the latter alternative it may act as an effective means of preventing a worsening of the existing situation and may partly even help to improve it gradually in the future. These effects will, however, become apparent only slowly over a series of years and a limit for the future cannot help towards reorganisation, if the existing situation in itself is extremely unsatisfactory. A minimum holding or a floor has not been used in any state in India yet for bringing about immediate reorganisation of production units in lands included in units below the minimum size.

Consolidation as practiced in India affects powerfully the internal organisation of a holding, though not usually its total size. The process of consolidation may lead to some saving in the land surface used for such purposes as boundaries or roads and may thus enable formation of a pool of land for specific common purposes. But the saving effected in this way is not likely to yield substantial acreage for distribution among existing holders. Cooperative farming or cooperative village management are yet chiefly in the stage of thought. What little has been taken is experimental and nowhere has any legislation been formulated of contemplated which bases itself on the formation of cooperative farming units for at least a part of state policy relating to land management.

It would be clear from the above account that the present programme of land reform in India does not bear to any significant extent on the structure or size of the unit of agricultural production. Excepting the attempts at consolidation of holdings, which affect the internal organisation of the unit, no other feature of existing or contemplated legislation is important in this context. The ceiling and the floor to the cultivated holding, in the few states in which they exist, are thought of essentially as regulating future activities, and their importance for immediate reorganisation must be considered negligible.

Legislation on land reform is essentially a matter for governments of states. Therefore, the programme is usually framed in the context of particular problems of each state. The activity of the National Planning Commission affords the chief occasion and instrument for integrating policies of different state government and for formulating a common Indian policy. Considerable importance, therefore, attaches to the views regarding land policy contained in the first Five Year Plan. It is difficult to piece together in a consistent whole the ideas contained in the chapter on Land Policy of the plan. The problems of the policies are divided into two aspects; Land Management and Land Reform. It is evidently considered that there is some conflict between the two; for, it is said that "Land Policy should include both elements but should maintain a balance between the two" (p.88).[1] The abolition of intermediaries is accepted unequivocally, though it would appear from later sections that all land owners are not considered to be intermediaries in relation to all land surface not cultivated directly by them. There is a similar lack of clarity in respect of the distinction between the cultivated holding and the ownership holding in other instances. Without entering into a discussion of these peculiarities of exposition, the main ideas of the Planning Commission may perhaps be summarised as follows. A limit for future acquisition of holdings or for resumption of personal cultivation is favoured. The limit proposed is three ties that of a family holding, which is defined as equivalent to a plough unit or work unit. In relation to existing large cultivated holdings it is laid down that the principal consideration should be whether or not such management is in public interest; and public interest, in this context, is somewhat narrowly interpreted. It is to be judged on the single criterion as to whether the management fulfils the standard of efficiency necessary to raise production in conformity with the plan. It is only when the large cultivated holdings are palpably inefficiently managed that land management legislation would give the appropriate authority power to take over the land 'in excess of the prescribed limit' and to arrange for its cultivation. Here again, there is evident confusion between the two possible reasons for taking away any piece of cultivated land from a person cultivating it; one, that the holding is too large and the other, that it is inefficiently managed. The most curious feature of this part of the exposition is

1. For ease of general ready reference the quotations have all been given from the People's Edition of the First Year Plan.

the optimistic forecast that the proposals would provide for a large measure of redistribution of land belonging to substantial owners.

The suggested land reform policy is again not uniform; it is evidently to change with size of land owning. The most important result is that the tenants of small and middle owners are recommended much less protection than the tenants of large owners. As middle owners are defined as owners of land up to three times the family holding the field of tenancy protection is thus seriously narrowed. The main reason given for the maintenance of a large class of tenants-at-will in this way is that otherwise movement of people from agriculture and rural areas into other occupations and towns may be seriously checked. It is fortunate that most state governments have paid little attention to this recommendation in their legislative programme.

In relation to the landless worker the main concrete reference is to the *Bhoodan*[2] movement. "It offers the landless worker an opportunity not otherwise open to him" (p.92). This can be only interpreted as meaning that the state itself considers it neither necessary nor possible to do anything for him. It is added that the problem of the landless worker must be considered in terms of institutional changes which would create conditions of equality for all sections of the population. The essence of these changes is described as a system of cooperative village management. Cooperative village management is referred to in the Plan in other context also. For example, after having formulated the important and unexceptionable proposition that 'the basic condition for increase in agricultural production is increase in the unit of management of land" (p.93) this also is said to be possible only through cooperative management at the village level. With regard to the small and the middle farmers again it is recommended that they should be encouraged and assisted to develop their production and organise their activities on cooperative basis. With this emphasis on cooperation it would be expected, that cooperative organisation or cooperative village management would be described in specific detail and a programme sketched out for establishing it. The following

---

2. Voluntary contribution of land.

extract summarises the ideas of the Planning Commission on cooperative village management.

'Broadly speaking, however, we envisage that the village *panchayat* should become the agency both for land reform and for land management in the village. In the first place it should be the body concerned with the management of land taken over from substantial owners, and also of village waste lands. The leasing of lands by small and middle owners should also be done through the *panchayat* and not directly. In this way the village *panchayat* may be able to provide cultivating holdings of economic size, at any rate for landless cultivators. The exercise of these functions would naturally lead on to the wider conception of the cooperative management of the entire land of the village and the undertaking of activities for creating non-agricultural employment in the village.'

It is not necessary to comment on the above except to note that it does not deal with the problem in the comprehensive manner that might have been expected, that it is overoptimistic and also that the working model presented is obviously unsuitable and inadequate in most respects. Moreover, no concrete programme is provided for progress in the direction of cooperatives village management even according to ideas contained in the Plan, beyond the ₹50 lakhs provided for study, training and experimentation. It would not be unfair in the circumstances, to suggest that the Plan proposes little of importance in relation to land management and that we have at present no effective programme in contemplation for dealing with what the Planning Commission itself recognises as the basic problem viz., that of 'the increase in the unit of management of land'.

Not only is the subject treated unsatisfactorily in the Plan but also the subsequent activities of the Planning Commission and the Central government do not indicate that it is currently held in importance or receives any special attention. The progress report of the Plan for 1953-54 notes that the government of India constituted in May 1953 a Central Committee for Land Reforms consisting of members of the Planning Commission and some members of the Union Cabinet. It is, however, not indicated what policy decisions, if any, have been taken by this Central Committee and no advance in ideas or concepts over those contained in the Plan is anywhere recorded. The Community Projects are supposed to be the special field of the Planning Commission and great reliance is placed on them in relation to agricultural development in the future. Not only do these Projects pay no attention to land reform or land distribution but even experimentation in relation to land management appears to find no place in them. The complete absence of any attention to it in the programme of the community projects is evidenced by its not having been necessary to devote any attention to the subject in the Evaluation Report on the first year's working of these projects. And the later report of the Evaluation Organisation, "Community Projects-First Reactions" contains enough evidence to show how the project authorities completely ignore land management and land reform problems.[3] In regard to cooperative organisations we have the following comment in the Evaluation Report. "While in the very initial stages of the formulation of community plans this aspect of mutual dependence between community development and cooperative organisation was not so explicitly stated, during the course of the year increasing emphasis was placed by the Community Projects Administration on promotion of cooperative organisations" (p.39). The cooperatives to which increasing attention is reported are still the credit, purchase and sale and other organisations and not experiments in increasing scales of land management. The putting forward more recently of agricultural extension as the panacea for all rural ills is, perhaps, a reflection of the same attitude it appears to be considered by the Planning Commission, as by many foreign experts, that all that is required to increase agricultural production in India adequately is to arouse enthusiasm and to transfer techniques. The problem as to whether conditions in the field are such as to favour generation of enthusiasm and the acquisition and continued practice of new techniques does not evidently need prior consideration.

3. See, for example: "This is a very poor village and most of the villagers are illiterate. Majority of the families are landless and land is their greatest felt need. The community programme being mainly related to agricultural improvement has not touched the basic need of many (majority being of non-land owners) and has not been appreciated by them. Due, however, to the VLWS contacts and propaganda, many know about the programme and there is nothing in the project which would make them hostile or apathetic towards it". *op.cit.* no.1: 72.

Finally, it is reported that in some states the Central government, presumably acting through the Planning Commission, was responsible for persuading governments to modify their original ideas regarding immediate operation of ceiling on holdings. All in all, one gathers the impression that while in its theoretical formulation the Planning Commission may recognise the existence of the problem of land management it is not ready to give this recognition any immediate of concrete form.

For a statement of integrated thinking on the subject, the steps which have been fairly reasoned out and which appears to have influenced to some extent, the action of governments, during the last few years, one must have recourse to the Report of the Congress Agrarian Reforms Committee. It cannot be said that this Committee was not impressed with the importance or the urgency of the problem. Writing at the time it did, the Committee was especially impressed with the need for an immediate and complete abolition of intermediaries. It, therefore, gave a great deal of attention to that problem and argued effectively the case in favour of the abolition. For the same reason, it intended to treat the problem of the reform of the unit of agricultural production as subsequent to the effective abolition of intermediaries. Therefore, though its approach to this problem is outlined fairly fully the Committee does not make recommendations in regard to it that are as specific or definite as those made in relation to the former, more immediate problem. The general approach of the Committee turns on three separate definitions of the holding which it formulates and which it calls respectively the basic, the economic and the optimum. The basic holding of the Committee may well be interpreted as a watered down or a 'realistic' economic holding, In the words of the Committee the basic holding is the holding 'Smaller than which would be palpably uneconomic from the point of vie of efficiency of agricultural operations.' The economic holding, the Committee defines in terms of the standard of living afforded by it to the cultivator and the employment that it provides to a family of normal size and at least a pair of bullocks. The Committee realised that the economic holding so defined sets a standard with which it would be impossible to operate in India unless the large majority of peasant family farms are to be denied independent existence. It, therefore, provides for a lower operating unit called 'basic,' the criteria for determining which are for obvious reasons somewhat vague. The 'optimum' unit of the Committee, which is the same as the general concept of the ceiling, is defined as three times the size of the economic holding. It is difficult to argue out a case for any particular figure as the proper multiplier in this instance. The important point to notice in the context is that the Committee is definitely against mechanised capitalist farming and that it recommends the operation of the ceiling on holdings so as to make impossible holdings which could be worked of managed and supervised by an owner only when mechanised. The 'optimum' of the Committee is thus the largest unit which an operator can manage without the farm falling within the category of capitalist mechanised farming. The basic of the economic holding are both concepts which are highly relative especially when based on a measure of the standard of living. The actual or aimed standard of living in any country must have some reference to the resources and the national income potential of the country. Whatever the problems involved in defining the various units of holdings the Committee's approach involves the use, as generally suggested, of the two limits which may be called the floor and the ceiling respectively.

The Committee does not, however, work out in detail a land management policy based on these two operative limits. It states clearly that the ceiling should be effective and lands above the ceiling should be redistributed. It also holds definitely, that lands below the basic holding should not be allowed to be cultivated independently and that they should be generally pooled into some type of cooperative productive organisation. The committee, however, nowhere gives consideration to the crucial factor in an operative programme, that of relative dimensions. What number of peasant households or what proportion of land is at present included in holdings below the basic, what proportions are likely to be available for redistribution by the operation of a ceiling and how the total land surface to be redistributed is to be reconstituted into effective operative units; these are the basic data in relation to which alone the large number of administrative and legislative problems connected with any policy based on the ideas of a ceiling and a floor could be solved. In the absence of such detailed calculations the Committee accepts or at least acquiesces in the prevalent notion that the vast bulk of existing units would continue to operate independently.

It is the merit of the Committee's report that it indicates an approach to the problem which in Indian conditions is both economically feasible and is likely to be acceptable politically. As pointed out above its great pre-occupation with, logically, the earlier problem of abolition of intermediaries did not, perhaps, allow it or induce it to work out the practical programme in detail.

Barring the report of the Congress Agrarian Reforms Committee there is no comparable official or semi-official document which contains a reasoned discussion of this problem. It is understood that some foreign experts have in recent years also advised the Government of India and the Planning Commission in relation to problems of land reform. The reports of these experts are not readily available to the public and it is not therefore possible to say what they contain and what the reactions of government have been to the recommendations that might have been made. It is however, known that a number of foreign experts, especially American experts, lay a great deal of store by the independent peasant unit of operation. There is, on their part, usually a considerable distrust of state effort and they entertain a generally sceptical attitude towards cooperative effort. It is in their tradition also to frown on any State assistance or direction to the building up of cooperatives. The notion of a 'collective' is, of course, totally repugnant to them.

As indicated above, the Planning Commission appears content to operate with the existing unit of agricultural production and does not propose to change in any radical manner the organisation of land management and operation. The Congress Agrarian Reforms Committee also formulate its actual programme very largely in terms of the peasant farm and it does not appear to consider that a programme of reorganisation involving large numbers of families and a considerable land surface is insistently called for. The experts appear to talk almost exclusively of the peasant unit and their ideas of land reform are confined mostly to abolition of intermediaries and dealing with large estates. In the circumstances, it becomes necessary to examine existing conditions carefully, especially with a view to throwing light on the strength and efficiency of the independent peasant farm as the unit of land management and agricultural production.

It is generally known that the size of the large bulk of farms in India is very small and that number of them cannot be called family farms in any valid sense of that term. However, no attempt is usually made to indicate with figures the dimensions of the problem. I shall, therefore, note certain salient features of the situation very briefly. My concern is with the unit of land management, the cultivated holding, and data relating to ownership of land are not relevant to my purpose. The data required are those relating to size, scale of operations and of investment, receipts and employment etc., from farming of the independent farming units. Through courtesy of the authorities of the Reserve Bank of India, I am enabled to use data contained in their forthcoming publication- the Report on the Rural Credit Survey. The data were collected from a stratified sample of 15 cultivating families in 600 villages in 75 districts in India and relate to the year 1951-52. In the context of the subject of my address the most revealing figures are those relating to the reported value of gross produce. According to the Survey a little over half of the cultivators reported a value of gross produce of farm business lower than ₹400 for the year. There was possibility of some under reporting or under estimation in these values. However, the most generous allowance for this factor would yet leave an estimate of more than half of the total cultivating families as falling in the class of value of gross produce of less than ₹600 per year. The usually cited figures of acreage of cultivated holdings present difficulties of interpretation because of the large variety in size of holdings and quality of land from region to region. The value of gross produce, though its detailed significance may no doubt vary from area to area, gives yet sufficient indication of the size of the total business in comparable terms.

This then is a useful starting point of the description, that half or more of the farm units i.e., independent units of land management in India may have a gross produce of farm less than ₹600 a year. The next step is to assess the relative importance of farming activity to the farm family. In the total number of cultivators reporting less than ₹400 of values of gross produce during the year two divisions were made representing those with a value of gross produce below ₹200 and those with value of gross produce between ₹200 and ₹400. The former formed slightly more than 29 per cent of the total cultivators and the latter group 21 per cent. Those reporting value of gross produce and those in the latter class reported total farm expenses that were on an average only about ₹60 less than that the average value of gross produce. The average reported cash receipts from sale of crops and fodder in the two classes were about ₹20 and ₹70 respectively and both classes reported cash farm expenses that were more than ₹50 on an average than the average cash receipts from sale of crops and fodder. Making all allowance for errors in reporting, etc., it is clear that the cultivators included in these groups earn little, if any, net cash income through their farming activity and the main advantage derived by them from farming is some contribution in kind to family living.

This is emphasised by the data relating to the relative contribution of the different sources of cash income to the total cash receipts of the cultivating family. Here, the data used are those relating to the upper and lower half of cultivators divided, per village, in relation to the size of their cultivating holdings. Because of the possibilities of some small holdings yielding proportionately greater values of produce and vice versa the division by value of gross produce. In the main, however, cultivators in the lower strata have much lower values of gross produce than those in the upper.

To illustrate the importance of farm business to the family, I cite the data relating to the sources of cash receipts of families during the year. The average cash receipts per cultivating family of the lower strata during the year was about ₹310; of these the largest single source was cash wage receipts amounting to ₹133 or nearly 43 per cent of the total. Value of crops and fodder sold was only about ₹108. Sale of milk and milk products represented a little over 7 per cent of the total cash receipts and carting and remittances each contribute between 4 and 5 per cent of the total. Perhaps even more illuminating in relation to the size of operations of the individual cultivating family and the role of farming in it are the corresponding data for the upper strata family were ₹594, less than double the average cash receipts of the lower strata family. That these families were mainly concerned with farming is evidenced by their receipts from sale of crops and fodder amounting on an average almost to ₹400. Sale of milk and milk products are relatively of the same importance for the upper strata as for the lower strata, representing 7 per cent of their total cash receipts. But the most significant feature is the amount of receipts from cash wages and remittances. Remittances received by families of the upper strata amounted to nearly ₹25 per family and cash wages to over ₹83. The proportionate share of cash wages in the receipts of the upper strata was much less than in that of the lower strata, but the difference in absolute amounts was much less striking. The cash wage receipts of the average family of the upper strata were more than 60 per cent of the corresponding receipts of the family of the lower strata and the amount of remittances received by the former was almost double that of the latter. The remarkable fact is that taking wages and remittances together as representing outside labour employment that cultivating families had to undertake, the average of such employment in the average upper strata family is not much less than that in the overwhelming importance of receipts from wages for families of the lower strata.

The picture may be elaborated by reference to other data which relate to all the families in 600 villages. The grouping of families in this case is by each tenth of decile, in the list of all cultivators arranged in order of the size of their cultivating holding. Of the total borrowings of even the middle four deciles more than half represented borrowing for family expenditure items; for the last three deciles the corresponding proportion was almost 60 per cent. The capacity of the average cultivating family to undertake capital expenditure was obviously extremely limited. The average expenditure undertaken on all items such as bunding, reclamation, irrigation sources, implements, etc. including expenditure on repairs, maintenance and replacement was about ₹22 for families of the lowest three deciles and ₹51 for families of the middle four deciles; the corresponding amount was ₹311 for families of the first decile. But if this expenditure is taken together with expenditure on purchase of livestock by the respective groups of deciles and the total calculated on a per acre basis the expenditure incurred actually increases as one goes down the group of deciles. This means that while the total outdays of the families of the middle and lower deciles are small and are known to be inadequate their burden in terms of per acre costs rules high.

It is not necessary to labour the point further. What I want to emphasise is that the size of farming business of at least half the cultivating families in India is such that it is futile to consider them as independent units of land exploitation in any plan for a developing economy. It is irrelevant, in the light of the data cited above, to talk in terms of family farms or economic holdings. Even the definition of basic holding of the Congress Agrarian Reforms Committee cannot cover these units. Therefore, a land policy for a developing economy must face up to the serious problem of the reorganisation of these units. A vague recognition of these facts is seen in the general comments made in various contexts by numerous experts and Committees on the 'non-creditworthy' or the 'marginal and sub-marginal' groups. This recognition must become more explicit and must lead on to a realisation of the inability of any supply or credit reorganisation to deal with fundamental defects of the small size and turnover of the existing basic units and of the large numbers involved in any scheme of reorganisation.

Before going on to further consideration of the problem it is necessary to pay attention to an important factor which has emerged during the last few years. I refer to the movement for *Bhoodan* initiated and actively led by Acharya Vinoba Bhave. The movement is so original in its conception, so novel in its methods, and so revolutionary as to its objectives that special effort has to be made to understand it and to place it in its proper perspective. I begin the discussion by presenting summary account of what I consider to be the essential features of the movement. Though I have read some of the original writing on the subject and also aware of the limitations on my knowledge of the subject and can only hope that the summary presented by me is fair and that I have not misinterpreted or misrepresented the essence. The movement may be said to owe its origin to an overwhelming feeling of revolt at the gross social injustice existing in our society and a deep compassion for the disadvantaged classes in India. It embodies an attempt at initiating a social revolution such that these submerged classes, of whom the numbers and proportions are extremely large in the Indian countryside, obtain some immediate relief and a ray of hope for the future. The chief initial propositions are that rural society in general and the landless rural classes in particular are heavily disadvantaged in India today, that rural society is being continuously exploited by urban classes, that in terms of political power, economic strength or social influence, it gets a very rough deal and that unless the balance is rectified there can be neither peace nor prosperity in the land. The immediate problem to which the *Bhoodan* movement addresses itself is that of the landless labourer but, as the extension of the movement in directions such as *Sampattidan*[4] or *Jeewandaan*[5] shows it is an all-embracing movement directed at the total problem of social injustice. The main objective of *Bhoodan* is to provide for each family in rural society with at least a small piece of land to cultivate. This is aimed at as the only way of giving every family some means of livelihood and the feeling of a stake in society. Rural reconstruction as envisaged by the *Bhoodan* movement is also directed towards encouraging rural industry and attaining comparative self-sufficiency. It is not clear whether these are considered to be primary objectives, as having values in themselves, or as methods of rehabilitating rural society. The doubt arises because Vinobaji seems to have made clear on various occasions that he does not object to an urban structure as such. In his opinion it is possible to have a proper division of labour between urban and rural society which is just and non-exploiting one. He does not also apparently object, in principle, to the use of machinery. He objects, in a measure, to its present use because it is evidently the instrument of exploitation, on the one hand and concentration of power on the other.

4. Voluntary contribution in money.

5. Voluntary contribution in labour.

The programme of action which the *Bhoodan* movement sets out is that of asking for voluntary gifts of land from present owners. It is made clear that the word *dan*[6] is not used in the sense of simple gift or charity. Starting from the definition *dan Samvibhag*[7]: that *dan* is proper or just distribution, the gift is sought as voluntary redress of present injustice. The process seems to be conceived of as follows. The awareness of distress of others may arouse sympathy, compassion, daya[8] and such sympathy joined to reflection may bring awareness of the existing stage of social injustice and one's responsibility, in part, for it. Such awareness would induce action to redress the injustice to the extent possible for oneself; and this is the basis of the general demand made of everyone of a fifth or sixth or some similar proportion of his land possessions. The land so collected forms the pool for redistribution. It is realised that the voluntary movement may not necessarily be sufficient. Legislation may ultimately come in, but legislation is thought of as the ultimate expression of a popular will, the culmination of a popular movement. To arouse public conscience and to obtain for redistribution of land general support, to create in its favour what is called *janashakti*[9] a power, a strength, or a force, arising from among people themselves; this is expected to be one of the results of the movement; it appears at least to be one of its primary objectives. On the manner of redistribution itself, there seem to be no dogmas, no hard and fast rules. The essence is that any pool obtained in a village must be redistributed according to the wishes of the people of the villages. Vinobaji personally, it is said, would emphasise the role of cooperation. For him the preferred general pattern would be widespread cooperative cultivation with a kitchen garden for every family. He is also supposed to favour both a ceiling and a floor for holdings. However, sensing perhaps that a voluntary movement will not be called upon to deal with general problems of redistribution, he appears content with each small piece of land acquired for a landless

6. Voluntary contribution.

7. Voluntary contribution for equitable distribution.

8. Mercy.

9. Mass power.

person as a step forward in curing social injustice, creating an element of stability and making possible the reconstruction of rural society.

Such, in brief, is the account that I have gathered regarding the objectives and programme of the *Bhoodan* movement. It now becomes necessary to subject these to some scrutiny. It is a task which I approach with some diffidence and with all humility. In Vinobaji we have not only the greatest but also, perhaps, the only force that has acted in independent India for dealing with the problem of concentration of economic and political power, which is at the root of all evil in our society. Such an effort made in such a bold, all embracing manner and involving such vital reinterpretation of our traditional values must receive the instinctive sympathy of all Indians. At the same time our intellectual tradition enjoins on us the responsibility of examining whether the premises of the movement are correct, its objectives appropriate, whether its programme leads us in the right direction and whether there are no repercussions of it that will deflect us from the desired goals.

Examining the premises from which the movement starts it is possible to show that they are well founded. The facts on the cognition of which the movement is based can be evidenced from statistical data. The data of the Rural Credit Survey can, for example, be drawn upon to support the premises in two important directions. The first is that of the existence of land hunger accompanied by the fact of extremely low sizes of holdings of land. The extremely small size of the holdings of what might be called the middle and the small farmer are emphasised in the data over and over again. These, however, are matters of common knowledge. The more remarkable support to the premises is brought out by the indirect evidence obtained during the Survey on the existence of land hunger. This was made clear through replies to questions asked to cultivators regarding their credit needs and the direction in which they would use credit if they obtained it. A very large proportion of the credit requirements, especially of cultivator amongst the lower strata i.e., the lower half of the body of cultivators arranged according to size of cultivated holdings, was for purchase of land. The data attain special meaning when they are compared with the actual expenditure which had been incurred by the reporting families on purchase of land in the year in which the Rural Credit Survey had been conducted. The average actual expenditure of the lower strata was ₹12 as against their reported credit needs for purchasing land which averaged at ₹333. The limited purchases were due both to non-availability of credit as well as to non-availability of land. It does not follow that if credit had been available the purchases would in fact have been more. If everybody wanted to purchase additional land and there was no new land available, net purchases of land could not be made by everybody, whatever the credit facilities. At the same time the replies emphasise the great urgency, the high priority above everything else given by members, particularly of the lower strata, to purchase of additional land. Land hunger is thus clearly exhibited in the Rural Credit Survey statistics.

Other data, supporting the premises, available from the Rural Credit Survey are those relating to the economic position of the non-cultivating classes. The non-cultivators of the Rural Credit Survey included all classes other than cultivators being defined to include all those who cultivated even a small plot of land. The non-cultivators of the Rural Credit Survey thus comprise traders, moneylenders, artisans, labourers, all those who had no cultivating holding of their own. Though a mixed class, it was, in the main, formed of the artisans and the labouring classes. According to the Rural Credit Survey data the average performance whether in level of debt, of repayment, of borrowing, of family expenditure, of capital expenditure, of financial investment, of the net balance of capital transactions or any other indicator of economic well-being of the non-cultivators on an average, was worse than that of even the last three decades i.e., the last 30 per cent of the cultivators arranged in accordance to the size of their cultivated holding. These are average figures for a somewhat mixed class. It can be realised from the data how the labourers and the inferior artisans, in particular, is much worse even a small plot of it, must appear to the landless as leading almost automatically to improvement or possibility of improvement which must remain beyond their hope till they obtain possession of this essential means of livelihood.

The premises on which the *Bhoodan* movement rests can thus be supported both by statistical data and methods of the movement are justified. Here again it is possible to argue that in a static society in which no technical change is going on an increase in population would require a redistribution of land such as is sought by the *Bhoodan* movement. There are a number of single agricultural societies in which periodic redistribution of land has been practiced as a means of adjustment between the resources of society on the one hand and the needs of groups of families or of the individual family on the other. The argument may be extended so as to apply to a society in which the progress of total technical change was overshadowed by the growth of population. Conditions in India during last century approximate to this picture. Since the time we have had comparatively reliable data, say 1880, a trend has been in evidence towards increasing numbers depending for support on agricultural activity which, in turn, was based on a comparatively non-expanding area of cultivated land and on techniques which were changing only very slowly. In the earlier stage it was not only a case of larger numbers being supported by land but even larger proportions of the total population depending on it. In more recent decades this trend has stopped and the growth of urban population has been somewhat faster than that of rural population. Comparative data in regard to occupational distribution of the population cannot be set out because of the many defects, in this regard, of the Census data. However, even though in recent years there might have been a slight decline in the proportion of the people depending on agriculture and a definite trend towards more rapid urban growth it yet remains true that the total absolute numbers of those depending on land and agriculture is steadily increasing and continues so to increase. The consequent adjustment that has taken place through the recent decades in India is that of a progressive diminution in the size of the average holding through successive subdivisions and the creation of a larger and larger class of the completely landless. The *Bhoodan* movement may be said merely to look these facts and to insist that while the existing situation lasts the landless must not be completely deprived of the primary means of sustenance i.e., some land and equipment, and that society must look at this problem as a problem essentially of social justice. Given the existing situation in India and given the present apparent powerlessness of the Government either to initiate rapid economic development in general or to devise any logical solution to the land problem in particular, the *Bhoodan* approach cannot be dismissed as either irrelevant of ill-directed.

The possible results, of the movement affects large areas, will obviously be a partial reduction of the average area of large holdings, a partial reduction of even the area of the average or below the average holdings and the creation of a large number of extremely small new units. Prima facie all these results may be called unsatisfactory than the existing situation or than the situation that will develop with existing trends. It may no doubt be unsatisfactory to create a large number of very small new units of holdings, But if the only alternative to the creation of these very small new units of holdings is for the landless to continue to exist as they do without alternative occupation, without sufficient sustenance and with no hope or zest in life, can it be said that the creation of these new small holdings is an evil or that it is not an improvement, however small, over the existing situation. A small reduction in size of the average holding or in size of the large holding is no evil, looked at in this manner. If finally it is realised that the movement desires specially a transfer of land from the largest holders to the smallest or to the landless and that it accepts gifts from the average or below the average holder as merely a token, as contributing to the creation of *Janasakti*, the whole will be placed in its proper perspective. The above argument necessarily proceeds on certain assumptions regarding technical change and economic development to which attention must now be directed.

The first assumption is that the cultivating possession of even a small plot of land brings about a radical change in the economic position or outlook of the landless and that even small additions to diminutive holdings have a large actual and psychological effect. Where employment in agricultural labour is extremely uncertain and the level of wages very low and where rural artisan industry is chronically depressed, a plot to employ oneself with and on which to grow some part of one's food supply, may conceivably bring about a transformation, economic, social and psychological. Further, a wider distribution of land may rectify the existing distorted balance of terms of exchange between classes; the possibility is indicated by the reported opposition, for fear of not obtaining sufficient agricultural labour, of the larger landlords to the *Bhoodan* movement.

The second assumption is that of the almost complete divisibility of land surface as a means of production, or to put it in alternative terms, the assumption that consideration of economies of scale can be ignored in agricultural production. This, no doubt, is a large assumption but it would come naturally to the protagonists of the movement who are used to thinking in terms of a simple non-mechanised technique and a simple structure of economic organisation. And it may be argued that for the large bulk of agricultural producers in India actual conditions are not far different from the assumed simple technique and organisation. For the larger part of the crops grown, the methods of cultivation, the degree of intensity, etc. are all actually at a uniformly low level and the per acre product does not appear to differ widely as between small plots and large holdings. The resource most wastefully employed is, in these circumstances, labour; but we are reasoning about a situation where labour has, after all, no alternative employment. Finally, it may well be asked whether the existing policy of the State in India with its neglect of the problem of land management and emphasis on agricultural extension is not itself tantamount to ignoring economies of scale. If propaganda aided by a little reorganisation of credit and supply is supposed to achieve the desired results in the existing distribution of size of holdings, the same achievement ought not to be difficult with a somewhat more even distribution of the same.

If one grants these assumptions, and I have argued that they are tenable and are also implicit in the ruling policy of the State, considerations of social justice must take high priority over those of economic production. The *Bhoodan* movement, in the circumstances, is not only completely justified but also remains, in face of the indifference of powerlessness of the State, the only instrument of the desired social transformation. This is not to say that if a plan of rapid economic development was under way and an appropriate land policy had been evolved, it would still be necessary to adopt the immediate objectives and programme of the *Bhoodan* movement. In fact, one could point to certain aspects of the movement which may, in that event, hamper the working of a plan of development and to certain repercussions of it which may create grave difficulties in the way of future progress. Before I leave the subject, it is necessary to pay brief attention to these.

It is assumed that a plan of rapid economic development and an appropriate land policy will be conceived in terms of advanced techniques and comparatively complex forms of economic organisation ; such a plan and policy will also not find it possible to ignore economies of scale. It follows that a movement which increases the number of diminutive holdings in the country will act in a direction which will be contrary to that of the plan. The *Bhoodan* movement has, further, no bearing on, and indeed ignores the problem of, the rational layout and utilisation of the land surface. Both these are bound to be important parts of a plan and the movements emphasis away form them may prove a handicap. Supporters of the movement may reply that what are logical features of the programme in existing circumstances need not persist and need not prove obstructive when circumstances alter. If the movement itself leads to the State undertaking a plan and a policy which result in increased and more diversified employment, wider security and greater equality for rural society, it will have largely succeeded in its ultimate aim and need not necessarily pursue existing programmes.

There is one possible repercussion of the *Bhoodan* movement which I greatly apprehend and in regard to which the above would not prove and adequate reply. This is that of increase in intensity of the sentiment of ownership of possession in relation to land. No doubt, the *Bhoodan* movement is itself based on denial of rights of private property, especially in the means of production. Its slogan in relation to land is *Sabai Bhhomi Gopalaki*.[10] But its actual programme results in investigating rights of ownership of land in a vast number of people. Even its appeal for gifts is couched in terms of a more equal

10. All land belongs to State.

distribution of ownership of land; so that those who give do not, through their gift, subscribe to the doctrine of the social ownership and use that would logically follow from their basic philosophy except when they obtained gifts of, say, the entire lands of village. Granting this, the fact remains that the main result of *Bhoodan* would be creating a large body of new very small holders and of generally intensifying the sentiment in relation to the ownership of land, its social prestige and its economic value. Taken together with steps such as that of the abolition of intermediaries and provision for tenants to acquire ownership of land, the total situation, in which landownership was not largely diffused in vast areas of the country and large bodies of peasants had not vested rights in the lands they cultivated, is being rapidly changed to one, in which most of the cultivated land surface will be vesting in a very large number of small holders who are likely to hold to their newly acquired rights with great tenacity. I am not arguing against steps such as that of the abolition of intermediaries which I hold as essential preliminaries, but only in favour of trying to avoid the undesirable effects, in terms of popular sentiment, resulting from it.

It would be unwise to ignore the dangers that might develop out of the existing situation. *Bhoodan* owes its origin to Communist activities in Telengana and it might be relevant, in this connection, to cast a glance at the pattern of Communist history. It shows consistently, in the first instance, a movement towards redistribution of land to win popular support and arouse popular enthusiasm, followed at no distant date, by a move towards socialisation, collectivisation and other programmes which all necessitated wresting lands away from the peasants in whom they had been recently vested. The latter step has invariably created a grave crisis. It is held that the Russian Revolution passed through its most critical phase at the time when the first large movement towards collectivisation was made and Communist countries of South and East Europe have, in the post-war period found collectivisation to be the toughest problem with which they had to tackle. It is unnecessary to enter into the justification or logic of Communist strategy. I merely suggest that Communist experience presents a grave warning to which we must pay heed. There is nothing in the Indian situation today, if we conceive it in terms of a rapidly developing economy, to require or to justify a further diffusion of the ownership of land. At present the jurisdiction of the State in regard to plans of distribution of land surface and its utilisation is widely recognised. It seems to me to be highly unwise to disturb this situation and to create conditions under which efforts of the State to create a new pattern of land management units may meet with active opposition. If this happens, the state in India, with its much smaller capacity to deal with such opposition may find the problem well-nigh insoluble. I feel strongly that it is the special responsibility of the leaders of *Bhoodan* to pay attention to this aspect, as their avowed emphasis is on the overall social or collective view and as, whatever they do, or say, its influence on popular sentiment is likely to be much greater than even their concrete achievements.

Before proceeding to a discussion of the main ingredients of a land policy appropriate to a plan of a development I may recapitulate why I consider the *Bhoodan* approach unsuitable to the context. *Bhoodan* tries to attain comparatively equal distribution of income through a similar distribution of the means of production. It conceives the problem in terms of a simple economy of family producers. How this cannot apply in case of a more elaborate economic organisation may be made clear with reference to the concept of *Sampattidan*. *Sampattidan* is supposed to be the counterpart of *Bhoodan* in the sphere of urban economy. But the problem of concentration of power in the sphere of urban economy cannot be dealt by any attempted simple transfer from large holders to small holders. In the urban economy the organisations that hold the key positions, to whom the greatest surpluses accrue and who have great economic strength and wield political influence are not simple divisible units like land. These organisations are complex integrated units which represent transport systems, power generation, international and internal trading on a large scale, credit and financial organisations, large mechanised units in all types of industrial activity. Bits of these cannot be separated and given away; the problem of concentration of properties and power in private hands, in this case, can only be dealt with by transferring the ownership of the integrated units to organisations-new or old-which represent social and collective interest. *Sampattidan* in the circumstances cannot result on diffusing or socialising control over concentrated means of production. It can only result in some giving away of current income and can thus only become charity in the worst sense. Agricultural production in a plan of development will necessarily be concerned more and more with advanced techniques and must resort to more complex forms of economic organisation. Therefore, in this case also considerations of distributive justice must be kept separate from those of organisation of production. The production pattern must be determined by the available proportions of productive resources and by the possibility of increasing absorption of advanced technique. Economic and social policy can separately deal with the problem of distribution. Land management units must be shaped in accordance with the needs of the production pattern.

There appears at present general agreement on the nature of the problem and the main approach to its solution. The following statement by the Prime Minister, Pandit Jawaharlal Nehru, may be taken as representative of this:

> As agriculture is the principal occupation of the great majority of our people, it must be the first concern of the State. The abolition of the zamindari system has been the first reform and this must be expedited. But it must be remembered that this by itself is no solution of the problem. Even before this abolition a very large proportion of land was self cultivated. An addition to it, without any further reforms, will not help much. The small subsistence farm makes progress difficult. We have to think, therefore, and think soon, of other and further steps. There should be a diversion of a part of the agricultural population to other occupations. There should be a development of cottage and small-scale industries. But essentially the problem of agriculture needs cooperative cultivation and the application of modern techniques. This does not mean necessarily mechanising agriculture all over India, though some degree of mechanisation is taking place and is desirable. But there is no escape from some form of cooperative cultivation, if we are to make agriculture progressive.[11]

11. Quoted in H.D. Malviya, *Land Reforms in India* (1954), pp.91-92.

Adopting the general approach indicated in the quotation it is possible to frame a land policy utilising only the concepts and methods that have already been widely discussed and partially adopted. These are the floor and ceiling to holdings, consolidation of holdings and farming cooperatives. I assume it to be accepted that the very small holding can find no place in the production organisation of the future. None of the problems of inadequate or disproportionate equipment, of backward technique and inefficiency, of underemployment and wasted labour, of poverty and exploitation can be solved as long as the bulk of farming units are small.

The concept of the floor is required to define the lower limit to the size of holding. The ceiling is not equally, obviously, justified from the point of view of productive organisation. For, the very large units may also be efficient in production. The main reason for adoption of the ceiling is socio-political. Among all resources, the supply of land is the most limited and the claimants for its possession are extremely numerous. It is obviously unjust to allow the exploitation of any large surface of land at the hands of single individuals unless other overwhelming reasons make this highly desirable. Further, in light of the available supplies of land, labour and capital it would be undesirable to encourage capital intensive methods of production in regard to which the very large holdings may have special advantage; without intensive use of capital the very large individual holding operated with hired labour is not specially productive. Moreover, whatever the economies of larger scale management, they should, in the congested state of our countryside, accrue to collective or cooperative bodies of cultivators rather than to an individual family. It may be noticed in this connection, that even the United States of America (US). Farm Ownership Farm programme contemplates creation of such farms by subdividing large holdings and recognises both a ceiling and a floor to the farm ownership. On the other hand, barring a very small number of exceptional cases, as that of certain types of plantations or orchards, it is unlikely that the reduction in size of a very large holding to a holding of moderate size will bring about any diminution in per acre productivity. Such diminution had been reported where land surface, included in large commercial estates had been broken up into very small holdings. But this procedure is not contemplated by us. A ceiling is finally necessary if social policy requires immediate redistribution, to some extent, of the land surface.

I shall not devote much attention to the problem of the determination of the floor or the ceiling. All the terms used and definitions given in this connection, such as that of the economic, the family, the basic, are capable of wide variations in interpretation. The determination of the limits, I consider, as essentially a matter of practical policy. If the concept of the floor is properly applied, its fixation gives the numbers of the uneconomic holders who cannot continue to exist as independent units. The floor will, therefore, be determined in relation to the ability of the state to pool the holdings of uneconomic holders and to form them into some sort of collective or cooperative holdings. The ceiling on the other hand, has reference to the area of land that can be made available for redistribution. The greater the area desired for redistribution the lower will be the ceiling.

Considerations of production economy do not seem to indicate any particular figures or proportions in this context. However, data from the Rural Credit Survey may be set out to illustrate the relative proportions involved. These data relate to all the cultivated holdings in the 600 villages selected for the survey in 75 districts. The cultivators were arranged, for the purpose, in order of the size of their cultivated holdings. When so arranged it was found that the first 10 per cent or decile of the cultivators held more than 25 per cent of the total land surface in almost all districts and held more than 30 per cent of it in 51 of the 75 districts. The first 30 per cent or three deciles taken together held more than 50 per cent of the land in almost all districts and in 48 districts they held more than 60 per cent; in no district was this proportion larger than 85 but in 6 it varied between 75 and 85 per cent. The middle four deciles held between 25 and 35 per cent of the total cultivated land in the large majority of districts. The holdings of the last three deciles included less than 10 per cent of the total cultivated land in the majority of districts ; but in 27 out of the 75 districts they held between 10 and 15 per cent. Taking the broad division of cultivators into the upper half and lower half the relatives size of their holdings of total land surface would be approximately 3:1.

The above taken together with the data relating to occupational distribution of families should give an idea of the overall dimensions of the problem. We may assume that about 60 per cent of families in rural India are cultivators in the sense of operating some cultivated land and that of the non-cultivators at least half depend for their support on agriculture and land. This gives about 80 per cent of rural families as interested in land management units and policy. We may, on the basis of data set out before, consider about half of the cultivators as having unsuitable units for independent land management, the proportion cannot at a minimum be put as less than a third of the cultivators. If we take the former figure the cultivating families together with the landless interested in reorganisation, will form about half of the total number of families in the countryside and, if the latter, about 40 per cent of them. Working with the data for existing distribution of cultivating holdings and assuming a transfer, because of the operation of the ceiling, of the order of 5 per cent or the total cultivated land, we have the following figures of the extent of land surface and number of families affected by reorganisation; if, half the units cease to be independent about 50 per cent of rural families and 30 per cent of the cultivated land; and if, about one third of cultivators are so affected, about 40 per cent of the rural families and about 15 per cent of the total land surface. These calculations are so broad as to be almost notional and they have been indulged at this place only to give some idea of the dimension of the problem. The floor and the ceiling will vary from region to region in relation to the quality of land and size of holding. However, their definition will also necessarily depend on the ability of government to undertake schemes of reorganisation.

The two fundamental steps in reorganisation of land management units in India are: (i) redistribution of the land surface and (ii) formation of the larger consolidated units out of the pooled resources of the uneconomic units and the surplus available for redistribution. The first step is an essential preliminary in almost all programmes of land improvement and agricultural development. Its importance is universally recognised in connection with the consolidation of fragmented holdings. It is, however, necessary from other points of view also. One can view the process of consolidation in a wider context as part of the process of a rational layout of the total land surface for agricultural and other utilisation. The operation should, therefore, be undertaken as that of laying out areas, sections or fields for total land utilisation and not be considered merely as that of mutual exchange of fields or fragments. Full advantage of the redistribution of surplus land could also not be had without this process of redistribution of land surface. The operation of the ceiling would yield small and large plots of land dispersed over a wide area. Some measure for redistribution would be necessary to utilise these appropriately. Therefore, for consolidation of fragmented holdings, for pooling together land included in uneconomic holdings and for utilising surplus land available for redistribution in a similar manner, a general redistribution of land surface must be undertaken. As pointed out above, it would obviously be wise to conceive the process as involving also a radical layout for total utilisation. This task is no doubt of vast dimensions. But tasks of somewhat similar dimensions have been undertaken in other countries and the total work indicated above is not more complicated or larger than the process of consolidation undertaken currently by many state governments in India. As I visualise it the first stage in the process will be that of determining the general layout and the second that of locating in this layout the independent farm units now fixed in location and made impartible, and the cooperative estates or farms.

The formation of larger units out of pooled land and other resources will have two aspects, one compulsory and the other voluntary. I have assumed that once the floor has been defined independent units of farm management, smaller in area, will not be allowed to exist. Obviously the enforcement of this must be by some sort of legislation. The measure of compulsion may vary. In the early stages the step may be initiated, as in some programmes of consolidation, only on the motion of a minimum number of families involved. However, as the development programme advances it must soon be generalised. The abolition of the independent existence of smaller units can only be subsequent to redistribution and consolidation. Also, before it takes place steps must be taken to formulate the basis of the new units and to equip them suitably. The formation of these new units out of the total pool may be essentially voluntary. It may be found useful to have floor and ceiling limits for these consolidated units also, so that in relation to their own scale of operations they might be neither too small not too large. The number of such units that could be formed in a village would be broadly determined by the ceiling and floor limits; within these limits the formation of particular units would be the result of voluntary groupings.

It has been assumed above that the landless families who depend on land and agriculture would be provided for in the new consolidated units. It would be unnecessary, even unwise, to bring into the organisation rural families who have distinct non-farm business and can depend on them. But as the process of reorganisation will materially reduce the number of permanent farm servants etc., and as one of its objective is to improve the condition of those dependent entirely on casual labour, the landless must be made members of the consolidated cooperative farms. It is not necessary for me to enter here into details relating to conditions of entry, the capital share equivalent granted to each for the land and equipment that he brings into the surplus resulting from the operation of the ceiling is specifically the share of the landless; such provision alone can act as inducement for voluntary admissions of the landless into cooperative pools. It would be desirable also to provide for the voluntary accession of the independent farming units to cooperative farms.

Both these tasks, that of redistributing the land surface and the formation of a small number of cooperative farms in each village is immense in extent and complexity. I would argue that not only are they a *sine qua non* of any programme of land reform but also that they are not beyond our capacity, if a sincere and concentrate effort is made. Consolidation is already generally accepted as a necessary part of the programme by everybody. The formation and proper functioning of cooperative farms is undoubtedly less generally accepted as part of universal policy and is somewhat more difficult. It is, however, equally necessary. I shall, without elaborating them, briefly state a number of reasons in favour of this proposition. In the first instance, without cooperatives, collectives or state farms an economic reorganisation for operation of the existing numerous small units is impossible. Secondly, except as members of a cooperative the small farmer can never be in a position to avail himself of technical, financial or other external aid of which he stands in need, more than anybody else. The findings of all studies, whether e.g., of the Rural Credit Survey or the Evaluation Organisation are that the bigger man gets the greater profit out of everything, government loans, cooperative finance or other financial or technical assistance of any type. One need not go into the reasons for this state of affairs; it is, however, clear from the studies that the scattered, weak units cannot really be helped effectively unless they come together. Also, as long as there is no consolidation of these weak units the balance in rural society will always remain against them. Any close study of the effects of recent land reform legislation reveals large variations in its results, dependent chiefly on the strength of tenants and smaller holders in a region or locally. In a large number of instances legal protection and other devices prove constantly fruitless because of the strength and ingenuity of the strong. It is sad to record that large deflection of original intentions, brought about by the strong, has been reported even in the working of the *Bhoodan* movement. The moral is that unless the weak acquire economic strength by joining together they could never stand up and get full advantage of state policy and state legislation. The formation of a small number of cooperative farming societies of the smaller holders and the landless in each locality is a necessary step in this direction. It will be noticed that I consider redistribution as only a part of the process in the formation of the large cooperative farms. I cannot see any virtue in merely tenancy cooperatives which perpetuate the smaller units of management. Leaving scope for allotments of kitchen gardens would be welcome but the main cropping operations must be in terms of the large units. The experience elsewhere, such as with the Ejido in Mexico, emphasises importance of this.

I do not think that widespread formation of such cooperatives is an impossible undertaking. In the initial stages the progress may be slow but once the movement gathers momentum the field could be covered fairly quickly. Wide extension of the activities of State governments in rural areas together with efforts made by the central Government in directions of Community Development and National Extension already provide comparatively large staffs in the field; the work of this staff would come to have real meaning and purpose only if it is linked to a programme of the formation and operation of such cooperatives. Instead of chiefly conducting propaganda they would find in these cooperatives numerous local units which could prove centers of demonstration and experimentation. All governmental information and propaganda, aid and assistance could be directed towards and routed through these cooperatives should combine the features of the programmes envisaged in the draft outline of the Five Year Plan for the Registered Farms and for the Cooperative Farming Societies. Instead of receiving very little or nothing of government protection and assistance this step would ensure that the most disadvantaged received, as they should, the highest priority and the greatest assistance.

Further, a land management policy which retains only a limited number of small and medium farms which can operate economically as independent production units and brings all other families into comparatively large cooperative farms is likely to lead to a better utilisation of the labour of rural families. The element of under-employment about which a great deal is written will be made clear under such an arrangement and the extra labour uncovered could be made available for programmes of either capital works construction or for rural industrial development in a systematic manner. The rural industrial plan will then be visualised in two clear divisions; one chiefly of domestic work directly supplements family living for the independent farm families and the other of industrial production for sale, local or otherwise, organised within the farming cooperatives.

I should like to emphasise that all that I have said is intimately related to a plan for rapid economic development. If such a plan is not in operation a vigorous land policy is hardly likely to succeed. It is now universally recognised that creation of employment in other spheres is the only remedy to a large pressure of population on land from which India, for example, suffers. Most programmes of improvement of technique in agriculture do not increase employment has chiefly to be in spheres other than agricultural. The minimum target of an employment plan for India has to be cessation, at the earliest possible moment, of the increase in the number of families dependent for their living on land and agriculture. Unless the number of families dependent on agriculture is stabilised there is little stability for measures of land reform. The prevention of sub-division or the creation of impartible holdings will be effective only if, in successive generations the number of families on land does not increase. The large cooperatives farms, even with allied industrial employment, could also not effectively absorb a continuously increasing number of families.

I realise that I did not leave myself sufficient time to elaborate and explain my own concrete proposals regarding the structure of a land policy. It is perhaps unwise, anyway, to compress too much in one statement. Therefore, in concluding I shall content myself with reiterating the following basic propositions. That in relation to land management we are following, in this country today, a policy mainly of drift; that a programme of economic development requires more positive approach in respect of agricultural productive organisation and units of land management; that a very large number of existing cultivating holdings are extremely unsuitable for functioning as independent units in a programme of agricultural development and that a rational layout of the land surface for its proper utilisation and a programme for the creation of large consolidated holdings are essential ingredients in any agriculture and land development policy. I may add finally that it would be impossible to deal with this problem except in term of bold steps and very large and strenuous effort. However, these would be no more radical and strenuous than those undertaken by many other countries in analogous situations, as evidenced, for example, by the Mexican Agrarian Revolution which redistributed land in favour of the Ejidos or the transformations brought about in countries of South and East Europe both during the inter-war years and in the period after the Second World War.

Indian Journal of Agricultural Economics, Vol. 10, No. 1, March 1955.

15

J.J. ANJARIA

The Role of Agriculture in a Developing Economy

THE central problem of the Indian economy at the present stage is economic development, and this has to be secured through simultaneous and balanced action in all sectors of the economy. The objective is to enable the community to produce more per head of population and per unit of resource and this necessitates a re-adaptation of the economic and social structure to facilitate and accommodate the new techniques and modes of production. There are, it would appear, two possible approaches to the problem. One proceeds from a narration of overriding social objectives and of the need to remodel the institutional framework in the light of these objectives. The other regards the attainment of economic—or shall we say material—ends as a matter of priority and postulates a relative indifference or averseness to institutional changes. It is, I should think, unnecessary—if not misleading—to pose the issue in terms of a choice between these alternatives. The urge for economic development is itself a manifestation of more deep-rooted social urges, and plans of economic development can rarely go far unless they take account of and meet these underlying urges. The problem at each stage in the process is to ensure that economic development and institutional change keep in step, so that each supplements and promotes the other rather than become a drag or a damper on the other.

I wish to ask myself today what role agriculture must play in Indian economic development and along what lines the transformation we are seeking in the economic and social conditions of the rural population could be attained with maximum advantage. This is a very broad topic. It has many facets and ramifications. Naturally, I shall confine myself to only a few of them.

May I start by recounting the salient facts of the economic structure? Of the total population (in 1951) of 36 crores, 25 crores depend on agriculture and some 11 crores on non-agricultural activities. Of the latter, some 5 crore live in rural areas. The rural sector of the economy thus comprises about 30 crores of people. The occupational pattern of the country reflects the overwhelming dependence on agriculture. Of the total estimated working force of 14.3 crores in 1950-51, 10.3 crores or 72.3 per cent were engaged in agriculture. Factory establishments and mining occupied only 2.7 per cent of the working force; small enterprises, including construction, some 8 per cent; and the remaining 9 per cent was engaged in other services. A major—and obvious—desideratum from the point of view of development is a change in the occupational pattern away from agriculture and in favour of industry, trade and ancillary services.

There has been, it would appear, little change in this pattern as a result of the developments that have taken place in the First Plan period. Industrial output has increased by almost 50 per cent in the last five years, but there has been little increase in industrial employment. Mining, railway, construction and urban trade have presumably absorbed fairly large numbers, but all this has made little impression on the occupational pattern as a whole. Agricultural production has increased by about 20 per cent in the last five years and although this must have reacted beneficially on the quantum of work available in the rural sector, it could hardly be expected to reduce visibly the surplus of manpower on the land. Judged in terms of the occupational pattern, the Indian economy has yet to feel the impact of expansionary forces. The process of evolving a more balanced occupational structure must necessarily be spread over a period of years, but it may be stressed that if the real income levels of the farming community are to be improved it is imperative that the number of those who share in the net proceeds of agriculture diminishes substantially. Technological considerations permit this and social considerations point the same way. It is, only, thus that the economy as a whole can be made dynamic and progressive.

It is easy to see that economic development requires a continually expanding output of agriculture. The incomes generated by investment create a larger demand for food. More raw materials are needed for the new factories, and circumstances may arise which necessitate some exports of agricultural produce to secure the imports of machinery and capital goods required. The increase in agricultural output cannot come from new land, of which there is none or very little. It has to be obtained through more intensive and more efficient cultivation. The present crop yields in India are low even as compared to countries like China and Japan where the unit of farming is small. There is little doubt that for many years to come no increase in the working force in agriculture is needed, and it would be most desirable if any increase could be prevented. In due course, as the development process gets accelerated, there could even be a reduction in the absolute numbers engaged in agriculture. The point, for the time being, is that given more water, better seed, fertiliser, livestock and implements, the country can produce far more food and far more of raw materials with a diminishing proportion of the population remaining on the land. As real incomes increase, the community spends an increasing proportion on goods other than the basic staples that agriculture produces. If, under these conditions, incomes per capita in the agricultural sector are to keep pace, there must be a steady diversion of people from agriculture to other occupations. More and more, therefore, the adequacy of our plans must be judged in terms of the opportunities they open out for employment in non-agricultural activities.

When the First Plan was formulated, the problem was, of securing an early increase in the output of food and raw materials. Schemes of agricultural development, including irrigation, were, therefore accorded the first precedence. With irrigation was associated power, and in the case of railways, there was a certain minimum programme of rehabilitation which had necessarily to form part of the plan. In the structure of investments that thus came about, there was little scope for a direct emphasis on creating a pull on the agricultural sector via increased demands originating from the rest of the economy. The First Five-Year Plan was admittedly a first step, although 'the need for initiating a process of development that will form the basis of the much larger effort needed in the future' was also kept in mind. The results attained have been satisfactory. The shortage of food and raw materials has been overcome, and although it is difficult to relate the increases in output in all cases to the investments undertaken, there is little doubt that the production potential of agriculture has increased. The major irrigation schemes undertaken in the First Plan period will have added 7 million acres to the area irrigated; another 16 million acres will be provided with water on completion of these schemes. Minor irrigation works will have brought about 10 million acres under irrigation by the end of the Plan period. The use of fertilisers and better seed and the adoption of improved agricultural practices have also been spreading. The Community Development and National Extension Programme have helped in making the rural population receptive to new ideas. With the further extensions of these programmes envisaged in the Second Plan, the prospects of a continuous and sizeable increase in agricultural production in the years to come must be adjudged as definitely good.

It can reasonably be claimed that the priority accorded to agriculture and rural development in the First Plan was right. It is usually the relative inelasticity of agricultural output that limits the growth in other sectors and so in the economy as a whole. In the countries that started early on their industrial career, the availability of raw materials from colonies was a favourable factor. This made it possible for them to concentrate on high-productivity, developments in manufacturers and to avail themselves of 'increasing returns'. Basically, however, improvements in agricultural productivity and a release of part of the labour force from agriculture were an essential precursor of the industrial evolution. Where, as in India, more food and raw materials are needed even for keeping up per capita consumption, improvements in agricultural productivity have to receive prior attention. Nevertheless, the crucial feature of the development process is a much more rapid growth in the non-agricultural sector and these sectors claim inevitably an increasing proportion of the investment resources available. In the Union of Soviet Socialist Republic (USSR), in eastern European countries and in China, the bulk of new investments have been made in industries, and the increases in productivity needed in the agricultural sector are sought to be obtained through direct investments of labour and through capital formation promoted by means of cooperatives or collectives. In the First Five-Year Plan of China for instance, of agricultural investment proposed, some 68 per cent is to be devoted to industries and lower, but only 9 per cent to agriculture, including irrigation. The lower priority given to agriculture appears already to be causing difficulties. The output of grain is reported to have increased by only 3.4 per cent in the first two years of the Plan as against the target of 17.6 per cent over the five year period. But, what we are interested in is not whether the priorities in investment in a particular country or in a particular set of circumstances were or were not right but in the underlying theory that rapid development demands heavy investments in the secondary sector. The limited impact of our First Plan on employment and the sharp fall in agricultural prices in 1952-53 also point the same way. This is not to say that agricultural and industrial output can always expand in such balance that the terms of trade between agriculture and industry remain unchanged. It is, however, important to recognise that in a developing economy, the secondary sector has to grow fast. In a paper read recently at a meeting of the Working Party on Economic Development at Bangkok, Dr Kuzushi Okhawa has brought out the fact that in the period 1872-1942, while Japan's national production increased 12.6 times, primary production increased 2.4 times, secondary production 47.5 times and tertiary production 21 times. The average annual rate of increase of primary production was thus 1.4 to 1.7 per cent. Japan, of course, depended heavily on the exchange of manufactured goods for primary products from abroad. The Six Year Plan (1954-1960) which Japan is now implementing envisages a higher rate of increase in production in the primary sector—3 per cent per annum—than has been observed in the past, but still primary production is expected to increase by 20 per cent, while secondary and tertiary production is to go up by 50 and 28 per cent in six years' time.

One need not multiply instances. Whether one views the development process from the point of view of increasing aggregate output or from the point of view of increasing employment opportunities, very large increases in output and investment in sectors other than agriculture are called for. Agricultural production has to increase in absolute terms mainly through a rise in productivity rather than in employment; the other sectors have to expand both output and employment. Investment-output as well as investment-employment ratios in these latter are generally very high, so that unless savings on a big enough scale are available in the system, it becomes very difficult to provide for sufficient increases in output and employment in them, and this, in turn, recoils on the agricultural sector which has to continue to bear the excessive pressure of population on land.

These observations are of special relevance in the context of an employment-oriented plan such as this country is now thinking of. They point to the conclusion that the creation of new employment opportunities depends upon the resources that can be made available for investment in construction, in industry—large scale as well as small scale—and in transport, trade and tertiary services. The Second Plan proposals that are now being formulated envisage a considerable expansion in the industrial sector, some 19 per cent of the aggregate expenditure proposed by Central and state governments being devoted to it, as compared to only 8 per cent proposed (the actual being lower) in the First Plan. The relative shift in priorities may be seen from the fact that the tentative allocations being considered assign as much as 57 per cent of the proposed outlay to power, industry and transport, while agriculture including community development and national extension gets about 12 per cent and social services, housing and rehabilitation about 20 per cent. Briefly, the step-up being considered is large in respect of industries and transport, but only moderate for agriculture and for social services. On present estimates, it appears that with the investment on the scale and pattern, that look feasible, an addition of a couple of million to the labour force in agriculture might be unavoidable.

It is of interest in this context to consider briefly the approach which Italy is exploring in her Ten Year Development Programme. This programme aims at maintaining over the 10 years an average annual rate of growth of 5 per cent in national income. It proposes that the natural increase in the labour force of about 2 million should be absorbed in occupations other than agriculture and that a further 2 million jobs be created also outside agriculture to take care of the existing unemployment and under-employment in the country. For securing these results, the development programme differentials between investment in what are called 'impulse' sectors and investment in industrial and tertiary sectors are worked out separately in terms of certain estimates of capital required per worker. Government's action is to be concentrated on the impulse sectors, while the provision for the capital requirements of the other sectors is to come by way of private investment. What emerges on the whole is a programme of net investment amounting to 18 per cent of national income on an average over the 10 years, as compared to 14 per cent in the year prior to the building up of economic overheads. The increase in investment needed in the industrial and tertiary sectors between 1954 and 1964 works out at 300 per cent. It is also envisaged that investment in the impulse sectors should decrease progressively as the development process spreads itself out, so that greater resources become available for the industrial and tertiary sectors. In part, this phasing reflects the fact that the programme outlined relies heavily on the private sector except in the matter of social overheads, but it does bring out the fact (assuming of course that the calculations made are correct) that for creating employment on the scale required not only has the rate of saving (gross) to be as high as 25 per cent of national income but also that investment in industries and the tertiary sectors had to increase from 30 per cent of the total in 1954 to 40 per cent of the total investment in the economy by the end of the 10 year period.

I should not go too far from the problem I started with. I wish only to stress the point that since the role of agriculture has to be fitted into the overall programme of development which necessarily means a differentiation of the economic structure through investments elsewhere, investments in agriculture proper have to be, in the main, labour-intensive and that for sustaining aggregate investment the high levels required, savings have to be generated both in the agriculture and in the non-agriculture sectors. Perhaps, I should dwell a little more on the latter aspect.

For many years, agriculture in India has remained relatively static, and it is only since Independence and particularly after the adoption of the First Five-Year Plan that substantial funds are being directed to agriculture and rural development. For many years to come, the public authorities will have to continue to devote large sums to the building up of the infrastructure of development in the interest both of agriculture and of industry, and as the situation stands in India, a large part of the industrial development programme itself has to be financed by the public sector. All these investments are usually capital-intensive, and as long as public savings are limited, the state has to necessarily depend upon what private savings it can draw upon and channel to itself. As agricultural economists, we should desire that as large a part of the public funds as possible should be devoted to the development of agriculture and to raising the real incomes of the agricultural classes, but I think we cannot but stress at the same time the most urgent need for the agricultural sector itself to finance as much of its development programmes as possible from out of its own unutilised or under-utilised labour power and from the new incomes generated by development. An effective development programme for an under-developed economy has necessarily to have as a major component the direct utilisation of the idle manpower in the countryside. The agricultural sector does this 'saving' in any case; the underemployed or unemployed are being maintained out of somebody's earnings in that sector. If, in building some of the overheads of development, like dams, roads, railway lines and the like, this saving potential can be utilised—with, of course, some supplementing by public authorities, the initial push to the economy could be given more easily. The task, thereafter, would be to see that the larger incomes resulting from the increased productivity thus brought about are not all spent on consumption and a reasonable proportion is utilised for further capital accumulation. This point is well recognised in our plans, and a good beginning has been made in the direction of mobilising local labour and resources through the community development programmes and local development works. It is difficult under democratic processes to bring about the same mass mobilisation of manpower as has been possible in totalitarian countries, but if available financial resources are to be deployed in larger measure for developments outside the agricultural sector—in the interests of the agricultural population itself—the greater the degree of success that can be attained in this direction, the better, obviously, will be the results.

I shall not venture on an estimate of the rate at which the various sectors of the Indian economy ought to expand if the resulting pattern is to be well-balance. The rates and which demand can increase in respect of existing products and the pace at which new demands may arise must be regarded as largely a matter of guess. When, therefore, we speak of a doubling or tripling of national income over a period, we are speaking of an abstraction, and it must be confessed that the elasticities of demand and supply in various sectors and sub-sectors and the scale of investments needed to secure given increases in output are in the nature of only broad judgments. But, I should imagine that something like a 4 per cent annual increase in agricultural output on an average should be needed for the next few years as against an increase of 5 per cent in national output, Should this happen, and should also the working force in agriculture increase mentioned earlier, by about 2 million, the net output per employed person in agriculture would go up by about 18 per cent over the five year period.

Per capita productivity in agriculture is at present roughly 2/5th of what it is in the rest of the economy. In this sense, agriculture is a 'depressed' industry. But, since it contributes almost one-half of the national product and employs some 70 per cent of the population, it is not amenable to the kind of treatment that would be adequate for a depressed pocket in an otherwise flourishing economy. Maintenance of agricultural prices through support measures, fixation of minimum wages for agricultural labour and similar measures involving a transfer of incomes from other sectors can at best be temporary, alleviating measures. They have their place in the scheme of preventing economic distress due to some temporary imbalance in the economy or in a scheme of preventing exploitation of those whom the mere fact of numbers and the accidents of social tradition have placed at a disadvantage. What is needed as a long-run and lasting solution is, however, an increase in production in this sector and a smaller number of persons to share the product. In regard to the former, that is, the increase in production, the outlook, as I have said earlier is satisfactory, since it is obviously the responsibility of the state to build up the necessary overheads, and no one will deny that in this matter the central and state governments have gone forward with determination. The latter suggests the need for special orientation of industrial programmes and the use of industrial techniques that would make it possible for the agriculturist to turn to industry. The fact that at present with wages of factory workers in India around 2½ times the average earnings of an agriculturist, there is no great exodus to the cities shows, in part at any rate, that the real value of the higher money incomes earned in urban areas is not rated proportionately high by the rural people. If industrialisation were to mean more and more urban concentration, then the costs in terms of the higher money rewards needed to attract labour would be obviously too high. There are, of course, other grounds on which a wide diffusion of industry is desirable, but there is an economic case for it in the rural or semi-urban areas. If it is a decentralised pattern of economy that we are seeking, the answer must be found in terms of a proliferation and diversification of economic activities in the rural areas. It is only thus that we can draw upon the large reserves of manpower the country has.

So far, I have made no explicit mention of the relevance of population growth to economic development and to the problems of sectoral growth. I think the position was summed up well in the Report of the Planning Commission on the First Plan. "If population is growing at a certain rate, the total national income output has to be raised at the same rate merely to maintain existing standards of income and consumption; this means that not only will so much less of further additions to national income be available for ploughing back into investment but a part of what is ploughed back will be taken up by capital equipment required for maintaining per capita incomes constant in the following period. ...Given a situation in which shortage of capital equipment rather than of labour is the main limiting factor in development, a rapidly growing population is apt to be more a source of embarrassment than of help to a programme for raising standards of living." The Registrar-General of India has in the Census Report 1951 dwelt at short length on the problem of agricultural productivity in relation to population growth. His calculations suggest that the increase in agricultural productivity required to meet the needs of the increasing population would be of the order of 21 per cent by 1961, 37 per cent by 1971 and 54 per cent by 1981, and he has suggested a ceiling on population at 45 crores. One could take a more optimistic view of productivity. It may also be that considering that we are already 38 crores, nothing we can do now can probably hold the population at 45 crores. Nonetheless, an increase in numbers makes the problem of employment more difficult and limits the possible improvement in living standards. Since, as we have seen from experience, transfer of population from agriculture to the other sectors has necessarily to be spread over a period of time, an increase in population will either add to under-employment in the rural areas or result in overt unemployment in the urban areas. In either case, the outcome is highly undesirable. Is it permissible, one wonders, while talking of planned development to think still in terms of unplanned population growth?

I must now turn to the set of problems that go under the name of reorganisation of the structural relationships within agriculture. The misfortune of agriculture has been that there have been too many instrument of production at his disposal; his holding or farm is, in most cases, too small to permit efficient cultivation. But, further, from whatever he produces, he has had to pay a share to somebody from whom he obtained the right to cultivate and he had also to incur heavy costs for any loans that he could obtain from the local moneylender. Attempts were made in the context of comprehensive planning that the problem is being viewed—I was about to say 'attacked' but I suddenly recollected Prof Gadgil's address to the Society last year—in a comprehensive way.

On rural credit, I need hardly say much. In the reports of the Rural Credit Survey Committee, we have a picture of the present situation—and how unsatisfactory it is—and of the measures needed to institutionalise credit facilities and make them available in ampler measure and at reasonable cost. If as much as 93 per cent of the total borrowings of cultivators come from private agencies, and if, as the report records, a large proportion of these borrowings carries a rate of interest as high as 25 per cent—interest rates of 50 per cent and above also not being unknown—all the moneylending legislation that has been undertaken so far has been futile. The cooperatives provide only 3.1 per cent of agricultural finance. Their operations are concentrated in a few States. A large proportion of the agricultural population has remained outside their purview, and they meet only a part of the credit requirements of members. The cooperative organisation has been, in terms of its achievements so far relatively to needs, a virtual failure, and yet cooperation remains, to use an old phrase, the only hope of the rural masses. As the report puts it, "The problem is not so much one of reorganisation of cooperative credit as of the creation of new conditions in which it can operate effectively and for the benefit of the weaker".

These new conditions call for state assistance and participation not merely for rural credit but for a cooperative production-cum-marketing structure for the rural economy. The Reserve Bank has, in this connection a crucial role to play. Central banking, in an under-developed economy, involves much more than keeping the economy on an even keel. Its function is to assist new institutions actively and to influence the pattern of investment by directing funds to the growing sectors of the economy. As the Committee has approached the problem, reorganisation of rural credit becomes a part of the wider problem of building up a cooperative commonwealth, a scheme of things which 'joins together in a common purpose the united strength of the Indian village with the united strength of the Indian state'. It is possible that some will feel sceptical as to the organisation remaining 'cooperative' if it is state sponsored and State-assisted, for current thinking has been that the state can either be force or it can be charity. But, then, what the authors of the Rural Credit Survey are asking for is not merely a change in the cooperative structure but a transformation of the state itself.

If the reorganisation of rural credit has to be related to the role and responsibility of the state for promoting economic development, this must be no less so in the case of what is called land policy. On this subject the Planning Commission had suggested a fourfold approach in its First Plan Report: (i) abolition of intermediaries, (ii) tenancy reform, (iii) a ceiling on land, and (iv) cooperative farming. As to (i), progress one could say has been satisfactory. The process of abolition of intermediaries had commenced even before the Plan was launched, and it has now been virtually completed. This reform raised difficult financial and administrative problems. For a time, they got bogged down in legalism, but these difficulties have been overcome. Fortunately, there was little social resistance to reckon with; the intermediaries whose interests have been acquired had, by and large, no organic connection with the land. One load has, so to say, been taken off the back of the cultivator. But, it must be noted that while the compensation payments are being made and the cultivator continues to pay the same dues to the State as before, there is no increase either in private or in public saving on this account. The increase in public saving will start after a fairly long period when the compensation payments have all been made. This if one may say so, is the price one pays for democratic and orderly change.

As to (ii), that is, tenancy reform, the basic idea is the same as in the case of (i) just discussed—a redistribution of incomes or transfer of ownership rights from those not working on the land to those doing so. Regulation of rents, which is one item in this programme, has had a long tradition in India, and in many States action has been taken in pursuance of the Planning Commission's recommendations to limit rents by law to $1/4^{th}$ or $1/5^{th}$ of the produce. But, like the legislation to restrain usurious lending, this has been far from effective, simply because the right to cultivate land and make some income out of it is, under the present conditions of land scarcity and absence of alternative occupations, something for which the tenant is prepared to pay up more than he need to under the law. He may cut into his subsistence earnings in doing so, but he can rarely afford the expense and trouble of seeking the protection of the law; and, after all, can a cat look at a king? The answer to the problem must, I think be that the State should place itself directly in touch with the tiller of the land, recover from him the legally determined rent and make the same over to the landlord after deducting the cost of collection. Another item in the programme of tenancy reform is limitation on the landlord's right of resumption. On this, what we have at present is little short of a debacle. Personal cultivation, for which resumption is asked for, is hard to define. It seems to facilitate only the conversion of the tenant into a hired labourer. This is not to say that there are no cases of resumption for genuine personal cultivation, much less could one argue that there could be no such cases. Various states have passed legislation in this regard, but no clear principles of action seem yet to have emerged, and there is not adequate machinery for implementation. Meanwhile, there have been large scale evictions, and action that was intended to benefit the tenant has, in practice, made his position more precarious. In a matter of this kind, which involves many complexities, solutions are not easy to suggest. But, it would appear that the thing to do might be to permit resumption on a portion of the land, subject to fair consideration representatives of all the interests on land approved of the proposal. The economist can only say that the less people come back to the land, the better, but there is, of course, the practical problem of balancing the claims of different classes or interests, for which the voice of the people, that is, the local community, may well be the nearest approximation to the voice of God.

Then, there is the vexed question of a ceiling on land. There is little difference of opinion as to the desirability for fixing a limit on future acquisitions, for larger individual holdings can only come at the cost of smaller ones. The case for a ceiling on existing land holdings is a great deal more difficult. The argument that there is so much land hunger which must be satisfied does not take us far. Hunger, after all, has no very specific content; what is wanted is work and the wages of work. Moreover, it is doubtful if there is land enough to satisfy all the hunger and one cannot but feel sceptical about any long-term satisfaction or social stability arising out of the distribution of small bits of land to claimants who have little of the other ancillaries needed. If one were arguing this matter in strictly economic terms, one would say that if there are some large-sized farms and well-to-do farmers in the country, and if they conform to certain standards of efficiency, it would hardly serve any purpose to disturb them. If their incomes are large, it might be argued, this is a matter for the tax authorities; the agricultural income-tax could be made as progressive as desired. This would be in keeping with the treatment of large incomes in the non-agricultural sector. According to the Rural Credit Survey Report, of the total gross capital formation in the agricultural sector, nearly two-fifths was by the 'big' cultivators and about two-thirds by the 'large' cultivators. A ceiling on land holdings may affect capital formation adversely. These arguments have weight. There is, on the other hand, force—and may be wisdom—in the contention that some redistribution of land cannot be avoided if only as an earnest of the state's desire to move towards as egalitarian society. Land in a densely populated country with very limited opportunities for other types of work in the countryside is, after all, in a special category, and it is not impossible to reduce the size of large farms a little without bringing down the standard of cultivation and the incentive to work and save. If a policy of fixing a limit on incomes in the rest of the economy is accepted in principle and action towards its implementation is part of a plan for a progressive approach to equality, there could be little objection to a beginning being made with land; it must be mentioned, however, that the application of the ceiling idea to all income or all forms of ownership may come up against many practical difficulties, not the least of being the limitations to action in a mixed economy pattern. If on a balance of considerations, ceilings on land holdings are decided upon, it would be desirable to earmark the land thus becoming available for cooperative utilisation rather than for distribution to individuals who cannot probably hold them for any length of time or cultivate them effectively.

The Planning Commission indicated in its First Plan Report the lines along which cooperative farming should be encouraged, and went on to suggest a broader framework of cooperative village management aimed at reconciling the various interests in the rural community including the landless labourer and promoting the organisation and development of the entire resources of the village from the point of view of the village community as a whole. The experience so far with cooperative farming has been small and not encouraging, and cooperative village management can hardly be regarded as anything more than a distant ideal. We are in the same sort of situation which the Rural Credit Survey depicts in respect of cooperative credit. Whatever the experience or record of achievement so far, there is little hope of success for the future unless a structure of cooperative organisations for production, marketing and credit can be developed fairly rapidly and extensively. The individual farm is too small a unit for planned land use, and the prospects of giving every farmer an economic size of cultivation unit are remote even on the basis of large additions to the irrigated area and more intensive cultivation. If agriculture is to make its effective contribution to the development process by producing more, selling more and in turn buying more, and if, further, it is to be organised on business lines with proper provision for capital formation and the adoption of newer techniques and improvements, there is no alternative but cooperative organisation and development. Will cooperatives succeed? Can the rural population be organised quickly for the purpose? How are the problems of administration and management to be solved? Is the necessary supply of trained personnel available, or can it be created with the requisite expedition? These are questions which immediately cross one's mind, and economists, by training and tradition, are apt to be sceptical. All they can do is to reiterate the well-known fact that however important and expedient redistributive measures in the field of agrarian reform may be, the ultimate strength of the economy has to come from an efficient ordering of the resources available and from reorganisation of the productive unit itself on technically efficient lines. Cooperatives are needed for enabling the farmer to avail himself of the economies of scale and they are valuable as increasing his 'countervailing' power. Perhaps more important still is the role they can play in assisting the process of capital formation. It may be that before full-scale cooperative farming with pooling of land is feasible, a great deal of experiments with looser forms in the shape of mutual aid teams will be necessary. There is probably a case for 'companionates' in this field. And here, again, a great deal of sponsoring and assistance from the state will be necessary.

Finally, however, I must add that the solution of many of these problems depends not on mere institutional change but on the underlying spirit which gives meaning and purpose to changes in form. We hear of hundreds and thousands of cooperatives being organised in other countries within the space of a few years or even months. These cooperatives are intended to serve as agencies for organising the people for work; they promote the necessary accumulation for further development, and represent an essential link in the chain of planning agencies. Whatever one might say about the processes by which this kind of economic organisation is brought into being and maintained, its achievements in terms of developmental effort are impressive. Each country has to evolve its institutions in response to its own needs and traditions, and the key to success lies in proceeding step by step, continually adapting and readapting ideas and institutions, trying them out on the ground, leaving room always for fresh experiments. Institutions are, after all, a means; the end is the development of human personality. There is a challenge inherent in a situation of under-development, coupled as it often is with social inequities. Democracy will fulfil its purpose to the extent that it organises itself effectively for meeting this challenge.

Indian Journal of Agricultural Economics, Vol. 11, No. 2, April-June 1956.

16

D.G. KARVE

Economics of Rural Development

IF we were to scan the occupational data supplied by the decennial censuses we would find little cause to justify a feeling that the number or proportion of our population dependent on agriculture is being reduced. Whether we take movement of population or the pace and pattern of economic development there is nothing to suggest that a major socio-economic process of rationalisation of agriculture and of modernisation of the techniques of industry has yet gathered a strong momentum. The number of persons employed in agricultural pursuit is near two-thirds, and the share of agricultural production in the national output is near one-half. As compared with this preponderance of agriculture in respect of employment and output the share of capital investment in agriculture is meagre, being only about 30 per cent of the total. If the experience of agricultural development made by other countries is any guide, it would be obvious that productive investment of capital, per acre of cultivated land and per employed person, will have to be substantially increased if a process of rapid rise in productivity of agriculture by the adoption of such measures as minor irrigation works, oil conservation and employment of improved standards of cultivation is to be promoted. This is an aspect of the programmes of agricultural improvement which has not been worked out with enough care, and hence the provision of funds intended for agriculture in the national plan appears to be meagre, especially if the pattern of progressive cultivation included in the national extension programmes is to be adopted by a majority of farmers.

Under the Second Five-Year Plan it is proposed to bring all the villages of India under the national extension scheme of rural development. Forty per cent of these villages have to be under the more comprehensive plan of community development. The total number of villages in India exceeds 500,000. Even allowing for a twenty per cent deduction on account of de-habitation, smallness, double counting etc., it would be fair to presume that not less than 400,000 villages are to be the scene of a programme in which every family is expected to participate in a plan of improved farming which would be devised to suit its conditions. Participation in a developmental programme along these lines would, apart from other things, require long and medium term investment, very much in excess of the figures of ₹25 crores and of ₹50 crores for long and medium term credit respectively provided in the Second Five-Year Plan. Most of the funds needed for agricultural development under the scheme of extension must come through institutional channels, as no other agency exists which will provide the funds either on the required scale or on economical terms. Great as is the paucity of current credit, greater still is the shortage of development funds. If the programme of agricultural progress envisaged by the plan, which has been recently augmented by a quarter, is to be really implemented, a much larger amount of long and medium term investment than is provided in the plan would be needed.

The importance of providing adequate developmental capital to agriculture is obvious. When it is realised that the process of industrial revolution of the entire economy of the nation has to start from the industry of the people, the truly national industry, namely, cultivation of land, it will be seen that starving agriculture is not only bad economy for farming, but it is also an evil turn done to manufacturing and other industries as well. Without the support of abundant and cheap supplies of food and raw materials, there is no prospect of organising an economical and productive programme of industrial development. Nor would the products of industry sell except in markets created by growing incomes among the non-industrial sections of the community. The desired balance between agriculture and industry will be attained not by alternating emphasis on the two in one plan and the next, but by keeping a continuing harmony between the needs and potentialities of both. The allocation of only 12 per cent of developmental outlay of the Second Five-

Year Plan to agriculture and community development, as against 16 per cent of the First Five-Year Plan, can have no justification at all. In any case it would not be realistic to expect that the enhanced targets of agricultural production in the revised Second Five-Year Plan can be realised with a lower percentage of allocated funds. Increase in productivity, as well as of total production, of agriculture commensurate with the effective demands of the non-agricultural sector must be treated as a corner-store of our national economic policy. Out of agriculture's increased productivity will arise not only the current supply of food and raw materials, and the demand for non-agricultural produce, but eventually also the investible surplus of capital needed to activise later stages of economic progress.

In one sense the importance of modernisation and scientific progress of agriculture is even greater than appears in an accountant's stock-taking. The habits of mind which make for successful and progressive industry are not built in a day nor would their possession by a few people suffice to improve the tone of a whole economy. In fact concentration of enterprise in a few people leads to a corresponding concentration of wealth. The social injustice and economic wastefulness of such a state of things have already been recognised in our national constitution. More important than an equal distribution of consumption goods is the equal distribution of productive opportunities. In an old ruralised economy seeking to make rapid progress on the road to industrial development emphasis must be placed on bringing opportunities of progress to the mass of the people, who are rural and for the most part agricultural. A change in the outlook, habits and occupations of the cultivating classes will contribute to the material as well as moral progress of the community as a whole. The measure of progressiveness, comfort and civic virtue which a nation claims to possess must be real in terms of the actual experience of the large mass of its people. In India, anything which the average cultivator has not an opportunity of experiencing by his own effort can hardly be claimed as a dependable index of the improvement of the nation as a whole.

It was, therefore, in the fitness of things that the First-Five Year Plan gave prominent place to agriculture, irrigation, community projects and similar items in other departments of economic life having a bearing on the progress of the principal industry of the people. In an old, big, populous and monsoonish country like ours a programme of agricultural transformation has to be technically sound as well as massive in its impact, and it has to be continued for a long while. Resources of capital and organisation have to be concentrated on the job till the transformation is complete. The achievements of the agricultural part of the First Five Year Plan were not disappointing so far as they went. Irrigational works, large and small, made normal progress, Community projects got off to a good start, and above all the rain gods were on the whole propitious. These were encouraging, but by no means completely reassuring circumstances for the planner. A large part of the programme of minor irrigational works had to face physical and technical difficulties of an almost insurmountable character. Even where the irrigational part made normal progress, schemes of actual utilisation, and especially of promoting progressive levels of cultivation could not be easily adopted. Legal, financial and organisational difficulties cropped up on a scale not anticipated in the plan. In the community projects a lot more of initial effort had to be incurred than was allowed for. At their best the projects touched only a portion of villages and of cultivators. The technical and scientific contents of the programme of extension were as a rule too generalised to be a practical benefit to individual farmers. All these and other deficiencies were gradually noticed and plans were made to remove them.

This process of slow initial impact, and of trial and error, was unavoidable. Lulled into complacence by the favourable results of two or three monsoons, the planners tended to overlook the sobering implications of the slow and halting progress of their agricultural programme. For long term achievements in raising productivity on a mass scale there was need to invest much further effort and resource than were provided in the programme, the technical implications of irrigational works had of course to be studied. But of even greater economic import were the limits on actual utilisation of land and water. Systems of land tenure, programmes of appropriate crop planning and provision of institutional finance and technical assistance could not be suitably rearranged in time to make the best use of irrigational facilities. Even now not much advance has been made, and it is of the utmost importance for the success of the plan as a whole, and of its agricultural portion in particular, that more resources should be channelled into these items. Any feeling of complacence, such as that which is implied in a rapid increase in the formal setting up of agricultural extension organisation without adequate technical and financial strengthening, or that in setting up targets of agricultural production without corresponding programmes of agricultural extension and institutional finance, will be fatal to the success of the plan. Recent happenings on the food front are not to be treated as passing events. They only show that the weaknesses of our existing agricultural structure are very deep seated. Unless large scale and sustained efforts are made to remove these defects the foundations of a major plan of economic transformation cannot be said to have been laid.

Nor is the importance of agricultural planning to be judged in terms of physical production alone. The republican constitution of India lists certain social values as paramount in the structure of relations to be sustained by the authority of the state. Among these values avoidance of concentration of economic power, and equalisation of possessions and of opportunities have been given a very high place. It is not open to any loyal citizen of the state and least of all to an organised governmental authority, to order its plans otherwise that in keeping with this scale of values. Where eighty per cent of the people live in rural areas and over two-thirds live by agriculture, the regime of equalitarian and progressive social arrangements must be established firmly in rural pursuits generally, and in agriculture in particular. The programmes of rural reconstruction outlined in the First-Five Year were largely based on recognition of the system of values contained in the Constitution. In the implementation of those programmes the same zeal for maintenance of the prescribed values has not been uniformly sustained.

Thus, the land reform measures which are basic of the whole structure of agricultural planning have not been pursued with the necessary vigour in all parts. If our national planning is to have any goal at all, it must be to reach over a period of time, say of twenty years, a stage at which the rate of economic development on the one hand, and that of the movement of population on the other, are so adjusted that a steady rise in per capita income is assure. This course of action would provide conditions of fair employment from which further stages of more and more productive employment can be reached. Pending the attainment of this first landing stage on the road to rising progress, employment for a large part of the population will have to be found in agriculture. This is not possible without a wide distribution of opportunities for cultivation of land. The constitution of the country does not treat private property in the means of production as in itself an undesirable thing. On the contrary, with specified exceptions, it guarantees the right of property. In the circumstances, the only way in which the large majority of the rural population can be provided with work opportunities, in a society where economic concentration is to be avoided, and equality of opportunity is to be established, is by redistributing land on the widest possible scale. As this is to be done by legal process it is only natural that land for redistribution, which is to be requisitioned from private owners has to be paid for on terms considered to be socially fair. The great merit of the *Bhoodan* movement consists in this that the social evaluation of ownership of agricultural land beyond what one actually cultivates with the assistance of one's own family drastically reduced, thus strengthening the hands and facilitating the way of legislative land reformers. The physical quantities of donated land made available through the *Bhoodan* movement is also not negligible. In spite of the strict injunction of the Constitution, and of the specific provisions of the First Five-Year Plan, in large parts of the country no appreciable progress has been made in regard of deconcentration of the possession of land as an instrument of productive activity. In some other parts where a frame work of law has been created in support of accepted ideas of land reform, actual formulation and execution of the law are being pursued in such a way as to defeat the basic purposes of decentralisation and of equality of opportunity. In fact, there is reason to apprehend that second thoughts which do less than justice to the values contained in the constitution have begun to assert themselves in some quarters. Unless a strong movement for land redistribution is kept up at a high level of public support, there is serious danger of the very basis of agrarian democracy being too loosely laid to support the structure of a socialistic state.

The Constitution of India as it stands does not sanction a collectivist solution to the problem of distributing land as a means of production among the larger number of people who, in the absence of any other means of employment, must depend on land for their occupation. Considering the large number of agriculturally employed persons, and the limited area of cultivated land, it follows that any scheme of redistribution would leave the average "holding" too small to be economic. As the technological standards of cultivation improve the size of an economic unit is likely to show a material increase. Confining ourselves to such standards as may be incorporated into our extension programmes during the next twenty years, it would be seen that even for the immediate future in many cases, for economical use agricultural land will have to be pooled in much bigger blocks than would be available to individual holders under any scheme of equitable distribution. A programme of joint cultivation thus occupies a central place in our schemes of agricultural development. The implications of joint farming, however, deserve very careful consideration. It should be clearly stated that the only kind of joint cultivation, which it would be proper for authorised developmental agencies to propagate, is joint cultivation which can be confidently expected to have a better economic prospect that would be available for individual cultivators assisted by institutional aids which can be offered to them through cooperative channels. Any sponsoring of joint cultivation which is not designed to secure to the farmers themselves concrete and sizeable economic benefit would be inconsistent with the purposes of democratic planning.

Wherever joint cultivation is likely to yield appreciably better economic results one may confidently advocate its adoption. A suitable educational programme in support of such step should be undertaken without delay. As normally an extension to community project organisation would exist in areas selected for joint farming, educative and demonstration programmes connected with it can be easily incorporated into the overall extension programme of the area. Experience of extension obtained during the last four years proves beyond a doubt that proposals for innovations either in behaviour or in institutional activity which can be demonstrably proved to be in the interest of the cultivators are accepted without undue resistance. The educative effect of successful acceptance by one group rapidly spreads to other groups in the same and in adjoining villages. Beyond intensive educational and organisational activity no more social or legal pressure is normally necessary. Joint cultivation involves the readjustment of legal rights, not only of the individual farmer but also of his family. It is, therefore, natural that some legal provision for the incorporation of joint farming will have to be made. A programme for the creation of joint farms based on firm calculations of economic benefit and accompanied by normal extension education will be easily fitted into democratic functioning of community life in our rural areas.

Two things regarding joint farming are obvious. On the one hand the process of fragmentation and subdivision of holdings has gone so far that there should on a priori grounds be considerable scope not only for consolidation, but also for joint cultivation. Even where land reform legislation is combining some consolidation with re-distribution of lands to tenants and agricultural labourers the average size of farms in many areas is expected to be less than what, with available means of cultivation, can be considered as optimum. On the other hand, in the comparatively hard and trying conditions of Indian farming the incentive of direct personal responsibility for the product is so essential that any weakening of this, not counter balanced by appreciable advantage on the technological or organisational sides, is likely to prove injurious to the objective of greater production. That joint farming, in the initial stages, would make for more conspicuous unemployment is a possibility which must be taken into account before large sized programme of joint farming is undertaken. Joint farming especially cooperative joint farming, represents a higher form of organisation which for its success depends on favourable economic conditions and on adequate training and experience.

The extension organisation has first hand appreciation of the physical organisational and psychological situation prevailing in each block. It should not be difficult for extension staff first to spot the most eligible places where to initiate a programme of joint framing and then to make the demonstrational and promotional effort to establish societies for the purpose. One obvious indicator would be the prevalence of a large number of definitely uneconomic farms. It should be very easy in this situation to plan and to demonstrate a more economical and productive grouping and reorganisation of members' resources including land. Even here it would be worthwhile to proceed initially on the pilot principle. Another situation in which economists and planners could combine to initiate a more radical programme is the existence of many farms of a size which is much larger than can cultivated by a farmer without the aid of hired labour. It would be an elementary step in land reform in a society which describes itself as socialistic to restrict individual holdings to a size which can be self-cultivated in the genuine sense, that is, without the help of hired about. All land in excess of such ceiling should be available for redistribution. For a long time and in most places the number of landless agricultural labourers or very small holders having a claim to share in the disposable surplus would be so large that it would not be possible to allot enough to watch to make an economic unit. In all such cases while a share in the right to hold land should be distributed, the use of the land to be economic should be organised in a cooperative society. Here, again the advantage of joint cultivation can be easily demonstrated, and in the absence of a vested interest a more productive regrouping could be easily brought about. Once the benefits of joint cultivation are established the normal extension and democratic process may be left to convey even to the more substantial farmers the real lesson of joint farming.

The extension organisation provides a machinery which may be utilised for the purpose of identifying the areas where cooperative effort among the several classes of farmers can with advantage go on to actual joint cultivation productive use of land by participating in a progressive programme of agricultural development is obvious. In fact, it is not difficult to demonstrate in favourable situations that not only their interest, but also their duty, consists in participating in the national programme of agricultural development through cooperatives. The real justification for a land reform measure is that in its absence the resources of the community, land, labour, capital and organisation, are not put to most productive use. If after redistribution of land standards of cultivation were to fall or were to remain avoidably stagnant the reform measure would not contribute to the welfare of the community. Even where as a matter of social policy land has been withdrawn from one owner and conferred on another the community's interest in its use is paramount. Both the act of revocation and reconferment have the same justification, *viz.*, social interest which includes the legitimate interest of individuals. It was never as true, as it is today in regard to all means of production in India, that the State is an active partner in ensuring, and if necessary guiding their best possible use for the community as well as for the individual.

The state sponsored movements of agricultural extension and community development stem from this realisation of the new opportunities as well as the new responsibilities of the Indian people. The small farmer, who is the corner-stone of the edifice of our new rural life, is not likely to command by his unaided effort those resources of technique, capital and organisation with the help of which his business of cultivation is expected to prosper. The national extension movement is a timely step in providing by community effort, state as well as local, those resources which in a capitalistic economy can be had only by the rich. State initiative and State partnership in a popular movement of community action in the pursuit of a better, fuller and more progressive life is the substance of the community project and national extension movements, it is most appropriate also that provision of resources is being planned through cooperative institutions which draw in proper proportion on public as well as private resources, on state resources of enterprise and organisation, is to perpetuate the regime of rural stagnation and capitalistic exploitation. To assume state responsibility for direction without popular initiative is to inaugurate an authoritarian regime which is bound to fail economically as well as politically. In a democratic state which is planning to build a progressive economy on welfare principles an equitable partnership between the state and the individual farmer can alone help to realise prosperity without the loss of freedom.

Several proposals designed to regulate the possession and use of land as also the manner of channelling public resources into the operations of individual farmers and their groups, have occasionally evolved protests on the score of individual liberty. One would have thought that by now it would be obvious to everybody that the enjoyment of freedom cannot be real to all citizens without ensuring conditions of equality of possession and of opportunity. Any action taken in pursuance of a programme of equalisation would be in support of individual freedom. This should be specially obvious in the Indian Republic where the purposes of a welfare state are inscribed into the Constitution, and where programs in support of the same are being formulated through the Planning which is compatible with a welfare and planning state, and not the negative one of lack of restraint on each, irrespective of its effect on the real enjoyment of freedom by all.

One would feel that this is too obvious to merit a repeated mention. And yet it has been my experience within the last year or two to be recipient of almost a personal remonstration from at least two persons holding responsible positions who felt that discouragement of small village cooperatives and State participation in the capital of cooperative societies somehow constituted a shocking violation of individual liberty. More specifically, both of these gentlemen felt that the doctrine of individual liberty propounded by John Stuart Mill as the basis of civilised freedom and of democratic government was somehow violated by the proposals contained in the policy of ensuring by state participation in farmer's cooperatives that the means of positively benefitted by land reform and agricultural extension reach the under-equipped section of cultivators.

There are elements of absolute truth in the mature thought of all philosophers, but the specific expressions and illustrations used by them naturally reflect the limitations of their own time and environment. One need not, therefore, be accused of comprising with one's regard for liberty if one prefers a varying expression or an altered field for the effective establishment of freedom in other times and in altered circumstances. Mill himself clearly distinguished between the doctrine of political liberty, in so far as it consists in lack of restraint on holding and propagating opinion and in pursuing modes of behaviour, on the one-hand and social relations arising out of vital interests affecting individuals as well as the community on the other. For instance, Mill specifically excluded consideration of the doctrine of free trade from his discussion on liberty. As is well-known Mill not only recognised that there are limitations on the beneficence of *laisseze faire*, but he went so far as to justify formation of commercial corporations by the state in backward countries. Almost everywhere he favoured state interference in the spheres of education and of family planning. What Mill objected to were restriction and interference which hampered the universal urge to exertion and development. Regulatory and positive acts of the state which expanded the area of enterprise and progressive activity would be welcomed by all, and especially by Mill, as essential bulwarks of liberty.

It is somewhat symptomatic of the only partially emancipated stage of the Indian people that people in responsible, almost strategic positions, should still quote Mill in support of a policy of apparent voluntarism which only breeds hospitalistic exploitation. It is to be hoped that the new generation will be more untrammelled in its endeavour to build an equalitarian and progressive social order. In designing and guiding the rural sector of that order the agricultural economist has an important part to play. In its comparatively brief career of separate existence agricultural economics has already achieved considerable progress and recognition. No small amount of credit for this achievement is due to the Indian Agricultural Economics Association, not only the general problems connected with the condition of peasants, but specific problems such as land reform, credit, marketing and farm costs have been intensively studied by members of the association. Thanks to the coordinated efforts of the research Programmes Committee of the Planning Commission, the Agricultural Economics Committee of the Indian Council of

Agricultural Research and the Economics and Statistics directorate of Food and Agriculture Ministry of the government of India, a large measure of purposeful research in the economic aspects of agriculture is proceeding in a number of places. This research had no doubt served a useful purpose in making agricultural economics more specific and more creative.

In view of the urgent needs of agricultural planning and development it is desirable that this trend towards enlarged and specific studies should be carried much further. Some aspects of agricultural policy have a justification based on much broader ground than economic. But the whole programme of getting the best return out of agriculture, becomes more and more important each day. All aspects of economy, especially the rural economy which have a bearing on this problem need intensive study. On account of the continental nature of our economy any study to be significant and fruitful has to be carried out in appropriate regional environment. Coupled with the tendency towards greater and greater specialisation, this would mean that unless studies in agricultural economics will not be able to make its full contribution to the understanding and solution of the problems of Indian agriculture.

In the absence of sufficient analytical and practical studies undertaken by our own scholars there is an occasional tendency among policy makers to rely too trustfully on some generalised claims and suggestions based on foreign experience. I would agree that it is not always the absence of such indigenous material which make some of our policy makers more receptive to foreign than to Indian suggestion. Probably this attitude is a vestige of foreign domination which can be expected to vanish only as local achievements become more substantial. Here indeed is a challenge to Indian economists generally and to the agricultural economists in particular. There are several aspects of urgent national policy which in the absence of convincing findings of local research are being decided on purely impressionist suggestions, often drawn from tendentiously reported foreign experience. For instance, the decision on promoting a programme of joint farming, in so far as it is based on economic grounds, assumes a number of economic propositions about the optimum scale of cultivation under varying conditions. Such conditions are not only physical and technological, but they are also organisational. That conditions of soil, crop and technique will influence the size of the optimum unit of cultivation, whether under individual or joint cultivation, is easy to understand, though even here the full economic implications of the various physical and technical factors involved need to be brought out much more specifically than has been done hitherto, if practical policy is to be more rationally guided. But the various implications of organisation for joint services such as credit, supply, marketing, technical advice etc., have to be studied in detail to decide as to which of these must be made available to joint farming units, rather than to individual members of farmers' cooperatives, so as to be most effective. In regard to the form, size and activities of the cooperatives themselves a constant study to assess not only usefulness generally, but specific economic advantage in terms of the costs involved, is necessary. In the absence of such study planning is apt to degenerate into an authoritarian sponsoring of ill-assorted measures.

Like other sectors of the economy, agriculture has now to be studied in a developmental context. The position of agriculture has to be viewed in the perspective of a long term objective for the economy as a whole. The contribution by way of employment, production and investment which agriculture would make to the plan, and the contributions by way of capital, supplies and demand which agriculture would expect from the other sectors have to be worked out by agricultural economists as their share in plan making. In the past while other sectors of the economy have evoked both interested and analytical participation by experts during and after the formulation of plans, agricultural economists as such have made only a modest contribution. This is due primarily to the fact that agricultural economics in India is still mostly concerned with problems of organisation, status, and latterly methodology of farm survey and cost studies. Data collected by these studies will no doubt be valuable in constructing an agricultural plan for the country. But unless economics becomes development-minded and economists study the business aspects of alternative forms of land utilisation under varying conditions of technology, crop pattern and organisation not enough rational planning is possible. The factual data, analytical tools and planning experience needed to achieve the requisite degree of competence cannot be developed within a short period. By all accounts in other developed countries also, capitalist as well as socialist, development of agricultural economics has taken a long time, and even today for analysis as well as planning agriculture continues to be on a much less firm ground than, say, industry and finance.

The process of carrying economics nearer to the production and sale aspects of agriculture which has already commenced has to go a long way so that agricultural economics may really claim to be economics of agriculture. Some division of labour seems to be indicated in this respect. While scholars studying agricultural economics in universities might with advantage attend to problems in which field and operational condition are not so important, a more intensive study pursued in the economics departments of agricultural colleges. The number of agricultural colleges is small and not all of these have a well staffed economics department. There is a real need for increasing the strength and equipment of economics faculties of all agricultural colleges.

The Food and Agricultural Ministry which has now established a well deserved leadership in organising intensive studies in several research centres, may exert itself further to ensure through central assistance the creation and augmentation of an economic research section in all agricultural institutions. The large number of agricultural extension organisations, including their training counterparts, must also include agricultural economics units. Unless extension becomes more and more economic-minded, and in its turn economics become more and more extension-minded both will tend to be superficial and on the whole fruitless. Agricultural economics can least afford to be cloister study. It must be very near the field and the *mandi*. In large measure it must be part of the study of agriculture and of the developmental efforts made for it through extension organisation.

What is termed agricultural extension in other countries has only a limited application to Indian conditions at present. Extension is essentially an educative process of carrying to the practical user the most advanced lessons of scientific research. The success of this process presupposes a developed research organisation at one end, an active, well-equipped and integrated community life at the other. At present, not only are our research achievements too modest and too general, but they rarely have a directly economical angle on their results. In researches connected with a few commercial crops some coordination between economic and technical aspect has occasionally been attempted. But nowhere has economics, i.e., calculating inputs and outputs, and studying the business aspects of the utilisation of new methods, been incorporated into that body of research knowledge which extension seeks to carry to the actual cultivators. There is urgent need to develop all agricultural research, but the need to develop economic research as part of developmental research is all the greater. In the rural areas themselves general education, as also growth of habits of progressive action through institutional channels have to progress considerably before all the efforts at organised development can gather enough momentum to constitute an agrarian revolution. While agricultural economics as such has not much active part to play in promoting this change, it has to realise that mutually complementary movements in fields like transport, industries, education, community development, etc. have a very close bearing on agricultural progress. In fact development can never be sectional. It is an integrated process where urban and rural, industrial and agricultural, in fact economic as well as the broader cultural movements all move forward and strengthen one another, the pace of one conditioning the pace of others.

There is hardly any issue of agricultural progress, as indeed of the whole pattern of rural development, which does not raise fundamental questions of social policy. Land reform undertaken with a view to giving to the agricultural worker his share of the means of productive employment , and establishing him in such a relationship to land as would maximise his incentive to efficiency is only the beginning, though an essential one. Even this, however, presupposes acceptance of a certain concept of property and of distribution of resources which is fundamental to our constitution, and which it is obligatory on us to follow not only in agriculture but also in other spheres. This is not merely a question of formal consistency. Reforms in agriculture cannot move forward on rational and progressive lines unless the acceptance of common principles of social organisation is ensured over the whole field of economic activity. Thus the adoption of joint cultivation in suitable conditions involves an act of voluntary surrender on the part of an agricultural family, which has, as a rule, more man power than can be economically utilised with the assistance of available land. Howsoever, one may try to disguise the fact any rationalisation of the size of the farm and the use of available means of employment is bound initially to create in many cases a concentrated degree of unemployment in rural areas. Unless there is some programme for the rational use of this surplus labour, the overall social and economic effects of a forcibly rationalised agriculture may not be altogether beneficient. In other words, we cannot afford to be progressive and socialistic in agriculture, without being at least equally progressive and socialistic in other spheres of economic life. Most of the earlier attempts at a socialistic reorganisation of society, even these in some of the communist countries, have been at the cost of the tillers of the soil. One service which agricultural economists can at this moment render to their science and to their fellowmen is to state the overall conditions in which alone it would be fair to expect that agrarian socialism, even in its more democratic form of cooperativism, would prove practicable as well as beneficient.

Indian Journal of Agricultural Economics, Vol. 12, No. 1, January-March 1957.

17

B.K. MADAN

Agricultural Organisation and Finance and Some Elements of the Appropriate Price Policy for Agricultural Products

I

The process of development is invariably associated with a relative decline in the importance of agriculture in the economy. This is an objective which economic thinkers in the country had set for our economic policy as it has been taking shape in the years since Independence has as one of its foremost objectives to lessen the dependence of the population on the primary industry of agriculture and to expand significantly the avenues of employment in secondary and tertiary sectors. In several countries, the stimulus to industrialisation has often come from the profits of a prosperous agriculture and has been imparted either spontaneously as in some European countries or through deliberate state policy as, for example, in Japan. Often, however the spur to industrial development has been given by the state—as in many underdeveloped countries in recent years—to correct excessive dependence of the economy on a few agricultural products which are liable to large adverse turns in their fortunes with consequent unfavourable terms of trade for the entire economy of the country. In our country the movement for industrialisation has been stimulated and sustained by a variety of forces and elements. In the earlier phases, however, the prosperity of agriculture, following upon the great growth of exports and the spread of commercial crops, partly contributed to that progressive ruralisation of the country associated with the decline of indigenous handicrafts and, therefore, to a setback for the industrial sector. The accompanying prosperity in trade in due course laid the basis for the growth of large-scale industry in the country; and this was accelerated by the deliberate stimulus to industrialisation by government policy.

The question thrown up by a review of the historical process of a declining role for agriculture as an incident of economic development—as a matter of fact in other countries and an objective of policy in our country where the role of agriculture has not declined—is, what should be the place of agriculture in our development programmes? What is its claim to a significant share of the increased public investment that is planned? Our answer to these questions has to be consistent with the long-term objective to which I have just referred of a relative shrinkage over time of the primary sector. I submit that it is wholly consistent with this objective to accord a larger relative place to programmes of agricultural development in our present planning than is given by the current allocations in the Plan, through such readjustments as would be made from time to time in this flexible framework. The reasons for this view derive both from the possibilities of development of production in the agricultural sector and from its needs.

The relative possibilities of development in various sectors of the economy have, in general, to be considered in the situation of capital scarcity which forms the general context of our development. Industrialisation normally requires a high degree of capital intensity and involves some technological unemployment. In a country like ours, however, with seasonal and disguised unemployment of large dimensions, the basic problem consists in fuller and more adequate utilisation of the existing labour resources through new and more effective way of doing things not only with the help of more capital but with increased use of capital-saving methods. In economies where capital is scarce, emphasis has to be placed, in a special measure, on better rather than more tools, on improved methods and techniques and on stimulus to resourcefulness in invention or adaptation of devices which enhance the quality and effectiveness of existing capital in the economy. While these considerations have a bearing on the pattern of our industrialisation, the claim of agriculture for a larger share of attention in the development programme in early stages is that particularly in agriculture, improvements in productivity are often possible with small additions to capital investment, with better organisational methods and more effective application of known techniques.

Over and above the significant possibilities of expansion of production in agriculture, the need for such expansion is related basically to the advantage of providing an adequate base for industrial development in an enlarged and assured supply of food and raw materials, to the great scope for extending the substructure of economic development through completion of the basic facilities in the agricultural sector, like irrigation and village communications as well as for extending the internal investment on the farms, among other things, for adequate utilisation of basic facilities like irrigation and to the importance of retaining the initiative afforded by the changes set afoot through agrarian reorganisation and carrying this process to some expeditious conclusion. Two inter-related considerations remain to be added. Though the long-term objective is to achieve a reduction in the pressure of population on the land, and therefore, to bring about a considerable expansion in the opportunities of employment in the secondary and tertiary sectors this objective is to be attained consistently with the over-all objective of the best utilisation of all available resources for increasing the national output in all sectors of the economy with a view to securing the greatest rise in per capita incomes everywhere. Secondly and more importantly, this rise in per capita incomes cannot be left to be attained only through the creation of more productive employment outside agriculture. Agriculture is yet the largest sector of the national economy affording employment to two-thirds of the population. A rise in the incomes and living standards of this large bulk of the population through improvement of the productivity and efficiency of agriculture, not through a mere shift away from agriculture but directly through investments of capital and improvements in technique bearing on the day to day operations of agriculture, is called for. The effects of incomes in the national product and national income must be visibly reflected in the lives of the mass of the people who live on the land. The objective of policy must, therefore, remain simultaneous attention to development in the agricultural and other sectors, with agriculture receiving at the present stage of our development more adequate attention comparatively to other sectors than is given to it in the existing allocations. This seems to be all the more necessary in view of the revised and increased targets of agricultural production decided upon after the publication of the Second Plan.

The desirability or soundness of aiming at an increase in output in the agricultural sector comparable to the general increase in national income—27 per cent in agriculture as against 25 per cent in national income—is broadly indicated by the line of thinking I have advanced, though, in practice, both the increases are likely to be smaller. We speak here of expected increases; the variable factor of weather conditions is liable to cause divergence of the actual from the expected increase in national income or output from year to year.

We may turn now to consider some of the problems of agricultural organisation. It has for a long time been recognised as axiomatic that to ensure the basis of progressive agriculture and social justice in the rural community as well to release the energies and enlist the enthusiasm of all strata of the population for the success of planned development, one of the foremost as well as one of the toughest problems of the agricultural planner is that of land reforms and agricultural reorganisation. The various measures of land reforms may be broadly grouped into abolition of intermediaries, tenancy reforms, imposition of ceilings and cooperative farming. It is not my purpose here to review the progress made in respect of each type of land reform. What strikes even a casual student of the subject is the large hiatus between promise and performance, between the legislative provisions and their implementation and the consequent uncertainty and suspense that hangs over the issue of rights in land in a sizeable sector of agriculture with its deleterious effects on investment and production in agriculture and on the social climate for economic improvement programmes in the country-side. There is hardly any field where the need for definition of objectives, for articulation of measures in terms of the objectives and for implementation is greater than in regard to land reforms.

'Land to the tiller' was the goal for which we set out on the road of land reforms. The achievement of this goal has, however, been substantially compromised by the large-scale evictions of tenants for resumption of land for 'personal' cultivation. Personal cultivation has, however, been a misnomer for the continuance in this garb of cultivation through temporary tenants or labourers. Adequate studies have not been made of the extent to which in the country as a whole the objective of land reforms has thus been vitiated or of the extent and manner of implementation if the other provisions or of the expedients devised to maintain the status quo in regard to ownership rights to the detriment of the interests of the actual tenant or tiller of the soil. It is now recognised that enforcement of legislation for regulation of the money-lending business has been largely a failure, not only in our country but in many other countries where legislative efforts have been made in this field. The situation in regard to compliance with legislative provisions for land reforms is certainly not as bad as in regard to regulation of money-lending, but the real change of economic balance in the country-side has been much slower than was avowedly visualised in the laws regarding abolition of intermediaries and tenancy reforms. The same combination of forces so vividly portrayed in the Report of the Rural Credit Survey which is stated to have prevented the benefits of cooperation from percolating to the lower economic layers of society, seems to have been operating in avoiding any significant tilting of the scales in favour of the working but weaker groups of the rural population.

In regard to the basic objective of the first two categories of measures, *viz.*, to confer on the actual tiller of worker on land rights to its ownership of tenancy to enable full exploitation of land including the right to mortgage it and to raise credit against its security with reasonably assured prospect of enjoyment of the fruits of his labours, there should be no question of faltering or flinching in its steadfast pursuit. This would involve more determined action in the direction of adoption of more rigorous standards for limiting the resumption of land for personal cultivation, and extensive review of cases ejections of tenants, in the teeth of overt and covert resistance from the powerful combination of elements which would seek to minimise the incidence of land reform in actual application. The problem has to be stated in this form to appreciate the magnitude of the task involved in the transformation that remains yet to be achieved.

As regards ceilings on holdings, these apply for the present mostly to fresh acquisitions of land, the question of redistribution of existing holdings above ceilings equivalent to those for fresh acquisitions not having been pursued to a conclusive stage. Meanwhile, divisions of landed property have occurred with a view to escaping the incidence of ceilings, should they be imposed. To some extent, the uncertainty regarding ceilings may have impeded investment. In Uttar Pradesh (UP) it is reported that it has increased the difficulties of consolidation of holdings, owing to a reluctance to avail of the benefits of consolidation in respect of land which might later be taken away. There are also the problems of making effective arrangements for redistribution of land without impairment of efficiency of cultivation. In any case, it is important that redistribution should proceed with due regard to the capacity of the administration to cope with the attendant problems of arranging for equally efficient cultivation; this would involve increasing commensurately the facilities of credit, minor irrigation, seeds, equipment and other supplies in respect of the land to be redistributed. Taking everything into account the imposition of physical ceilings in terms of any multiple of an economic or family holding, such as have been considered, now appears to offer a less promising approach to the problem of unequal land distribution than it perhaps did earlier. There are two alternative suggestions which offer themselves in this connection. In the first place, more resolute implementation of the reforms in terms of conferring the substantial rights of land exploitation for utilisation on the tiller together with more rigorous application of the provisions for resumption for personal cultivation can largely obviate the need for a direct treatment of the problem of ceilings. This may be supplemented, secondly, by action on lines similar to those adopted in UP, where the legislature has very recently passed a law to levy a steeply graded tax on agricultural incomes intended to tax the produce of lands many of which would be above any ceilings, under a highly progressive schedule of rates. For the rest, it would become a question of the effective levy and collection of taxes; and evasion should be easier to prevent in taxation of agricultural incomes and property than of general income. In fact, the very basis of computation of agricultural income has usually some multiple of revenue or rent.

I come now to the last of the main categories of reform, *viz.*, cooperative farming. This measure of reform is essentially to be considered in terms of its potential implications and value; its actual scope so far has remained restricted. There is hardly any field, however, where the ground for full scale advance is so little prepared, or where the need for adequate preparation is greater. In the Second Five-Year Plan the consideration of this question commences with the remark: "There is general agreement that cooperative farming should be developed as rapidly as possible. The practical achievements in this field are, however, meagre." This introduction to the subject bears out what I have said regarding our state of preparedness for the reform. The Second Plan Report clearly depicts the trials and tribulations of the limited experiments in this field that have been conducted so far. Its appraisal of the progress or failures to date in this field is realistic. The Report states: "On the whole, little action has been taken in these directions. In most States, there are groups of individuals who have come together to form cooperative farming societies. A few of these societies have been successful, but many of them have experienced practical difficulties for which they have not always been able to secure the necessary guidance. The result is that after a time efforts which begin with enthusiasm are given up as failures." In conformity with the objective regarding which general agreement has been said to obtain the Plan Report goes on to suggest: "A close study in the Indian setting, of failures equally with successes which occur in efforts to develop cooperative farming, is likely to indicate directions in which effective solutions may be found".

The Plan recommends in particular the selection, as experimental or pilot projects, of a cooperative farm in every national extension block during the Second Plan period. Particular scope is stated to exist for cooperative farming in respect of holdings which might be below the basic or floor limits which, in terms of the land policy in different States, might be fixed; these holdings could only avail of the economies of large-scale organisation through being organised cooperatively into larger units. Also, surplus lands and other lands available in a village might be regrouped into cooperative farming units. Several of the cooperative societies at present working are made up of fresh settlements of persons brought together for rehabilitation on Government land, a circumstance which is more conducive to possible successful working than the voluntary pooling of old peasant farms. The plan recognises the very great and real difficulties of making a success of cooperative cultivation and refers to the need of promoting cooperation in non-farm activities as a step preparatory to cooperation in cultivation. In states: "Cooperation in credit, marketing and processing will also further cooperation in production." The necessity of widespread training in cooperative farm management including accounts and the human relations aspects are specifically mentioned in the Plan. In particular, the Plan recognises the claim of the existing societies for the first share of attention so that the example of successful working of old societies may stimulant formation of new ones. The Plan, however, states that "an attitude of experiment should be maintained in all matters of detail." The pattern of village development visualised in the Plan reflects this spirit experiment in the continuance side by side of an individual sector, a voluntary cooperative sector and a community sector. The inevitable gradualness of the whole process of development of cooperative village management is also recognised.

Those of us who are bred in the countryside know the farmer's passion for his own bit of earth in our country. And it seems difficult to assume that for a long time to come a sufficient number of cultivators will be found voluntarily willing to merge their lands and their labours, with implicit assurance that—difficult as the problem of promoting cooperation in lesser and perhaps immediately more fruitful forms are—the all-too-complex problems of providing for rewards proportionate to effort or contribution in the varied and heterogeneous operations of intensive agriculture will be solved to their satisfaction. Though I cannot state that I have had the opportunity of some intimate discussion with people closely familiar with their working in their respective countries, where farms are operated under the most favourable auspices, and have had some testimony of the stubbornness of the peasant, even when he is literate, about maintaining the separate identity of his holding and ensuring the fullest enjoyment of the fruits of his labours. Farming involves a multiplicity of operations to be performed by the same person—and not standardised, specialised operations capable of easy evaluation—a highly flexible time schedule according to variations or vicissitudes of weather or natural calamity—a farmer has often to go scurrying to his farm at dead of night on tidings of approaching flood, or storm or fire—and a normal seasonal cycle of work from 20 hours in the day at times (e.g., of preparation for sowing and harvesting operations) to no work at other times, during which, again, the forms of spare-time occupations related to the arm (as a producing unit) or home (as a consuming unit)—are diverse. All in all, farming activity is so discontinuous, farming effort so unequal between the indolent and the industrious and farming operations so far-flung and scattered that their adequate measurement in any country is a task of exceptional difficulty. Considering the fact that the vast mass of the people concerned are also illiterate, they are likely to look askance at attempts at comprehensive cooperation under which they are remunerated at periodic intervals on, what is bound to be, rough and ready assessment of their contribution to the fruits of cultivation; the discontent would tend to be greater the smaller the land contribution and the larger the contributions in labour. In such circumstances, the amount of direction or supervision required for a large-scale excursion into cooperative farming is likely to be beyond our resources of trained and skilled personnel—when this resource is altogether too scarce—for a long time to come. Training must, again, be afforded to the right type of personnel, to leaders from within the ranks of farmers, as the white-collar supervisor may not prove up to the far more exacting responsibilities in this type cooperation.

There is another relevant set of considerations at this point. The Rural Credit Survey Report has pointed out that cooperation has failed, but it rightly avoids the inference that the path of cooperation can be abandoned. So it says: "Cooperation must succeed," and has suggested an integrated programme of credit and marketing cooperatives, fostered and assisted by the state, to help the agriculturist. If after a half a century, cooperation in its relative simple forms—which, besides, are of easily demonstrable benefit to the cultivator—has failed the prospect of success in the most advanced form of cooperation, *viz.*, cooperation in cultivation by independent peasant farmers, is necessarily more dubious. It might be suggested that comprehensive cooperation such as is implied in farming has a better chance of succeeding than partial cooperation of the type that has been tried and has largely failed in the past. This argument holds to the point of suggesting that an integrated programme of cooperation in credit, supplies, and marketing with the substantial exclusion of the money-lender and the trader is much more likely to succeed than mere credit cooperation; which is also partly an explanation of why cooperation has not shown significant success to date in its major objectives, except in certain area. But the experience of the past should rather caution us against increasing our commitments in the field of cooperation and in favour of measuring our steps in terms of the programme of integrated development of cooperation in credit, marketing and processing.

As I have said above, however, the approach of the Plan to the question is characterised by an 'attitude of experiment' and the difficulties are probably fully appreciated in relation to medium and substantial independent farmers. The proposal has, therefore, been made particularly on behalf of very small farmers whose holdings fall below a minimum size. Though the rationale of this approach will be more easily appreciated, the difficulties of promoting cooperation among the smallest holders are likely in some ways to be greater, and the need for careful assessment of the experience of the farms which would be started as an experiment, would be all the greater.

The above observations, however, go to emphasise all the more the great need for a very considerable extension of cooperation in spheres other than cultivation including credit, marketing, processing, supplies, irrigation, farm equipment, etc., and even cooperation in crop planning and land utilisation. Indeed, cooperation holds such promise of a better life to the farmer that every avenue for its furtherance, every mode of its greater use, should be exploited to the full, the uniform objective being to achieve real and durable results by enlisting the interest and enthusiasm of the farmer. Scope for cooperative farming may also exist particularly among newly established farmers, or newly settled labourers and agricultural workers, or in respect of Government and community lands. The *gramdan* villages may, in addition, provide good auspices for the genuine fostering of cooperation in all forms including partly farming, as the voluntary character of association of the farmers in the new way of life and work will be assured. Further, the joint family may be recognised as an appropriate and promising starting point for joint farming. Thus a minimum size of holding may well be prescribed—necessarily varying from one state to another—any sub-division below which should not lead to disturbance of the pre-existing pattern of joint cultivation. Provided the limit of minimum cultivation unit is not fixed high, the requirement of joint cultivation could be imposed in respect of farms at present below the minimum size which might well be regrouped into large units on the same basis of quasi-family farms. The coordination of a programme of promoting joint cultivation through such measures with much more vigorous prosecution of consolidation of holdings may offer a line of solution for the highly fragmented and uneconomical cultivation. Some of the special inducement, facilities and subsidies which have been spelt out in the Plan as appropriate for fostering cooperative farming could be partly directed to such smaller exercises in joint cultivation, which would be used in maintaining certain minimum standards of economical and efficient cultivation.

At this stage, we may take note of a question in regard to the form of the proposed cooperative organisation. The Report of the Rural Credit Survey made a major proposal in this sphere—state partnership in cooperation. The decisive considerations in favour of State partnership are now too well- known to require restatement, but we have constantly to remind ourselves of the essentially catalytic character of the proposed state partnership. It is in the measure that cooperation, encouraged by state participation, becomes self-reliant, as one of the first fruits of the prosperity it helps to engender, that the success of state partnership will be judged.

That leads me on to refer to what appears to be a somewhat weak point of the programme of cooperative development. Naturally the targets of the programme pertain to loans—short-term, medium-term and long-term. These are to be financed both by increased internal resources and by considerably enlarged borrowing from the Reserve Bank. There is not, however, I believe, an adequate drive for increasing the resources available to the cooperative movement from within the sector of its operations. State-partnered cooperation, steadily widening its beneficent scope, should be the spearhead of the savings movement as it is of higher investment in the betterment of agriculture. This applies particularly to medium-term resources, which call for a special effort to increase time deposits in the cooperative sector. Insufficient attention to the resources side might eventually lead to excessive reliance on the ultimate money-creating powers of the central bank and is full of potential risks, inasmuch as lending by the central bank constitutes an addition to the volume of deficit-financing of the private—even if cooperative—sector, which we are liable to overlook as an element in the monetary outlook. True enough, such expansion of credit finances productive activity. So, however, does deficit-financing of the public sector. With increased supply of cooperative credit the money-lender's resources should be released to some extent—though not to a corresponding extent if the volume of investment in agriculture increases—and an increased contribution to the savings stream in the country-side should be rendered possible.

II

May I turn now to an aspect of agricultural economic policy in which I have been especially interested in recent months? This is the problem of agricultural prices, and in particular, the prices of food grains, which also constitute the largest element of agricultural production, more than three quarters of the cultivated area being under food crops. The question of agricultural and food prices has its importance from the point of view of the terms of trade for the agriculturist and the producer of food grains, and these have to be considered over a longer perspective than the last one or two years. The question has also to be considered in relation to general price and economic stability and in terms of the purchasing ability of the large body of consumers with cash incomes. Regulation of prices involves numerous administrative problems in the control of consumption and production. I do not propose to deal with all these questions here. As economists we are particularly concerned with some of the larger aspects of the problem. Two of these are: (i) the relationship between prices and production, and (ii) the implications of such relationship as might subsist between prices and production for price policy and for the extent and technique of such regulation of prices as might be considered necessary in terms of economic stability and social justice. There is very often a demand for fixation of minimum prices at fairly high levels to ensure high prices with a view to securing higher production; this is sometimes put forward as a solution for the present shortage of production in relation to demand. Apart from the fact that the agriculturist is shrewd enough to be guided by the actual trend of prices, and prices can be maintained at high levels even without fixation of any minima, it is necessary to bear the likely relationship between prices and production in mind in framing the objectives of a suitable price policy for agriculture. Now the relationship we seek is a complex one, and not all sectors of agriculture or classes of agriculturists have the same production response to a given price change. A broad distinction should be made between the reactions of a large farmer engaged in farming as a business, and the small-scale subsistence farmer working on the farm with his family, with whom farming is largely a way of life. If there is only one crop and no alternatives, it has been observed that the large farmer is induced to increase his inputs and expand his output when prices rise and vice versa with a fall in prices. In regard to the small family farmer, on the other hand, not only is no such direct correlation observable, but the observed experience of agriculture over a long period, including phases of very low prices like the Great Depression and very high prices like the period of post-war shortages, as well as phases of smaller variations in prices, shows an inverse, elasticity of supply with changes in prices. Year after year, during the early 1930s Arthur Loveday reiterated this in his *World Economic Surveys*. When prices fall sharply, as needs for cash are somewhat inelastic, the small farmer in trying to make up in volume what he loses in value, works harder and produces more, labour being the most important input in small-scale agriculture. A very recent illustration of this type of inverse relationship is furnished in the current price and production trends in a commodity where producer response is prompt and which, therefore, forms a particularly apt case study, *viz.*, milk production in the U.K. *The London Financial Times*, in a leading article on "Food Prices" in the issue of 27 November 1957, observes:

> There are two basic reasons for the fall in food prices...The reason which is more fundamental is that food production does not respond in the usual way to price changes. To the typical small producer, a fall in price means that he must raise output in order to maintain his income. To the consumer, even marked price changes do not seem to have any great effect on demand. The classic example recently was U.K. output of milk. A surplus of milk above the standard quantity leads to a reduction in the price paid to the farmer. Trying to maintain his total receipts, the average small farmer has raised his output-and the surplus—even further.

On the other hand, during the early post-war years of high shortages and large fallows in agriculture it was brought out that, with prices above a certain high water-mark, the small farmer did not consider it worth the trouble to himself and his family to cultivate the marginal acre. A certain inelasticity in the small farmer's standard of living in terms of additional consumption beyond a point makes him opt for a little less hard work when for certain times during the year he has to labour extremely hard, indeed, to make both ends meet. Times are changing and these responses, too, may be undergoing some slow transformation, but in the main remain unaltered. Of course, a level of prices which should be adjusted high following a sharp rise in prices would cease to be so as cost catch up. Again, the over-all result in any economy would depend upon the relative importance of various types of producers. Besides, the capacity of the farmer to undertake agricultural improvement and increase investment is affected by the terms on which he trades his produce. The long-term effects on production have, therefore, to be considered over and above the short-term. The terms of trade, moreover, are particularly relevant from the distributional angle, as low prices entail distress and hardship to the agricultural producers. My purpose here, therefore, is by no means to argue for the maintenance of relatively low agricultural prices in the interest of production but rather to bring out that the interests of production, on the other hand, also do not require unusually high or rising prices.

Indeed, over and above any effects on production, the prospect of rising prices is liable to have sharply adverse effects on available supplies, or market surpluses, by inducing the producer as well as trader to hold more of the produce longer with a view to benefiting from higher prices by selling later. This phenomenon has been revealed by several studies in the rate of market arrivals which have been conducted recently at the instance of the Foodgrains Enquiry Committee. The cultivator's response in the circumstances is but natural and proper. I refer, however, to the general tendency to greater stock building which is liable to accentuate the rise in prices with some possible stimulus to the type of producer's response that affects production.

The above observations would generally hold in terms of agricultural production as a whole. If prices of one commodity of group of commodities, say, foodgrains, rises relatively to other commodities, such a change in prices can definitely stimulate production through diversion of resources to the cultivation, in this instance, of more food rather than other crops. I would suggest, however, that the use of price policy for diverting resources from production of one crop to another is not, in general, to be commended in terms of the objective of maximising agricultural production as a whole. We will recall, how not long ago, grow more food campaign was supplemented by grow more jute and grow more cotton campaigns, and in due course it became a unified grow more campaigns for all crops. Our immediate preoccupation with the food problem should not lead us to overlook or minimise the paramount importance of an overall approach to the problem of production in the interest of achieving sectoral targets. The application of special incentives, and disincentives to the production of particular crops related, e.g., to the supply of fertilisers or other requirements of the cultivator, is liable to detract from the proper allocation of resources, insistent as the demands of one form of agricultural production may see for the time being. In fact, the cultivator usually applies a good corrective to any schemes of assistance which are linked to specific purposes by modifying the allocations to suit the interest of agricultural production in general.

I wish to refer at this stage to one particular aspect of the need to effect the largest practicable increase in agricultural production in general. This relates to the need to remove all disincentives to production, including production which might immediately contribute to swelling exports rather than domestic consumption. One of the scarcest commodities today is foreign exchange and production of commodities other than in respect of foodgrains should receive the fullest stimulus of all available influences, including exports, for its maximisation. To this end, the substantial elimination of all impediments to their exports is called for. Even if this should involve a rise in prices—with some consequent restriction of consumption—of certain articles which are of subsidiary importance in the family budget of the average consumer or form the raw materials of industry. For the stimulus to production that will result in the process and consequent encouragement to export, should, on balance, enable a fuller development alike of the domestic consumption and of our productive facilities in the near future than efforts to protect through export restriction the standard of consumption in specific sectors, e.g., in vegetable oils.

To revert, in conclusion, to the board objective of price policy, if I may take this opportunity to make a confession of faith in such a mundane sphere, the best atmosphere for sustained real development of productive investment and activity in our economy, including agriculture, is the avoidance of large undue fluctuations in prices and maintenance of reasonable price stability. Rising prices are not only not necessary for satisfactory development but are liable to impair the basis of sound and healthy development. A certain degree of rise in prices may be unavoidable, but to avoid it must remain the constant objective of policy.

Important as are the questions with which I have briefly dealt they lie upon the fringes of the agricultural problem, the positive content and core of which consist of the programmes of agricultural development, including irrigation, village communications, education, agricultural extension, soil conservation, etc., and of internal investment on the farms in improvement of land and farm equipment, besides specific programmes directed to promoting the welfare of agricultural labour and backward agricultural tracts and classes. The agricultural problem itself lies at the heart of the general problem of development of the economy. In many ways, it constitutes the problem of economic development. As such, it is much wider than the programmes of agricultural development as such; the whole programme and process of development is primarily an answer to the agricultural problem; this problem thus sets up the challenge of planning. It is a challenge as much to our organisational abilities and material prowess as it is to the moral fibre and spiritual quality of our people. The challenge has been taken up by the nation to find an answer within a measurable time which will be consistent with the fundamental democratic values on which the structure of our constitution and of the state are founded.

Indian Journal of Agricultural Economics, Vol. 13, No. 1, January-March 1958.

18

S.R. SEN

Agricultural Economics Research and Economic Planning in India

I

It is obvious that the development of agriculture can help the development of our economy from the stagnant to the progressive stage in three ways: (i) by increasing the gross national product, (ii) by supplying the physical surplus required by other sectors of the economy in the shape of food and raw materials, and (iii) by providing the economic surplus which constitutes the material basis for economic development. From the point of view of economic development, increase in production is, no doubt, essential but it is not enough. It must be accompanied by an increase of surplus and thereby of investment at a compound rate which should be well above rate of increase in population. It is true that in more advanced stages of economic development, the bulk of the surplus required may come from the industrial sector. But in an economy like ours, where nearly half of thc national income still comes from agriculture, a large part of the surplus required for economic development will have to come from agriculture until the stage is reached when industry is able to contribute a major portion of the national income. A programme for industrialisation makes a large demand for investment and there is also usually a fairly long time-lag between investment and output in the industrial sector with the result that invariably there are considerable stresses and strains in the economy during this period of lag. In most of the stagnant societies, where agricultural techniques are still somewhat primitive, the application of known improved techniques like the use of fertilisers, pesticides, etc., often leads to a substantial and quick increase in agricultural production. Until the entire field of agriculture in such a society is saturated with the application of such known techniques, the development of agriculture can provide, as it were, a bargain sector—a sector with a large unexploited potential which can produce the requisite surplus with relatively low investment and in a comparatively short time. It is true that this bargain sector provided by an under-developed agriculture cannot last for long but so long as it lasts it provides a very useful breathing space for such a society seeking rapid industrial development.

In practice, however, it is often found that there are a number of serious inhibiting factors which tend to hold up agricultural development at this very stage. In fact, it is one of the characteristic features of an under-developed economy that its agriculture has reached some sort of a quasi-equilibrium at a low level of productivity but the species of crops and animals are also found to evolve in such a manner as to suit only that low level of productivity. If any attempt is made to lift any part of this mesh of interlocking vicious circles, there is usually such a pull downwards from the other chains in the mesh that any sustained progress becomes almost impossible.

The real difficulty about stimulating a self-sustained development of agriculture in an under-developed economy, therefore, consists not so much in making one part of it or the other a little more efficient or in tightening a screw here or a bolt there but in extricating the entire agricultural economy from these vicious circles so that the state of quasi-equilibrium at the existing low level is disturbed and conditions are created for lifting the economy to higher and higher levels of productivity.

In the ultimate analysis, an increase in agricultural productivity can come mainly in three ways: (i) larger investment for building up the infra-structure of the agricultural economy, (ii) more intensive use of traditional techniques and (iii) introduction of improved techniques. The first is a rather costly and also slow process. A concentration on traditional techniques may show some improvement in the short period but it has a limited value. The adaptation of improved techniques practiced elsewhere and the discovery of new techniques, on the other hand requires considerable ingenuity on the part of the people concerned. Once, however, the ground is prepared for the introduction of such techniques, an ever-expanding horizon is opened up. In fact, a substantial and sustained increase in agricultural productivity becomes really possible when the economy is so geared that the search for and adoption of new techniques become a built-in process, as it were.

The various measures that can be tried to increase the volume of surplus available from agriculture can be broadly classified into four categories, e.g., compulsion, inducement, collectivist organisation and strategic stimulus.

As a system, compulsion is the simplest and has indeed the longest history from the Lycurgean polity of ancient times and the feudal economy of the middle ages down to the plantations of the 19th century and its economic *raison d'etre* was accepted by the philosophers of ancient Greece, medieval church fathers as also the economic writers of the confederate regime in the United States of America (US). The underlying principle was that normally the primitive man would not work unless forced to produce anything beyond his barest necessities. For he preferred leisure to work so that when a surplus has to be produced, some people had to compel others to do so. This was the origin of slavery or serfdom. Rack-renting by landlords and usury by money-lenders were only alternative forms of compulsion for making the farmers work hard and for squeezing out the surplus from them for the consumption of the non-farming sectors of the economy.

It was the pre-classical English economists of the 18th century who in dealing with problems of agricultural production and marketed surplus under conditions not very unlike those which contain today in many under-developed countries, led that attack against the system of compulsion. They emphasised that there was an alternative system, namely, the method of multiplying the wants of the farmers by popularising 'superfluities' (or 'luxuries' or 'equivalents') and encouraging exchange which was a much more efficient system than any form of compulsion. For, once the taste of the farmer for superfluities is developed, he becomes willing to work hard to produce an agricultural surplus with which to procure these commodities. Compulsion is no longer necessary. For a violent method of making the humankind laborious in raising food, there is substituted peaceful exchange. Formerly they worked because they were slaves to their masters, now they work because they are slaves to their wants. These pioneer economists explained that a system based on compulsion tends to stagnate in the long run, although it may show spectacular results in the short period, while that based on inducement gives more sustained results and leads ultimately to progress and prosperity. "In the first supposition," as James Steuart emphasised, "It is the head of the master which conducts the labour of the slave and turns it towards ingenuity; in the second, every head is at work and every hand is improved in dexterity." The pre-classical economists, therefore, definitely set their face against the idea of self-sufficient subsistence farming as also against the use of compulsion for increasing production and marketed surplus. They pleaded for the adoption of a system based on inducement ad exchange but they felt at the same time that it may frequently lead to serious imbalances and, therefore, assigned to the state the role of a overall regulator of the system.

Later the classical economists carried the case for free enterprise to the extreme when they minimised the role of the state and gave whole-hearted support to Adam Smith's dictum that "little else is requisite to carry a state to the highest degree of opulence from the lowest barbarism, but peace, easy taxes and a tolerable administration of justice; all the rest being brought about by the natural course of things." And this set the pattern for the development of agriculture in western countries during the 19th century.

The system of 'collectivist organisation' and 'strategic stimulus' were evolved as a result of dissatisfaction with the various shortcomings of the system of laissez-faire and also as a result of the desire in less developed countries to accelerate the process of economic development faster than could be normally expected from such a policy. In both, the state was required to play a much more active role than in the system advocated by the classical economists.

In collectivist organisation, as in Soviet Russia or China, the state takes the lead in cutting the Gordian knot by eliminating private enterprise altogether and organising agriculture on a collectivist basis and there is a strong element of compulsion in this system, somewhat reminiscent of that of Lyurgus.

In the system of strategic stimulus recommended by a number of modern economists for achieving some of the quick and spectacular results of the system of collectivist organisation without giving up private enterprise in agriculture, the state is expected to give special stimulus to agricultural development by using different types of investments as levers, as it were, for pulling apart the vicious circles which inhibit agricultural production.

In some respects the position taken by the advocates of the system of 'strategic stimulus' is not unlike that taken by the pre-classical economists. Like the latter, they agree that the state should protect the farmers against unfair competition and also against the vicissitudes of the market economy. They also emphasise that it should create the infrastructure which is essential for agricultural development. But they feel that these measures by themselves will not be enough to bring about such an accelerated development of agriculture as may lead to what is described by some of them as 'a take-off into self-sustained growth'. There is a view that the simultaneous and balanced development of a number of sectors is essential to provide the element of material support that alone will create the necessary demand for the newly produced goods and also to ensure the necessary structural balances on the supply side. If any under-developed country finds that its own resources are inadequate for this task, which is obviously a very difficult one, foreign assistance can be of great help in giving an initial 'big push' and enabling it to effect a 'successful take-off'. There is, however, another view that the most effective strategy in such a case is not 'balanced growth', namely, a transition from under-development to development through a chain of equilibria but "unbalanced growth", namely, a movement from under-development to development through a chain of equilibria. It is argued that since it is difficult to lift all at once the trap of vicious circles which is today holding down the agriculture of under-developed countries at a low level, the right policy should be to develop only such industries which are linked with the more vital sectors of the agricultural economy so that as these industries develop they pull up the linked sectors of the agricultural economy as well and thereby disrupt the chains of the trap which has been holding agriculture down. Such a policy of unbalanced growth may, no doubt, lead to stresses, strains and tensions in the economy, but it is held that these stresses and strains are the very forces which are needed for promoting rapid development.

It is obvious that like all generalisations, there is a element of truth in each of these theories, but there are also elements which are true only for particular cases and do not have universal validity. Moreover, it is important to recognise that any 'take-off' may not turn out to be a 'sustained take-off' reminiscent of Baudelaire's albatross. There may be a 'hauled take-off' comparable to that of a glider which just helps to raise the economy to a higher level but where further progress is limited. There is again the 'assisted take-off' as in the case of an aeroplane catapulted from an aircraft carrier where the initial push is only a precursor of far-reaching progress in future. Lastly, there is the "self-propelled take-off" of the space rocket which once it crosses the gravitational pull will not come back to earth again. It is obvious that the strategy for agricultural development has to be so devised that the take-off is not of the first two types and is at least of the third type, if not the fourth.

II

With this preamble about principles, we may now turn to the facts of the Indian situation. Prior to the last quarter of the 19th century, Indian agriculture was almost entirely of a subsistence type with more or less self-sufficient villages concentrating on the production of foodgrains; there was very little transport and trade in agricultural products. The surplus produced by agriculture was, therefore, very meagre. With the introduction of railways, opening of the Suez Canal, annexation of Burma and the policy of laissez-faire which the British rulers scrupulously followed, Indian agriculture can within the orbit of international trade. There was also considerable extension of geographical division of labour within the country. Burma concentrated on rice, Punjab and Sind on wheat and cotton and Bengal on jute while in the rest of the country comprising what now constitutes the Indian Union, there were large transfers of land from foodgrains to the production of oilseeds, tobacco and sugarcane. New crops and new varieties were introduced in Indian agriculture to cater to the new demand. There was, however, no appreciable change in the technique of production or in the efficiency of operations. The Second World War and the Partition cut off supplies of rice from Burma and Wheat, cotton and Jute from Pakistan and the geographical division of labour which had taken place during the previous seventy or eight years, had to be suddenly reversed at the cost of considerable stress and strain for our economy. The country was obliged to put a much greater emphasis on the production of rice, wheat, jute and cotton that hitherto and to sacrifice some of the advantages of geographical division of labour. Largely as a result of this, the index of agricultural production (base 1949-50) which stood at 103 in the quinquennium ending 1938-39 that preceded the Second World War declined to 97 in the quinquennium ending 1950-51 that preceded the First Five-Year Plan.

With the initiation, however, of the Five-Year Plans which prescribed a number of measures for eliminating or reducing the exploitation of farmers by landlords, moneylenders, etc., provided for much larger investment in agriculture than hitherto and set up a machinery extension work, there has been a definite upward trend. The index of agricultural production averaged 123 during the first three years of the Second Plan period (1956-57 to 1958-59). A part of this increase in production was no doubt due to the extension of cultivation of new areas and of multiple cropping but it also reflected a substantial increase in agricultural productivity.

These facts undoubtedly suggest that Indian agriculture is no longer stagnant. But in spite of a significant increase both in production and in productivity in recent years, the progress made so far cannot be said to have marked a definite transformation from a stage of chronic stagnancy to that of sustained growth. Even in the past, there have been periods of three or four years when there was a significant increase in production as well as in productivity but those trends were not sustained fo long and sooner or later there was a decline to the previous level of low productivity.

One important characteristic of the agricultural development during this period has been that it has not been gradual but has come in spurts as it were. It has also been very uneven so far as different parts of the country and different classes of farmers are concerned. Elements of growth and stagnancy are now continuing side by side in our agricultural economy. Sometimes and in some areas, the former get the upper hand which on other occasions and in other areas, the latter predominate. In this context it may be interesting to make a note of the experience of certain regions in the country which seem to have some bearing on the theoretical consideration referred to earlier. There are, no doubt, several factors which require to be considered in this context but I propose to draw here special attention to only two or three of these factors just to highlight certain significant points which are usually given insufficient consideration.

Let me first take Assam where large-scale plantations have been developed right in the middle of a peasant farming area. These plantations, which use modern and scientific techniques do not seem to have any significant 'demonstration effect' on the neighbouring agriculturists. For instance, the use of chemical fertilisers and pesticides, is very intensive in the plantations but the use made by the farmers of Assam is amongst the lowest in India. The average production of foodgrains in Assam during the Second Plan period has shown an increase of only 4 per cent over that for quinquennium preceding the First Plan. There seems to be a chasm between the planters and their neighbouring farmers in Assam which has not yet been abridged because the later have rightly or wrongly the feeling that the modern techniques followed by the former are either beyond their means or unsuitable for their purpose.

I may turn next to Chhota Nagpur where there has been a very substantial development of heavy industries and yet the farming practices and the yield per acre are among the poorest in the country. The average production of foodgrains in Chhota Nagpur during the Second Plan period has shown an increase of only 5 per cent over that for the quinquennium preceding the First Plan. There is very little use of modern implements and machinery or fertilisers and pesticides by the local farmers. No industrial 'ladder' seems to have developed over the last few decades which could bring step by step the steel, fertilisers and pesticides produced in the area to the small farmers even 20 miles away in such forms and for such uses as could induce and enable him to modernise his farming techniques. The industrial development is, as it were, an island—nowhere integrated as part of area development with the agricultural sector. The experience of Chhota Nagpur would seem to indicate that there is no automatic relationship between heavy industries and modernisation of agriculture at least until certain complementary developments have taken place.

A most interesting contrast is provided by recent development in eastern Uttar Pradesh (UP) and eastern Punjab. Before partition, the eastern parts of both the former provinces of UP and Punjab were much more backwards than the western parts—although eastern UP was in a relatively worse position than eastern Punjab. Since partition, there has been a rapid improvement in agricultural production in eastern Punjab while eastern UP continues to be almost as stagnant as before. For instance, compared to the quinquennium preceding the First Plan, the average production during the Second Plan period of foodgrains was 58 per cent higher in eastern Punjab as against only 0.1 per cent in eastern UP Corresponding increase in the production of all agricultural commodities was 65 per cent in eastern Punjab and only 4 per cent in eastern UP neither the development of trade nor that of large-scale industries shows a very different character in these two areas. While eastern UP has a larger number of sugar factories than eastern Punjab, the latter has a larger number of textile mills. There is no heavy industry in either of these two areas. The effort made by government for agricultural development has not also been so different in these two areas as to explain such marked difference in agriculture production.

There are, however, no important factors which seem to have a bearing on this difference. First, there are many more medium and small industries in eastern Punjab than in eastern UP Many of the small industries in eastern Punjab are power driven and they also represent a fairly large variety, e.g., manufacture of pumps, agricultural implements, jigs and drills, light electrical goods, bicycle, footwear, hosiery, etc., in UP they are largely of a cottage industry type which do not use very much of electric power, like weaving, leather and tanning, sericulture, *gur* and *khandsari*. It seems eastern Punjab has been able to develop, as it were, a connecting 'ladder' between large industries (some of which are situated even outside the state like steel and chemical factories in Chhota Nagpur) and agriculture through a chain of small and medium industries which eastern UP (and as we have seen earlier also Chhota Nagpur has not been able to do yet). In eastern Punjab, the fairly dispersed growth of power operated small industries has helped the people to acquire an ingrained habit of technological improvement which is proving a strong force for progress while in eastern UP the people have yet to develop such a habit. Secondly, there is a strong urge for a higher standard of living amongst the farmers of eastern Punjab which is not there to the same extend amongst the farmers of eastern UP Many of the farmers in eastern Punjab are migrants from western Punjab where they had got used to a higher standard of living. The displacement caused by partition faced them with a new challenge which was difficult but at the same time not quite overwhelming, partly due to the assistance that they received from the rehabilitation authorities. The result was that they responded in a manner which was a practical demonstration of Toynbee's well-known thesis of challenge and response. Although they had now a smaller area to cultivate compared to what they used to have in western Punjab, they wanted to go back to old standard of living that they enjoyed before partition and, therefore, not only did they put in more effort but were also anxious to try new techniques in order to produce from a smaller holding the income which would enable them to maintain the same standard of living that they enjoyed from a larger holding before partition.

It may be said that what has happened in eastern Punjab is a very special case and has no other parallel. But there are several other areas in the country where also there has been similar progress, although the motivation and the challenge may have been somewhat different. In fact, the Kaira district of Gujarat, which some of you may be visiting during this conference itself is another good illustration. A healthy farming community whose outlook has been broadened by a close contact with the outside world and whose opportunities have been enlarged by the establishment of a modern dairy industry, the introduction of mixed farming and diesel pumping sets and the setting up of a number of small ancillary industries has recorded a rate of agricultural development in recent years which compares favourably with that of eastern Punjab. The level of foodgrains production in Kaira district in the first three years of Second Plan period was 30 per cent higher and that of agricultural production 130 per cent higher than the average for the quinquennium preceding the First Plan.

Such differences are noticeable not merely as between different areas but also between different groups of farmers. In the same area, the best farmers are known to have produced yields per acre several times higher than those produced by average farmers. For instance, the average yields of rice per acre in different states vary between 400 and 1200 lbs while the highest yields obtained in crop competition vary between 3000 and 9000 lbs. Corresponding ranges for wheat are 300 to 1000 lbs and 2500 to 6000 lbs respectively. In fact, while the best in Indian agriculture does not compare unfavourably with the best elsewhere, the difference between the best and the average is much wider in India than in the technically advanced countries. This is both an index of the backward character of Indian agriculture and a measure of its potentiality for development.

III

What, then, should be the strategy for the development of our agriculture and an early exploitation of this bargain sector? It is obvious that in our special political and economic situation neither laissez-faire nor collectivisation can be accepted. Laissez-faire has been given a long trial and has failed. Any attempt at collectivisation is likely to create serious social difficulties, at least in the transition period, which may outweigh whatever economic advantage it may have. Moreover, in a country where the population pressure is so great and the land-man ratio is so low as in India, collectivisation is likely to accentuate the problem of rural unemployment at lease in the initial period. The element of compulsion associated with collectivisation is also not in accord with our political ideals. It would appear therefore, that our basic approach has necessarily to be a suitable form of strategic stimulation, namely, a 'big push' given by the state supplemented by certain strategic 'pulls' through the deliberate planning for a limited amount of 'unbalanced development' of both agriculture and industry. Care, however, will have to be taken that the 'push' is of the right magnitude and at the right points and the 'pull' is not of such proportion that the resultant stresses and strains are so severe that what is gained in the swings is lost in the rounds.

An essential condition for the 'big push' is that the 'decks' should be first cleared for subsequent operations. Adoption of an appropriate land policy, credit policy and marketing and price policy is an essential part of such clearing of the 'decks', the main objective being the removal of the various inhibiting factors which tend to hold up development. On the whole, it would seem that the broad policies in regard to these three items recommended in the First and Second Plans are in the right direction. Unfortunately, however, some of the basic recommendations made in the two Plans, e.g., security of tenure, agrarian reorganisation, supply of timely and adequate credit and stabilisation of the price level, have not yet been implemented effectively. In spite of general agreement on these broad policies, there has been a certain hesitancy in regard to some of the details and also considerable laxity in implementation, with the result that it is taking an unduly long time or the full benefits of these policies to be realised. The first thing that requires to be done is to make sure that these reservations, most of which are really of minor importance compared to the major issues involved are settled quickly by permitting a certain amount of local variation and flexibility in regard to details and the basic policies which have been accepted by the nation are effectively implemented as early as possible by simplifying the procedure and stream-lining the administrative machinery. Unless our farmers are fully protected against exploitation by landlords, provided with an agrarian structure which is conducive to efficiency, supplied with credit which is adequate enough for adopting all the new techniques and for making all the investment that are needed for achieving a higher level of production and also assured that if they take the risk of stepping up their investment, prices will not be allowed to fall to unremunerative levels, even the best of technical and administrative programmes for agricultural development will not produce the desired result.

The 'big push' that our agriculture requires can be given only if we have strong and efficient organisation for providing an intensive agricultural research and extension service and ensuring adequate supply of fertilisers, pesticides, improved seeds, livestock, modern agricultural implements and irrigation. Here too, there is no new principle involved. We have already taken steps to establish a national extension service. But compared to the depressed character for our agriculture, the various restraints that it is subject to and the large volume of work that requires to be done for accelerating its development, the extension staff that we have today is too thin in coverage and too superficial in quality, especially in the context of the present and poor level of development of people's organisations in our rural areas. A young village level workers, who is only a matriculate and has been given some elementary training in agriculture, is asked to look after as many as 1200 farmers and give them day to day guidance not merely in agricultural and animal husbandry techniques but also in farms planning, cooperative organisation, *panachyat* social service, etc., and we call it agricultural extension service. Now, this is certainly very inadequate compared to the work that has to be done in most areas. In all countries which have a progressive agriculture, the agricultural extension service is much intensive. For instance, in Scotland only 2000 farmers are served by one extension agent, in Holland about 400 and in Japan where holdings are small, usually less than 2½ acres, between 500 to 600. In addition to these agents maintained by government, private manufacturers of fertilisers, pesticides, agricultural implements, etc., also employ a large staff of salesmen and extension workers in these countries. In China, there is a well staffed technique popularisation centre within walking distance of every farm and the arable area per agricultural extension workers is about one-fourth of that in India. Moreover, there are a large number of members of the Communist Party who are whole-time paid workers of the cooperative farms and are available for assisting the agricultural extension worker in his day to day work. It is not commonly realised that much of the progress made by Chinese agriculture in recent years is due to the intensive extension work which is done by agricultural technicians. An eminent Chinese agricultural expert mentioned to me, when I was in China some time back, that 85 to 90 per cent of the recent increase in production in China is due to technical services and supplies and only 10 to 15 per cent due to the particular form of farm organisation which they have introduced. A time has come when we should recognise that it is not possible to give a 'big push' to our agriculture unless we have a large enough extension staff which is fully qualified for our job and is required to attend to agricultural work with single-minded devotion. The utilisation of millions of the rural unemployed and under-employed for development work, which should be a key feature of the economic plan of an underdeveloped country, cannot be also effectively done unless there is a proper organisation for mobilising and guiding them on a works project basis. If on account of paucity of funds or shortage of personnel it is not found immediately possible to cover the entire country any more intensively than is being done in the present community project areas, the least we should do is to supplement it in selected areas or on a peripatetic basis with a special 'task force' of agricultural extension workers which has been recently recommended by a number of eminent experts. After all, it is an elementary principle that if a certain mass has to be moved against a certain friction, a strong enough force has to be applied to start with although once the mass has been set in motion, relatively less force may be required to keep it moving. If, therefore, any backward agricultural area has to be developed, the first impact of the agricultural extension service must be very intensive, although once the process of development has started, the size of the staff can be reduced.

Even after the 'decks' have been cleared and a strong enough agricultural extension service has been provided for giving the 'big push', and accelerated development of agriculture requires that there should be sufficient provision for supplies of fertilisers, pesticides, improved seeds, implements, livestock, etc. as even the best of armies cannot win a battle without adequate ammunition, similarly even the best of extension services cannot enable the farmers to increase agricultural production if the requisite supplies are not available and the farmers do not have adequate savings or credit for securing these supplies. If the farmers in a particular area appear to be not working intensively enough, there are often very important economic, social and physical reasons of which a basic one is that at the low level of productivity which is the inevitable consequence of traditional techniques followed by them and the lack of supplies, the additional yielded that they can get from additional work of a most back-breaking kind is so meagre that for them it is not really worthwhile to put in any additional effort. It is a common experience that when a new technique is introduced which is more productive and less onerous, a farmer gets interested not only in working more intensively but also in all-round improvement in his operations. Once the conservatism and apathy of a traditional farmer are overcome through the introduction of one impressive new technique, he becomes ready to accept other new techniques too even if these may be less spectacular. Moreover, introduction of modern techniques, e.g., use of chemical fertilisers or Japanese type farm machinery to a subsistence farmer not only improves his productivity and makes his work less onerous but also brings him closer to the nexus of exchange and increases the marketed surplus. If we really want to rescue the Indian farmer from the mesh of vicious circles to which he has fallen a victim, it is essential that we should provide him with an agency which will promptly pass on to him the results of modern scientific research and set up an organisation through which better implements and machines, fertilisers, pesticides and other supplies will become available to him without difficulty and delay and improved seeds will be introduced promptly in every area and in such a manner that it may quickly replace the older and less efficient varieties.

Irrigation is no doubt equally important. But in our country we have tended to give in the recent past much more attention to the provision of new major irrigation sources rather than to the optimum use of the available supply of water. There are large areas in the country where the adoption of modern dry farming techniques and introduction of drought resisting crops could increase output substantially even on the basis of the existing moisture. In other areas a more economic method of using water should enable is to irrigate a much larger acreage than we are doing in most areas. Yet there is a mounting demand for new costly and time-consuming major irrigation projects which have become symbols of prestige, as it were, with some of the local pressure groups although in the very same areas projects for agricultural extension and for stepping up of essential supplies are suffering for want of funds. I do not suggest that the major irrigation projects which have been taken up are not worth having or that we should not have any more of these projects at any time in future. The point that I wish to make here is that we should go slow with the major projects for at least the next plan period and reserve these projects only for such areas where there is no alternative way of improving agricultural production.

All this is about the 'push' that requires to be given to our agriculture. But as I have mentioned earlier, the push will become more effective if we can also provide for certain strategic 'pulls'. This can be done by adopting a bold policy of geographical as well as functional specialisation both in our agriculture and in our industry. It is obvious that in view of the scarcity of our resources we cannot make a uniform progress on all fronts. We have necessarily to plan for a more rapid progress in a few fronts and to remain content with slower progress in other fronts. But we should be careful to select the fronts where we concentrate in the near future in such a manner that they help maximise the rate of economic development in the country.

From this point of view it is only rational that we should pay no need to the rather obscurantist demand for regional self-sufficiency in food grains. On the contrary, it should be our endeavour to take full advantage of geographical specialisation in crops and follow a policy of land utilisation which will seek to optimise agricultural production of the country as a whole and for all crops taken together. Our programme for agricultural development has to be built not around self-sufficiency in foodgrains in different regions but around the best crops rotation for each of these regions from the technical as also the economic point of view. Moreover, crop production should be taken up everywhere on a commercial basis. Even where foodgrains are grown, they should be grown as far as practicable as cash crop rather than as mere subsistence crop. Until we are able to bring our crop production fully into the nexus of exchange, we shall never be able to put our agriculture on a modern and scientific basis nor to increase the much-needed marketed surplus.

We should also adopt a deliberate policy of developing a complex of industries and diversifying our rural economy in such a manner as would help accelerate agricultural development. It is not true that development of any industry would automatically induce the development of agriculture. It is only the development of such industries as have a link with agriculture either because they draw raw materials from the latter or because they provide the supplies required by the latter which can help agricultural development. The impact of these two types of industries on agricultural is however, somewhat different, the former help to extend the market for agricultural produce and also introduce new varieties. But it is really the latter which help improve the efficiency of agriculture itself. But as we have learnt from the examples cited earlier, this impact is not automatic but is subject to a very important condition. This condition is that between industry and agriculture there should be some sort of ladder of which the different steps are within easy reach of one another. For instance, if we decide to develop a steel industry, we should also arrange to set up subsidiary industries which will produce agricultural implements and machinery from that steel and use all the techniques of salesmanship not only to make these implements available to the agriculturist but also to make him interested in using them. It is only through the ladder of a series of efficient medium or small industries between large-scale industry and agriculture supplemented by adequate transport and marketing facilities, that the stimulating influence of industrialisation can become really effective. It is in the absence of such a ladder that the farmers of Chhota Napur continue to be backward while the presence of such a ladder enables the farmers of Punjab to benefit from heavy industries set up in Chhota Nagpur itself. If the different steps of the ladder are not within easy reach of each other, the so-called link between industry and agriculture will remain there only in theory. It is true that given sufficient time such defects may get corrected but the object of planning should be not to leave such developments to chance or 'natural course of things' but to ensure that there is no undue delay or obstacle at any point.

The introduction of power—both electrical and diesel—in rural areas is a very powerful force for providing this linkage between industry and agriculture. As I have mentioned earlier, the supply of electricity in the rural areas of Punjab and the introduction of diesel engines in the villages of the Kaira district have helped agriculture not only directly by operating irrigation pumps but also indirectly by enabling a number of small industries to be set up in a well dispersed manner. These pumps and industries have not only provided the 'ladder effect' but have also made the local farmers more technically minded and, therefore, more progressive.

We should also take another lesson from what has happened in Punjab and the Kaira district. We should recognise that unless we take appropriate steps to create the necessary social conditions and in particular foster amongst our farmers a desire for a standard of living which is appreciable higher than their present level of living, they will not have the necessary incentive and urge for working harder, for taking greater risks and for trying out new techniques with a view to producing more and earning more so as to get the wherewithal for purchasing the goods and services for which they have a new demand. It is not, however, suggested that we should make them crave for a standard of living which is completely unrelated to their present level of living. In fact, if it is too unrelated and appears unattainable to them, it will not produce the stimulating effect that is expected. All that is needed is that like the "Holy Grail", the standard of living should be always a little ahead of the level of living and yet not so far ahead that it appears to be beyond their reach. It is only then that they will not merely have a desire to improve their present position but will also have the feeling that with a little enterprise and a little more effort it will be possible for them to reach that higher level.

In this context the organisation of an efficient hire-purchase (or instalment credit) system, preferably through cooperative societies, not only for producer goods but also for durable consumer goods may prove to be of considerable help, especially in those areas which are still in the stage of backward subsistence farming, both for stimulating production and for expanding the marketed surplus. Many of our farmers are not used to saving but they are used to repaying debts. The hire-purchase system will take advantage of this fact for inducing these farmers to economise the consumption of their own produce and step up their sales. In fact provision of all the credit that is justified by the production programme and the repaying capacity of the farmer and collecting it promptly in suitable instalments through a cooperative organisation is a very useful way of increasing the flow of marketed surplus. We should recognise that if our farmers are to take sustained interest in producing more and selling more, we must provide them with necessary inducement and take special steps to create an interest in them for such inducement where they may be lacking on account of their backward conditions. Even in Soviet Russia where collectivist organisation is such a powerful instrument for making the farmer work harder, a great stress is laid on the need for providing 'material interestedness' for the farmer.

The task of pulling our agriculture out of the rut into which it has fallen and setting it on the path of accelerated development is a very difficult and complicated job and involves very hard decisions regarding priorities, especially when we choose to do it in a short time and without using collectivist methods. No dispersal of efforts and no symbolic gesture should have any place here. When faced with a difficult job, some of us have an unfortunate tendency to take a symbolic action because a really effective action will strain our resources too much and then consciously or unconsciously create an impression that necessary action is being taken. Such confusion of the symbol with reality and the complacence that it engenders can have only one consequence, namely failure. We should appreciate that when any action is based on grossly insufficient preparation or effort or resources, it tantamount to symbolic action.

There are again others among us who are today pleading for a holiday from planning or for a small plan which will not require much effort on the part of the community. They seem to forget that there is hardly any choice left to us. There are three broad facts that face us today. An accelerated increase in population, a growing disparity between our rate of growth and that not only of western countries but also of some of our neighbours and further the urgent need to complete our 'take-off' within the next decade if we are to pass over the 'period of trouble' that is likely to overcome us otherwise. History tells us that no nation has made a real progress unless it has decided to take up a big enough challenge and respond to it manfully. In our present situation, it is only a bold effort that has a reasonable chance of success. Any timid effort will only mean a fruitless dissipation of resources. In a dynamic world where one cannot stand still and the only alternative to forging ahead is to fall behind, we shall be inviting disaster if we listen to these 'lotos eaters' and prefer a temporary respite because we may face some difficulties in our journey ahead. I do not by any means underestimate these difficulties. But I feel that the very backwardness of our agriculture—our poor yields—also provide our silver lining. If our yield per acre were already as high as that of, say Japan or Egypt, we would not have the same bargain sector (i.e., the same advantage in terms of unexploited potential, low investment-output ratio and short gestation period) in our agriculture that we have today. If we are able to exploit this sector effectively, it should not only relieve us of many of the stresses and strains which inevitably accompany rapid economic development, but also pave the way for a prosperous and happy future for our people.

Indian Journal of Agricultural Economics, Vol. 15, No. 1, Jan.-March 1960.

19

M.L. DANTWALA

Agrarian Structure and Economic Development

THE decade and a half since Independence has witnessed unusual spate of activity in the field of agricultural economics studies. There is hardly a field of country's agricultural economy which has not been enquired into and reported upon by a Committee or a Commission, or a Working Group, or through a Research Project.

The problem of rural credit has been extensively and comprehensively surveyed by the Committee of Direction appointed by the Reserve Bank of India (RBI) in 1951. It reports—in three volumes—contain, besides the recommendations on the entire complex of rural credit, marketing, processing and warehousing, most valuable data on several aspects of the agrarian structure. Early this year, the RBI published the Follow-up Survey on Rural Credit conducted in 1956-57 in 11 districts. The recommendations of the Rural Credit Survey Committee and policies based on (or deviating from) them have been discussed and debated by the State Ministers' Conference at New Delhi in April 1955, at Mussoorie in July 1956, at Mysore in July, 1959, at Jaipur in January-February 1960 and at Srinagar in June 1960. The problem of credit was once again investigated by the Committee on Cooperative Credit under the distinguished Chairmanship of Shri V.L. Mehta.

The country's food problem was examined by two Foodgrains Policy Committees (1943 and 1947) and by the Grow More Food Enquiry Committee (1952) under the Chairmanship of Shri V.T. Krishnamachari. In 1957, the Foodgrains Enquiry Committee under the Chairmanship of Shri Asoka Mehta again reviewed the problem and made valuable recommendations. Two years later (1959), a Team of eminent experts from the United States (US) sponsored by the Ford

Foundation made a comprehensive Report on 'India's Food Crisis and Steps to meet it'.

The Research Programmes Committee (RPC) of the Planning Commission financed an ambitious project of research on Farm Management. Though the primary purpose of the project was to discover a suitable methodology for farm management studies, it has yielded valuable material on the working of the farm economy in several regions in India. The ministry of Food and Agriculture is continuing these studies. The RPC also initiated studies of the working of the Land Reform legislation and the Economics of Irrigation Projects.

The Community Development programme has been evaluated by the Programme Evaluation Organisation in as many as seven annual Reports. In 1957, the Committee on Plan Projects' Team under the Chairmanship of Shri Balwantari Mehta made a comprehensive report (3 vols.) on Community Projects and National Extension Service. Two years later, Community Development Evaluation Mission headed by M.J. Coldwell reported on the problem.

Two Land Reform Panels were appointed by the Planning Commission: one at the time of the formation of the Second Five-Year Plan and the other is currently in session and its Report is expected soon. In 1952, the Indian Society of Agricultural Economics sponsored an intensive study of the working of the cooperative movement in one taluka of Gujarat. Its Report "Cooperation in Kodinar" is a pioneering study of the various facets of the movement. This was followed by another enquiry into the problems of Low-Income Farmers in Kodinar. The Report gives an interesting analysis of the impact of the cooperative movement on different classes of cultivators. In 1958, the Society organised Seminar on Cooperative Farming and produced a very stimulating Report. It has also published a Report of an Enquiry into Cooperative Farming in Gujarat. The enquiry was sponsored by the RPC. The programme Evaluation Organisation of the Planning Commission studied the working of 22 cooperative farming societies and its Report "Studies in Cooperative Farming" was published in 1956. The latest in the field of cooperative studies is the Report of the

Working Group on Cooperative Farming to which reference is made later.

The Agro-Economic Research Centres financed by the Ministry of Food and Agriculture have carried on many ad hoc surveys—e.g., Market Arrivals of Foodgrains, 1958-59 Season—besides their continuing study of the change in rural economy.

The oft-quoted Report of the Agricultural Labour Enquiry (Ministry of Labour) gives for the first time a detailed account of the state of employment and income of agricultural labourers. A second Enquiry was conducted in 1956-57 but its Report is not yet published.

The numerous rounds of the National Sample Survey (NSS) have yielded valuable information on several aspects of the rural economy and in this address, I have extensively used the data collected in its 8th Round on Size of Holdings.

This is neither a complete nor even a representative list of enquiries and studies recently undertaken in the field of agricultural economics; nor is there any need to provide such a list to a gathering of agricultural economists. What I wish to point out is that the professional economist has no reason to think that his sphere of interest has been neglected. In fact, he has been kept busy, if not participating in these enquiries, attempting to read and digest their voluminous reports. The university teacher is nearly breathless in trying to keep with the rapid flow of these reports.

This intellectual activity has been matched by activity in the field of policy making and enactment of legislation. While one cannot say that all the recommendations contained in these report and the findings of the research projects have been fully utilised, accepted and incorporated in the agricultural policy of the government, their impact on policy deviated from the recommendations, the latter provided a great stimulus to the thinking on the subject and the discussions and debates which ensued have been fruitful, though occasionally acrimonious.

Nor have the agrarian structure and institutions remained static. Whether as a result of the policy or otherwise, extensive and even radical changes have occurred in the country's agrarian structure and institutions, sometimes in directions anticipated and hoped for, sometimes in divergent directions.

In the sphere of land tenures, which is basic to all agrarian institutions, intermediary tenures like *zamindaris, jagirs, imams*, etc., covering approximately 173 million acres of land, have been almost entirely abolished. Agrarian relations tenants have been brought in direct relations with the state. Considerable areas of forests (20 million acres) and cultivable waste lands (21 million acres) have been vested in the government. An idea of immensity of this operation will be had by the figure of compensation payable to the intermediaries. The total compensation payable along with the accumulated interest is estimated at ₹635 crores, the total amount actually paid to-date being ₹156 crore.

Tenancy legislation has been passed (or introduced in legislatures) by practically all the state governments. Besides regulating rent and safeguarding the tenant from wilful eviction, the legislation provides for purchase of the leased lands by the tenants at reasonable prices. Rents have been scaled down from the traditional half of the gross produce to one-third and one-fourth and, in one case, to one-sixth of the gross produce—though in one or two states the level is ceiling on future acquisition of land in several states and Bills have been introduced for this purpose in all the remaining. In regard to the imposition of ceiling on existing holdings, legislation has been passed in Assam, Punjab (PEPSU area), Rajasthan, West Bengal and Himachal Pradesh and Bills have been introduced in Andhra, Kerala, Madhya Pradesh and Orissa legislatures. I am not suggesting that the intentions of the legislation have been fully translated into reality. As is well-known, there are many flaws and imperfections in the laws and their implementation has been glaringly inadequate. On several occasions in the past I have drawn pointed attention to this problem and have even observed that in the wake of tenancy reforms there have been more evictions than ever before. But, judging for the country as a whole and taking an overall view, I think it will not be wrong to state that the agrarian relations are today more rational and less iniquitous compared to the situation before Independence.

Reliable, up-to-date statistics regarding land holdings and tenurial status are not available. Some broad facts regarding the pattern of land ownership and use based on the NSS Report on Land Holdings (1954-55)[1] are given below.

The agrarian structure may be examined by reference to the rural population as well as the cultivation pattern. The former is fairly well-known. A little over one-fifth of rural households (23.09%) own no land. About one-fourth of the households have a small parcel of land less than one acre. Their aggregate share comes to 1.37 per cent of the total land owned by rural households. At the upper end, about one-eighth (12.8%) of the households have more than 10 acres each covering about two-thirds of the whole area. The top-most one per cent of the households owned more than 50 acres each and in the aggregate about 17.5 per cent of the total area. (First Report).

The above facts lead to the conclusion that "the distribution of land is extremely concentrated with a small minority owning most of the land". In the context of the scarcity of land in relation to persons aspiring for its ownership, such a phenomenon is perhaps inevitable. The NSS data indicate that in order to provide just two acres of land (irrespective of its quality) to all such rural households who do not possess any or less than two acres of land, it would be necessary to impose a ceiling on individual ownerships at 20 acres (again irrespective of the quality of the land). In other words, "a floor of 2 acres would be consistent with a ceiling of 20 acres." The pertinent question therefore is to enquire whether the small minority owns excessive—judged by reference to the technological optimum—areas of land and further how much of the total land is owned by the owners of such excessive areas. Since the NSS data do not take into account the quality of land, this aspect of the problem may be ignored for the present, though there is enough evidence to show that larger holdings contain relatively inferior land. Assuming that ownership and/or operation of land above 50 acres is considered excessive, according to

1. The NSS data pertain to all rural households as well as land holdings, agricultural as well as non-agricultural. Ownership comprises the right of permanent heritable possession, with or without the right to transfer the title. Household holdings included lands of all the 'usual' members of the household.

the NSS data, only one-sixth (17.5%) of the total land falls within this category. If the limit of excessive ownership is placed at 100 acres and above, less than 6 per cent is so owned.

Viewed from the point of cultivation, the picture we see has some unexpected features. As much as 85 per cent of land is owned and cultivated in holdings of more than 5 acres each. Assuming that a holding of 10 acres and more could be generally considered to be economic, under our technique of cultivation, as much as 66-67 per cent of the total cultivated land is being operated in units of 10 acres and more (Second Report). Thus, though about three-fourth of all rural households had either no land or less than five acres and should therefore be considered as uneconomic, at least two-thirds of the land was being cultivated in units which could not be characterised as uneconomic.

The NSS data reveal some interesting features also in regard to the nature and extent of tenancy. 81.35 per cent of land owning rural households owning 69.23 per cent of land do not lease-out any land; 16 per cent lease-out a part of their land; and only 2.65 per cent lease-out all their land amounting to 1.84 per cent of the total area owned by all rural households. Small as well as big land-owners lease-out land and though the percentage of leasing-out families increases with the increase in the size of ownership holding (44% in the ownership group of 50 acres and above), the percentage of such families in the group owning less than one acre was 9.42 and in the group owning between 2.50-4.99 acres it was as high as 22, nearly 70 per cent of the total leased-out area by rural households was contributed by those who owned and operated less than 10 acres and out of this 70 per cent, 36 per cent was contributed by those households whose per household area owned was just 1 acre. (First Report). For all-India, the percentage of operated area 'leased-in' was 20.34, about a half being contributed by non-resident owners. Land was leased-in by small as well as large operators; but more than 60 per cent of the total leased-in area was with those who operated by the group with more than 50 acres was taken on lease (Second Report).

The above analysis should explode the belief firmly held even in the highest quarters of policy-makers that those who lease-in land (tenants) are necessarily small farmers and those who lease-out are big absentee land owning rentiers.

Apart from the changes in agrarian relations and structure, significant developments have taken place in the functional organisation of the agrarian economy. I shall mention only two. At the end of June 1959, there were 1,83,000 agricultural credit societies with a membership of 12 million. They serve 3,43,000 (or 60% of the total) villages with a population of 110 million (or a little more than one-third of the population in the rural areas). If the credit requirements of agriculture are placed at ₹1,000-1,200 crores (against ₹750 crores estimated by RCS [Rural Credit Survey] for 1951-52) the cooperative sector may be said to be providing about 10 per cent of these requirements as against 3.1 per cent reported by RCS in 1951-52. The share of the cooperatives in the marketing of agricultural produce estimated from National Income Statistics and on the generally accepted assumptions regarding the market percentages of food and non-food crops, the share of cooperative marketing may be placed at less than 1 per cent for food crops and about 4.5 per cent for non-food crops. Cooperative processing has made some headway in manufacturing of sugar and it is estimated that during 1959-60 season, cooperative sugar factories manufactured 2.86 lakh tonnes of sugar, which comes to 11-12 per cent of national production. The Community Development programme which has been in operation for seven years serves about 3,60,000 villages with a population of little over 179 million. The coverage works out to 61 per cent of the total rural population. It is expected to cover the entire rural area by 1963.

We have been so long accustomed to think about the agrarian structure in terms of concentration of ownership and preponderance of tenancy that the impact of the change which is taking place since Independence is not fully appreciated. That there are still pockets of concentration and exploitative tenancy need not be denied; but, when the situation is viewed in its totality and for the purposes of broad national policies, it is necessary to take into account what has happened during the last decade and a half. The NSS data are nearly

5 years old and while it is possible that some deterioration may have taken place in the extent of uneconomic ownership and cultivation, there is reason to believe that in regard to the tenancy situation, there is some improvement.

II

The question we are interested in is whether we have now a pattern of agrarian structure and institutions which could be relied upon for the purpose of rapid growth not only of the agricultural sector but of the total national economy. More pertinently we may enquire whether when the programme based on current policies is fully implemented we shall have such a structure or a further radical alteration will be needed.

During the Third Plan, and also subsequently, the pace of economic developments has to be accelerated, and this will impose further demands on the agricultural economy. We talk of 'take-off' and self-sustaining growth. I presume if the national economy has to achieve take-off, its every sector will have to be geared for the task. Too often in the past we have thought of the take-off as an exclusive assignment of the industrial sector; the capital-base producing machines for the manufacture of machines. What this would mean to the agricultural sector is not fully thought out. An unprogressive agriculture cannot partner a modern self-sustaining industrial sector. The task of economic engineering involved in constructing communication locks for a system operating at such disparate levels—industrial production being carried on with the aid of electric if not atomic power and agricultural production, with plough and bullocks—is indeed stupendous.

If by posing the problem in this manner I have unwittingly aroused the expectation that shall be attempting to indicate the precise pattern of the institutional structure appropriate to the purpose of rapid planned economic development within the democratic framework, I should beat a hasty retreat. Not only I do not know the prescription; I do not even believe that there is any such single unique structural pattern which we must either adopt or perish. I shall merely try to describe what is being attempted in our country; the trends in our thinking on the question; policies already adopted or are likely to be adopted; and towards the end draw attention to some of the economic consequences of institutional change.

We may begin with a review of our policies on the agrarian structure and more particularly, the trends in our thinking as revealed in the evolution of these policies. The latter perhaps is more important for assessing the stability of these policies and removal of uncertainty which, rightly or wrongly, pervades the atmosphere today. Even this limited query can be pursued only on a sampling or an illustrative basis. Extension of the cooperative organisation to all rural economic activities is the main stay of our agricultural policy. The Prime Minister, inaugurating the International Cooperative Alliance Seminar on Cooperative Leadership, recently remarked that: "He would like to 'convulse' India with cooperation, to make it the basic activity of every village and finally, to make the approach the common thinking of our country".

There are several aspects of cooperative policy; but, here, we are concerned only with those which affect the agrarian structure. Cooperation is a very broad category and before any judgment is formed about its appropriateness, it is necessary to be quite clear about its content. It may comprise anything from the traditional forms of informal mutual cooperation to the highly integrated form such as the Kibbutzim of Israel. Even the Collectives and the Communes in the communist countries are referred to as 'higher forms of cooperation'. The distinguishing characteristics of these various forms emanates mainly from the relation of the cooperatives with the state. Understanding of this relationship is therefore essential for deciding how far the cooperative sector can legitimately be considered as an independent sector distinct from the state sector.

In India, the official as well as the non-official opinion has, after much deliberation, accepted the general principle of state partnership with the cooperative movement. It is, therefore, not necessary to go into the pros and cons of this general principle; the only question still relevant is to see how this principle has worked in practice. The point to note is whether the partnership has vitiated the autonomous character of the cooperative sector and weakened the cooperative element in it. The main element of this partnership is the participation by the state in the share capital of the cooperative societies, leading to the inclusion of Government nominees, with varying powers, on their Boards of Management. We may confine our observations to the state participation in the share capital of primary agricultural credit societies. Time and the occasion do not permit a detailed review of the developments even in this field of the animated discussion which took place and the evolution in our thinking since the principle was recommended in the Report of the Reserve Bank's Rural Credit Survey. We shall therefore merely note the latest position as indicated in the Ministry's letter dated 4 October 1960, which is as follows:

(a) The share capital by the state should not be more than the contribution made by the members of the Society.
(b) The State should participate in the share capital of a primary society only if 60 per cent of its member desire it and the Central Bank, to which the Society is affiliated, supports the proposal.
(c) State participation in primary societies should, as normal rule, be indirect, i.e., through apex and central banks. Direct participation is however, permitted for special reasons in exceptional cases. But even in these cases, the state should not nominate the directors to the Board of Management of the primary societies and if this could not be altogether dispensed with, the authority to nominate directors should be delegated to the central cooperative banks.
(d) The central bank, holding shares in the primary society, may nominate one-third of the members of the Board of Directors subject to a maximum of three.
(e) State contribution may continue from the initial stage for a period of 5-8 years during which the society is expected to reach the optimum level of share capital required; and
(f) The principle of retirement of share capital should be accepted.

The policy as stated above appears to be unexceptional. Only when one looks to the development in the field of cooperative processing societies than a somewhat different picture emerges. This is illustrated by reference to the cooperative sugar factories. When the First Five-Year Plan was prepared, the block capital requirements for a plant of 1,000 tonne capacity were estimated to be about ₹80-90 lakhs. The cooperatives society was required to collect from its members (cane-growers) about ₹10 lakhs as share capital. An equal amount was contributed by the state government with the help of a loan from the Central Government and subsequently through the National Cooperative Development and Warehousing Board. Rupees 40-45 lakh were advanced as loan by the Industrial Finance Corporation of India. When it was found that some of the small and medium growers were not in a position to command sufficient money for purchase of shares, an arrangement was devised through which the RBI agreed to provide loans up to a limit of ₹1,000 per member. The amount of loan is advanced to the State Cooperative Bank on the condition that the repayment of the principal and payment of interest are guaranteed by the state governments concerned. The RBI agreed to advance 100 per cent of the amount if the paid-up value of the share did not exceed ₹500 and if it exceeded ₹500, 50 per cent of the amount after meeting the full requirements up to first ₹500.

Subsequently, the cost of the plant and machinery for sugar factories went up, and it is now estimated that a 1000 tonne capacity plant would, on an average, cost about ₹115-125 lakh. This necessitated a revision of the state's share of contribution to the share capital of the sugar factories. In 1957, it was decided that the National Cooperative Development and Warehousing Board may contribute (through long term loans to the state governments) ₹15 lakh to the share capital of each factory. The Board's contribution was raised to ₹20 lakh and further still to ₹25 lakh.

About one lakh sugarcane cultivators are member of cooperative sugar factories. By no token can they be classed as belonging to the disadvantaged group of farmers. Till the end of June 1959, the State Government had contributed ₹4.60 crore to the share capital of the cooperative sugar factories, constituting as much as 40 per cent of their total paid-up capital. This long-term loan assistance works out at ₹460 per sugarcane farmer joining a cooperative sugar factory.

In the Punjab, as against ₹73.5 lakh contributed by the producer-members to the paid-up capital of cooperative sugar factories, Government had made a contribution of ₹82 lakh; and another ₹63 lakhs were contributed by cooperative institutions, some of whom had contributed their own entire share capital to the share capital of the sugar factories. In Assam, Government's contribution was ₹27 lakh against the contribution of ₹4.4 lakh by producer-members. In Uttar Pradesh, the contribution of the producers to the share capital of the sugar factories came to less than 25 per cent of the total. The borrowings of sugar factories at the end of June 1959, were ₹21 crore, to which the Industrial Finance Corporation and the Central Financing Agencies had contributed respectively ₹9.2 and 7.2 crore. Incidentally, it may be noted that out of the total amount of loans amounting to ₹84.61 crore sanctioned by the Industrial Finance Corporation (IFC)—as on 30 June 1960—₹18.88 crore (22%) represent loans approved for 32 cooperative sugar factories. Another ₹1343 crore were provided by the State Bank of India (SBI) and nearly a crore by government. The amount of loans given by the RBI to enable the small sugarcane grower to contribute to the share capital of his cooperative factory is not known. Whether so much of the scantly financial resources of the government and the central financing agencies—not to speak of the resources of the RBI, SBI and the IFC—should have been diverted for the purposes of cooperative processing of sugarcane by just one lakh of comparatively well-to-do sugarcane cultivators is a question which needs serious consideration.

One major aspect of cooperative policy pertains to the contemplated extent of coverage of the agrarian structure by the cooperatives. Though there is considerable ambiguity on this matter, it appears that the ultimate objective of the Indian policy is to bring all economic activities—credit, marketing, processing and farming—within the fold of the cooperative movement, and to cover them to cent-per cent extent. By implication, when the transitional period is over, there will be only one form—the cooperative form of organisation. Since compulsion in any form or shape is to be eschewed, other forms such as household or family enterprises may persist. Nonetheless, the aim which the policy will strive to achieve is that of a mono-pattern of cooperatives. Many other policy decisions flow from such an aim. This, therefore, is the crucial question which may be examined in some detail. Once again, we shall do so on a selective basis, taking up two fields for consideration—cooperative farming and cooperative marketing—as illustrative of the general trend in thinking.

The problem of land tenures was comprehensively examined by the Congress Agrarian Reforms Committee in 1949. The Committee recommended imposition of ceiling on large individual holdings, and cooperative farming for what it called the below-basic holdings; but the dominant agrarian pattern recommended was that of 'peasant farming on suitable units of cultivation under a properly determined scheme of rights in land'. By suitable units, it meant that farms must not be too large (and therefore subject to a ceiling) nor too small. This floor limit was not placed at the level of economic holdings judged in terms of income or plough unit, but considerably below it, in fact, at one-third of the economic holding. Two considerations were involved in recommending this new operative concept of small holdings. It was believed that farms which at present were some what below the economic level could be upgraded through both additions of small areas or extension of irrigation and the assistance of service cooperatives. But the major consideration was purely administrative or organisational. One cardinal test of the suitability of a land reform measure is its practicability. The Committee felt that the task of bringing all the existing uneconomic farms in the cooperative fold would be beyond the official as well as non-official administrative and organisational capacity. The recommendation of the Committee regarding the maximum size of the cooperative farm may also be noted. The Committee wrote: "To minimise the odium of coercion and to maximise the sense of individual freedom, we recommend that the farmers whose holdings are smaller than the basic may not be pooled into a single giant farm but may be allowed voluntarily to join in any cooperative joint farm up to the size of the maximum holding which we have put at three times of he economic holding. The idea is that the management of such a cooperative should be within the capacity of the farmers and they may not be reduced to mere autonomous in the whole framework." Two points may be noted: the Agrarian Reforms Committee desired that: (i) cooperative farming should be confined to the sector of palpably uneconomic farms; and (ii) a limit should be placed on the maximum size of the cooperative farms.

Subsequently, the Government policy seems to have deviated from both these recommendations as evidenced by the First and the Second Five-Year Plan Reports, though the statements contained in both these documents are highly ambiguous. The Draft Outline of the First Five-Year Plan introduced a rather vague concept of cooperative village management. One of its essential features was that "the unit of land management should ordinarily be the area of the village was to be regarded as a single farm." It also suggested that ownership of the land should be separated from its management so that it may be possible to use the land wholly in the interest of the community. The section on cooperative village management was, however, deleted from the final draft of the First Five-Year Plan. The only justification for the quotations from the Draft Outline is just to indicate in what direction the mind of the policy-maker was moving. Regarding cooperative farming, the recommendation of the First Plan was that "for reasons (mentioned above), it is important that small and medium farmers in particular should be encouraged and assisted to group themselves voluntarily into cooperative farming societies. The area under a cooperative farming society should not be less than the prescribed minimum. It is perhaps not necessary to prescribe maximum for a cooperative farming society".

The approach of the Second Five-Year Plan on cooperative farming was equally cautious and it recommended that the general aim should be to bring the below-basic holdings increasingly in the cooperative pools. It also suggested that such essential steps should be taken as will provide sound foundations for the development of cooperative farming so that over a period of 10 years or so, a substantial portion of agricultural lands is cultivated on cooperative lines. The ambiguity regarding cooperative village management continued and though the Plan said that even when a larger area or the village as a whole is the unit of management, for many years, the common unit of operation will be the peasant holding, the aim was stated to be to 'enlarge the cooperative sector until the management of the entire land in the village becomes the cooperative responsibility of the community'. It was further stated that "once the stage of cooperative village management is reached,....the distinction between those who have lands and those who are landless will lose much of its significance." If such esoteric statements regarding the Government's intention in the matter of the future pattern caused uncertainty, if not suspicion, in the minds of the peasantry in India, there need be no surprise.

In January 1959, the Indian National Congress, at its 64th session in Nagpur, passed a Resolution on the agrarian organisational pattern. Inter alia the Resolution stated that "the future agrarian pattern should be that of cooperative joint farming in which the land will pooled for joint cultivation." In the Resolution, there was no qualification regarding restricting the cooperative pattern to the small holders. A presumption, therefore, arose that the intention of the government was to convert the entire agricultural economy into cooperative farming. Even when attention was pointedly drawn towards such implications, no clarification was thought to be necessary on the plea that after all the development of cooperative farming was to be on voluntary basis and, therefore, those who did not wish to join could keep away. In June 1959, the Government of India appointed a Working Group on Cooperative Farming mainly for the purpose of examining the organisational and managerial problems likely to be faced by the joint farming societies. The Working Group therefore was not called upon to express its views on the policy of the development of cooperative farming. But it noted that their visits to various farming societies confirmed their belief that cooperative farming was an effective method and agency for improving the economic and social conditions, particularly for small and middle cultivators. As for the size of the cooperative farm, the Group expressed the view that it was not necessary to lay down a rigid minimum size. But as regards the maximum size, its view was that compact and homogenous units are likely to prove more successful because the chanced of friction and misunderstanding were reduced considerably. They, therefore, recommended that "the size of the farm should not be too large and it was not essential that the cooperative farm should cover the entire village. On the other hand, there would be an advantage in having more than one cooperative farming society in the same village of average size." The programme of development it recommended for the Third Plan was quite modest. It was recommended that 320 carefully planned pilot projects should be carried out in the next four years in the NES Blocks. In each block, 10 societies may be organised. At the same time, it hoped that as a result of the working of the pilot projects, the programme of education and training and widening of cooperative effort in general, the idea of cooperative farming would gain ground and that 20,000 new societies would come into existence by the end of the Third Five-Year Plan.

The recommendations of the Working Group were generally approved by the National Development Council in September 1960. it is, however, significant that the Government's policy decisions communicated to the state governments in their letter of 23rd September, 1960, while emphasising that for the purpose of special assistance from government a minimum size in terms of membership and area may be prescribed by State Governments, stated that no maximum size need be prescribed.

Very recently (November 20, 1960), an important statement regarding the agricultural pattern was made by the Union Minster for Community Development and Cooperation, presiding over the first meeting of the National Cooperative farming Advisory Board. For the first time, after the recommendation of the Congress Agrarian Reforms Committee and the persistent ambiguity of statements contained in the First and the Second Five-Year Plans and the Nagpur Resolution, it was stated that:

> A mixed patterns of agriculture in the country was envisaged. There would be three patterns of farming: firstly, there would be a network of cooperatives—big and small—in different parts of the country, carrying on experiments in cooperative farming; secondly, the existing better class of cultivators would continue and should be given the facilities of advanced technological methods and skilled services to increase agricultural population, and thirdly there would be a significant sector of the agricultural population, the majority of whom owned small holdings, some being even landless and who constituted the marginal cultivator. These marginal cultivators

could not afford scientific means of farming, but if they pooled their resources of land and labour, they could avail themselves of the latest technological methods and also resist the exploitation of the absentee agricultural landlords.

We are thus back to the position from which we started after the report of the Congress Agrarian Reforms Committee. It is to be hoped that the Government will stick to the policy as enunciated by the Union Minister for Community Development and Cooperation. If thus view had been incorporated in the Nagpur Resolution on Agrarian pattern, the political party map of India would have been probably different.

In the field of cooperative marketing, the targets of coverage have been relatively modest. But even so, serious difficulties have been experienced in achieving them. Not that the contemplated number of cooperative marketing societies have not been established. The Second Plan had envisaged the establishment of 1,900 marketing societies. 1,743 societies have been already organised and it is expected that the Plan target will be reached by 1961. But as against the expectation that these marketing societies would handle about 10 per cent of the marketable produce, their actual share would hardly come to 5 per cent. It is in such frustrating situations that the suppressed layers of thought in the mind of the policy-maker reveal themselves. A single example will suffice. The Ministry of Community Development and Cooperation had constituted a Working Group to look into the question of cooperative development during the Third Plan. The Group found that a large majority of marketing societies did not undertake marketing activities as such, but were distributing fertilisers and other agricultural requisites in addition to advancing loans to the members against pledge of produce. The general condition of cooperative marketing societies, the Group felt, was far from satisfactory. With a view to giving an impetus to development of cooperative marketing during the Third Five-Year Plan, the Group recommended the following measures:

(a) Ensure adequate supply of marketable surplus by taking agreements from members of primary credit and service societies and through grant of monopoly for procurement of foodgrains whenever state trading is introduced.

(b) Assured markets for sale by grant of exclusive rights of sale for the requirements of state governments, hospitals, defence organisations, etc.

(c) A minimum assured price for agricultural produce collected by the marketing societies.

(d) Contribution by the state to the share capital of the societies on a more liberal basis.

(e) Loans and grants for construction of godowns and subsidy for the managerial staff in charge of their own housing.

(f) Increase in the quantum and duration of subsidy for managerial and paid staff.

(g) Provision of grading staff and grading equipment.

(h) Loans for capital outlay for the purchase of (250) transport vehicles.

(i) Provision of special staff for auditing the supervision in the cooperative departments.

The remedies suggested are thorough and almost leak-proof. If supplies are assured, markets are assured, prices are assured, share capital is assured, loans and subsidies for warehousing and transport vehicles are assured, and if the managerial staff is provided with subsidies, the societies have hardly anything to do except to collect the financial assistance and the trading privileges provided under the scheme. It would be interesting to note that the highly successful cooperative cotton marketing society in Gujarat—a region in which the private trader is by no token a nincompoop—had none of these assurances and privileges. Are we barking against the wrong tree!

III

The choice of the agrarian structure for a country depends upon several factors such as the historical situation mainly in regard to the resource base—pressure of population, availability of land, etc.,—the perspective level of technology and the socio-political ideology within whose context economic development is planned. It would be beyond the scope of this paper to deal with this entire range of the problem. I therefore assume that taking everything into consideration, the choice for the policy-maker in India is between the self-employed household pattern and the cooperative pattern of agrarian structure. In all probability, it will be a mixed pattern, the only choice being in regard to the proportions of each. It may be noted that if the cooperative unit is small-sized, ideologically and organisationally, the distinction between it and the household form may not be sharp. The household form is already in existence. The problem, therefore, boils down to the introduction and extension of the cooperative form—in production, distribution, and processing of agricultural produce. As we have seen, this also is the accepted policy of the Government. I would, therefore, like to draw attention to some of the problems involved in the organisation and extension of the cooperative sector.

The policy of the extension of the cooperative organisations has two major aims: (i) to make the small units of enterprise economically more efficient; and (ii) to insulate them from the powerful economic and social groups. But very often, as we have seen, the policy assumes ideological overtones relegating the economic objective to a secondary place.

It is important to remember that the structure of the economy reflects the market complex and gets adjusted to it—though there may be a reverse process also. In a recent enquiry conducted in the Department of Economics of the University of Bombay it was found that the crop pattern was positively correlated to the distance from the market centre. The result of a regression study between the percentage of area devoted to crops for self-consumption and several other variables showed that the area devoted to crop for self-consumption increased at the rate of 2.17 points for every additional mile to the distance from the market place.

A change in structure of the economy will bring about a corresponding change in the factor markets both on the demand as well as the supply side. Theses changes may be in consonance with changes taking place in other sectors of the economy. For example, the pooling of small, uneconomic farms would greatly increase their capacity to utilise credit and improved input factors. If, simultaneously, the cooperative and other organisations are in a position to meet the increased demand, the change will have an all round beneficent effect. But as I shall presently show, the position may not be equally favourable in regard to other factors of production. This does not necessarily imply that therefore, the change should not be brought about, if it is thought otherwise desirable. It only means two things: steps may be taken in advance to meet the contingent situation and if on a careful appraisal we feel diffident about the same, we may decide to adjust the timing and moderate the magnitude of the change.

Substitution of an individual enterprise by a cooperative one raises the question of efficient performance of the entrepreneurial functions. As long as the activities involved are of a routine type, no difficulties may be experienced; but when they go beyond and involve competitive functioning, the quality of entrepreneurship—good judgement, ability to take decisions and bear responsibility—assumes great importance, as for example in the case of cooperative marketing. The salaried personnel, materially and emotionally uninvolved in the enterprise, is not particularly suited for such tasks. Importance of extensive training programme is emphasised in this connection. But it is a moot point whether mere theoretical training can impart entrepreneurial competence. It is perhaps believed that a straight-forward non-profit making organisation does not need this sort of competence and its display by individual enterprise is even characterised as unfair competition. The highly successful cotton marketing cooperatives of Gujarat exemplify the importance of the quality of business acumen.

The other probably more serious problem concerns the effect of the extension of cooperative organisation employment. As is well-known, both the production and distribution sectors of the agricultural economy harbour a lot of disguised unemployment, albeit at the cost of diminished efficiency and low incomes. Prima facie, there is, therefore a clear case for the rationalisation of this institutional structure. But, by the very hypothesis, such rationalisation must result in the break-up of the employment quantum into two distinct components: full employment and overt unemployment. To the extent the rationalisation process opens up new opportunities of employment, the unemployment component would be reduced. Thus, for example, the adoption of cooperative farming for uneconomic units may open up fresh opportunities of expanding employment through a programme of land improvement measures—well digging, bunding, levelling, reclamation, etc.—the adoption of which was technically and financially inhibited by the small size of the farms and their resourcefulness. But the magnitude of such additional employment potential, and more particularly its duration, should not be over-estimated. Cooperative organisation may also release some labour for non-farm employment. But this would involve the question of the relation of the 'relived' farmer with the cooperative farm. Will he be permitted to retain his interests in the cooperative farm? How will that fit in with the policy directive to the effect that 'membership of cooperative farming societies should be confined to those who are prepared to work on the farm or in its ancillary activities?

To revert to the problem of the employment situation on the farm itself, we should remember that the *raison d'etre* of cooperative organisation is the under utilisation or uneconomic utilisation (as indicated by the low marginal returns) of human and bullock labour and implements on small family farms. It should not be difficult to get rid of the surplus cattle and implements. But what does one do with surplus labour, in some cases further augmented by the admission of landless labourers to the membership of the cooperative farm? Will the work be equally distributed to all members (and adults in the family), irrespective of their contribution to the land pooled? If this is done, two problems arise. Labour will continue to be under-utilised and, unless the organisation is one resembling the family type with mutual understanding, pressure will develop either for provision of more work or payment for (less) work at current wage rates for a full day's work. By hypothesis, the former is not possible, while the latter would mean running the cooperative farm at a loss. Once the cooperative organisation becomes larger, the family type of mutual adjustments would become increasingly difficult and it will have to function as a proper business unit adopting the usual accounting norms. When this happens, the manager (salaried or a member) will be compelled to employ labour only up to the point at which value of its marginal productivity is equal to the market wage-rate. Though the possibility of increase in productivity as a result of technological improvement is not denied, this point is soon reached in labour-intensive enterprises. The crux of the matter is that corporate as distinct from individual enterprises cannot harbour disguised unemployment. The difficult choice for the national, social and economic policies is how far and when—the timing is as important as the magnitude—disguised unemployment should be brought into the open and broken up into it two components—employment and unemployment; and by implication accept the organisational, if not also the moral responsibility for the emerging situation.

In the marketing sector also the employment effect of cooperative organisation will be similar to that in the production sector. One of the avowed objects of cooperative marketing is to eliminate the long chain of middlemen. Personally, I do not accept that the length of the chain adds to the cost of marketing. It merely reduces incomes of the individual links, no doubt generating temptations to make up through malpractices. That apart, adoption of institutional marketing must function on the basis of fully employed honorary or paid staff. Instead of a long chain of disguisedly unemployed middlemen with no claims to minimum remuneration, we will have a minimum of fully employed salaried staff paid at Union rates, unless the Parkinson's Law, a civilised device for disguised unemployment, comes to the rescue.

It has become almost axiomatic that defects in the agrarian structure constitute a serious obstacle to economic development. What is however not equally well recognised is that these defects are not merely aberrations of a faulty social and economic policy. Aberrations, no doubt, there have been, and in plenty; but do these wholly explain the existence of the senile or "semi-feudalistic" framework? Many of these defects are birth-marks of backwardness and cannot be erased till the very backwardness which gave them birth is removed. The institutional framework, perhaps, provided the defence mechanism against the stresses and strains of backwardness. The institution of household enterprise in agricultural production as well as trade provided a mechanism for disguising the widespread unemployment in the economy. Even when ownership of land was concentrated in a few hands, it made no difference to the operative pattern; the land being cultivated by small tenant households. It is difficult to imaginc how clse the problem of excessive pressure of population could have been solved. Such adjustments, however, were not without its price in terms of economic efficiency and exploitation.

The change in the institutional framework derives its rationale from the economic and social drawbacks of the household structure. But there is ample evidence to show that attempts, democratic as well as totalitarian, made for changing the agrarian structure have never been an unqualified success. Where the methods adopted are democratic, the blame for unsatisfactory implementation is placed on the social and political bias of the ruling class. Where totalitarian method has been followed and a particular agrarian pattern has been forced upon the economy, in spite of the tremendous sacrifice of human welfare, the economic achievement has been doubtful. This fairly universal experience establishes at least a case for the re-examination of some of our basic premises. Defects in agrarian structure undoubtedly obstruct economic development; but it is only the economic development that can finally and fully remove these defects. There is perhaps not a single instance, contemporary or historical, of a socially and economically faultless agrarian structure coexisting with overall economic backwardness.

I do hope, what I have said will not be interpreted as a counsel of despair or a plea for the retention of the status quo. If this is not implicit in all that I have said, I should like to state unequivocally that I support the extension of the cooperative sector to all fields of agricultural economy—production, credit, marketing and to the limited extent, also to the field of processing. My qualifications are: firstly, the cooperative element in it should not become subordinate, making it hardly distinguishable from the state sector; secondly, cooperative organisation need not aim at eliminating the individual sector, rather it should act as a powerful countervailing force to both the individual and the state sectors; and thirdly, cooperative units should not develop into giant impersonal managerial organisations manned by a salaried class. In the ultimate analysis, the problem of the removal of the defects in the agrarian structure cannot be separated from the problem of economic development as a whole. To say that the former constitutes an obstacle to the latter is but a partial statement of the phenomenon, because it can be said with equal validity that absence of economic development constitutes an obstacle to the removal of defects in the agrarian structure.

Indian Journal of Agricultural Economics, Vol. 16, No. 1, January-March 1961.

20

V.K.R.V. RAO

Agricultural Production and Productivity during the Plan Periods

A Review of the Past and Some Reflections on the Future

AS we have now completed two Five-Year Plans and have already entered on the Third, it may be useful to look at the performance of the agricultural sector during the last 10 years, particularly in terms of productivity and by individual crops. The relation between the crop pattern and productivity is also worth examination. The difference that exist in inter-state productivity in agriculture also needs to be formulated. Finally, it would be useful to view the Third Plan programme for agriculture and see how far it attempts to diminish differences in inter-state productivity in agriculture.

I am deliberately talking of productivity rather than production, not only because the former is more important in itself but also because it is bound to increase in importance with our limited supplies of cultivable land and the rising targets that are being put forward for our agricultural production. Productivity is a difficult theme, both conceptually and in terms of measurement and any definition that one adopts is bound to meet with some objection. It is important to remember however that productivity is a physical rather than a value concept and describes the changing relation between output and one of the major inputs like land or labour or capital. For purposes of statistical convenience, I propose to deal with productivity in this paper in terms of the relation between output and land or in more familiar language, the yield per acre.

In calculating the figures of yield per acre either nationally over time or by the different states at one point of time, one is confronted by the difficulty that our agricultural statistics have been constantly undergoing improvement during the last 20 years both in terms of a more complete recording of the area under cultivation and of a more accurate estimation of output by the substitution of the traditional *anna-wari* method by the more scientific one of crop-cutting experiments. As this improvement is automatically taken into accounts in the recorded statistics of area and production year by year, it follows that a part of the increase or decrease, as the case may be, which a historical series shows is merely of a statistical character and does not reflect a corresponding change in reality. A somewhat extreme illustration of what this can mean is shown by the statistics relating to the *tur*-crop. "Thus, for the year 1949-50, the all India production estimate stood at one million tonnes. From the year 1950-51, Orissa, Assam and Mysore states, which did not provide any production figures for his crop in the past began to supply estimates. In Madhya Pradesh and Uttar Pradesh states, which are large producers of tur, the crop sampling method replaced the traditional method of estimating yield from the same year. The result of these changes was that the all-India production estimate of *tur* shot up to 1.69 million tonnes and has maintained that approximate level since." It became important therefore to devise some method by which to eliminate these purely statistical variations and build up a comparable historical series of agricultural production. This was done by the Ministry of Agriculture by adopting the ingenious device of linking up each year's production estimate for a crop with the preceding year's estimate through a second estimate strictly comparable with the latter and by using the chain method, constructed a time series of adjusted production estimates comparable with one another over the entire series. Details are given in an article by Dr V.G. Panse and Shri V.S. Menon on "Indian Numbers of Agricultural Production in India" published in the *Indian Journal of Agricultural Economics* in their April-June 1961 issue. The same article contains such corrected estimates of the index numbers of agricultural production for the years 1949-50 to 1958-59 crop-wise, two alternative sets of figures being given, one with 1949-50 as base and the other with 1956-57 as base. The index numbers of production given in this article have been brought up-to-date to include 1959-60 and 1960-61 in a note published in the latest issue of the Ministry's Agricultural situation in India (August 1961).

Before one can use these figures to estimate the change in agricultural productivity or yield per acre during this period, it is necessary to have corrected figures of area, taking into account purely statistical changes that are the result of better coverage (or inclusion in acreage figures of areas which were previously non-reporting). The Ministry of Agriculture has not prepared an index of corrected area under cultivation during the period and use of the published figures of area without adjustment will tend to underestimate the increase in the yield per acre, as large increases in the area under cultivation have been recorded merely on account of better coverage.[1]

When one examines the figures of additions to reporting areas, however, one finds that their significance is much less after 1949-50 than before that year. In fact, I was responsible in 1944 as Director of Statistics in the Food Department of the Government of India for obtaining from all state government ad hoc estimates of food production in non-reporting areas during 1944-45 and these were published by Government in 1945 in a special publication entitled Food Statistics in India. Subsequently, the Governments were asked to prepare regularly estimates for the non-reporting areas on an annual basis. Since then some improvement in the method of estimation of crop acreages was in areas which were either not cadastrally surveyed or where regular land records were not obtained, and this process still continues.

1. My attention was drawn to this defect in the original version of my Presidential Address. I have taken advantage of this to do fresh work on the subject and arrive at revised figures of increase in productivity of yield per acre after allowing for purely statistical increase in the figures of the area under cultivation during the period. This paper contains the revised data, which supersedes the original Table 20.2 in the Presidential Address as delivered at Pilani. It will be seen from text however that the difference this makes is not as significant as may have been expected. This is mainly because of the fact that the bulk of the statistical increase in area coverage has been completed by 1949-50, due to exclusion of Pakistan occupied areas of Jammu and Kashmir state.

Data on new reporting areas is available only till 1956-57, and these have been included in the published figures of area under cultivation, thus introducing difficulties of comparability till 1956-57. Figures of 1957-58 and after do not included any new reporting areas and thus do not present difficulties of comparability from 1956-57. The following Table 20.1 gives the figures of additions to reporting areas from 1947-48.

Table 20.1

Additions to Reporting Areas

(in millions of acres)

1947-48	20.0	1952-53	7.3
1948-49	4.2	1953-54	0.7
1949-50	91.0	1954-55	0.2
1950-51	22.0	1955-56	0.5
1951-52	0.7	1956-57	1.0

Thus the bulk of the additions to reporting areas or the purely statistical; increase in area statistics has been completed before the beginning of the First Five-year Plan in 1951-52, additions between 1952-53 and 1956-57 accounting for only 1.2 per cent of the total geographical area. Moreover, it must be remembered that the additions that this makes to the total area under cultivation is much smaller than the increase in the reporting area and is only of the order of 4 to 4.5 million acres. Thus, figures of area during the Plan period are not vitiated to any significant extent by the inclusion of new reporting areas, and the bulk of the increase during this period can be legitimately regarded as non-statistical or real and therefore admitting of comparability. Nevertheless, I have made the necessary adjustments in the area figures, as the figures I am using for estimating the increase in productivity are based on three-year averages for each year of the Plan period, the average used for each year including the data for the previous two years. Thus the average figure used for 1951-52 includes the data for 1949-50 and 1950-51 and 1951-52; and as significant increase in reporting areas took place during these two years, adjustment has to be made on this account. Moreover, significant increase in reporting areas also took place in 1952-53 and these also had to be taken into account in calculating the adjusted figures for 1949-1950, 1950-51 and 1951-52. It may be added that the total cultivated area thus covered was of the order of about 11 million acres, and it was shared between Rajasthan and Saurashtra in terms of states. And *jowar*, *bajra*, other millets, gram, other pulse, groundnut, sesamum and cotton in terms of individual crops.

I have made use of the adjusted data to derive figures of the corrected increase in the yield per acre during the first two Plan periods. I have taken the corrected index numbers of the area and production for the years 1949-50 to 1951-52, and of 1958-59 to 1960-61, in each case taking a three years average to make some allowance for the seasonal factor. Taking the average of the three years ending with 1960-61 as a percentage of that of the three years ending with 1951-52, I arrive at the following figures of the increase in area, yield per acre, and production which has taken place in different crops on a national basis during the 10 years of the first two Plan periods.

You will see from Table 20.2 that it also contains an aggregate estimate for all crops taken together. This is based on the figures worked out by the Directorate of Economics and Statistics of the Ministry as a fixed base quantity index, based on gross production of the 28 crops listed in the table,[2] with farm-harvest prices prevalent during the year 1949-50 employed as weight. It shows that while agricultural production as a whole has increased by 36.4 per cent during this period, the increase in the average yield per acre is only 18.6 per cent, the rest of the increase in production being accounted for by increase in the area under cultivation. It is also seen from the table that the largest increases in yield per acre has taken place in the case of foodgrains—gram 29.5 per cent, rice 27.4 per cent, *ragi* 23.8 per cent, *jowar* 23.3 per cent, *bajra* 18.6 per cent, wheat 17.8 per cent and maize 11.8 per cent—and of the fibres—cotton 20.7 per cent, mesta 15.6 per cent and jute 6.4 per cent—while in the case of the oilseeds, the increase in yield per acre has been low and is actually negative in the case of linseed. There has also been a noticeable fall in

2. In addition to the 25 crops listed in Table 20.2, the index also includes tea, coffee and rubber.

Table 20.2

Percentage Increase in Area, Yield Per Acre and Production in 1958-59/1960-61 Over 1949-50/1951-52

Crop	*Area*	*Yield*	*Production*
Rice	9.8	27.4	39.9
Jowar	6.8	23.3	31.7
Bajra	8.1	18.6	28.2
Maize	31.9	11.8	47.5
Ragi	9.9	23.8	36.1
Small Millets	- 1.4	12.1	10.5
Wheat	32.1	17.8	55.6
Barley	4.2	7.9	12.4
Gram	30.7	29.5	69.3
Tur	3.8	- 5.7	- 2.1
Other pulses	19.5	-7.8	10.0
Groundnut	25.5	12.1	40.7
Sesamum	- 9.0	4.6	- 4.8
Rape and Mustard	27.9	8.8	39.1
Linseed	21.7	- 3.8	17.1
Castor	-7.9	14.0	-6.4
Cotton	30.1	20.7	57.0
Jute	11.0	6.4	18.1
Mesta	60.5	15.6	85.1
Sugar Cane	24.7	9.1	36.0
Tobacco	14.0	- 2.9	10.7
Potato	40.2	10.6	55.0
Pepper (black)	17.2	5.0	23.3
Chillies (dry)	7.7	- 2.4	5.1
Ginger (dry)	13.6	19.3	1.8
All crops	15.0	18.6	36.4

the yield per acre in the case of pulses other than gram, *tur* showing a fall of 5.7 per cent and other pulses of 7.8 per cent. It will also be seen that percentage increase in the yield per acre crop-wise ranged from +29.5 to –7.8 for the 25 crops listed in Table 20.2, only 6 of the crops showing an increase higher than the average for all crops taken together, and as many as 11 crops having an increase of less than half of the general average. I suggest that these inter-crop variations in productivity are deserving of more adequate examination than they appear to have received so far either at the hands of government or of non-official agencies or individuals.

It is of course true that the yield per acre is bound to be different for different crops and cannot lend itself easily to inter-crop comparability merely in terms of physical output. Lower volume of output with a higher value per unit may mean a higher productivity per acre than a larger volume of output with a lower value per unit of output. When, however, both value and volume are low per acre, productivity is certainly lower. And it is a matter of deep concern to a country if its farmers have a crop pattern that favours crops with a low yield per acre without compensating high value per unit. That this is so with regard to India is a fact long known to students of Indian agriculture and its economics. Thus, e.g., while rice, wheat, maize and barley all have yields above 700 lbs, per acre, which is low enough in all conscience, and *ragi* has a yield fairly close to 700 lbs, *jowar* has an average yield per acre which is 35 per cent less, while in the case of *bajra*, the yield per acre is 60 per cent less. Taken together, their yield per acre is barely 400 lbs. These two crops occupy about 10 per cent of the cultivated area or nearly a third of the entire area under cereals and they recorded an increase of more than 5 million acres in the area under their cultivation in the last year of the Second Plan as compared to that in the year preceding the First Plan. This persistence, and in fact even expansion of crops with both low yield per acre and low value per unit of output is a matter for disquiet from the point of view of improvement in agricultural productivity. Whether this is due to technical factors beyond human control that do not admit of the desired change in the crop pattern or whether changes in crop pattern inhibited by marketing habits in the part of the cultivator, not only with regard to sale but also purchase of foodgrains or whether it is the result of failure to effect the required change in the pattern and volume of agricultural inputs are all matters that more properly belong to the field of agronomy rather than economics. All the same, the attention paid to the question of either altering the crop pattern or otherwise increasing the yield per acre by improving the quality of these low-yield crops does not seem to be adequate to a layman like myself.[3] With the increasing demands that economic development makes on our soil it seems to me that this is a subject that calls for more immediate and more intensive examination than it has received so far.

I would also like to dispel any illusion of complacency that may have been caused by the fact that our agricultural production has recorded an increase of more than 36 per cent during the two Plan periods or an average of 3.6 per cent a year. I have already shown that about half of this increase is due to an extension of the area under cultivation a factor that is not likely to be available during the Third Plan period and certainly not during subsequent Plan periods. Increase in productivity or the yield per acre, which is the only solid and enduring base for the required progress in our agricultural production is only 18.6 per cent over the period or an average of 1.9 per cent a year. It may be of some interest in this connection not only to draw attention to the much higher absolute yields per acre in Europe, North America, Japan and certain other countries, but also to the fact that the yield per acre in these areas has also been increasing and, in the case of some crops, at a much faster rate than in India during this period, even though these other countries had the statistical disadvantage of starting from much higher initial levels. By way of illustration, I would like to give you the following figures. In the North American continent, the annual yield per acre during the 3 years 1957-58 to 1959-60 has risen over that during the 5 years 1948-498 to 1952-53 by 23.8 per cent in the case of wheat, 27.8 per cent in the case of barley, 59.7 per cent in the case 33.8 per cent in the case of millet, and 41.2 per cent in the case of sorghum, which corresponds to our *jowar*. In the case of Japan, yield per acre during this period has risen by 19.4 per cent for wheat, 51.4 per cent for maize, 11.6 per cent for barley and 15 per cent for rice. During the same period, Burma has recorded a rise of 60.7 per cent in the case of

3. It must be admitted, however, that a part of the preference for *jowar* and *bajra* in spite of low grain yield would be their high yield in stalks and straw.

millets, Taiwan of 89.7 per cent in wheat, 34.7 per cent in barley and 23.2 per cent in millets and Thailand of 64.8 per cent in maize. In the case of Europe, the corresponding figures of increase have been 23.8 per cent for wheat, 27.8 per cent for barley, 59.7 per cent for maize, 33.8 per cent for millets and 41.2 per cent for sorghum. Figures for India of a comparable character as given in international statistics are 9.1 per cent for wheat, 4.0 per cent for barley, 20.2 per cent for maize, 8.1 per cent for millets, 23.9 per cent for sorghum, and 18 per cent for rice. The only major country with whose figures of the rise in yield per acre India shows at an advantage is Pakistan in whose case the corresponding figures are 4.1 per cent for maize, 5.4 per cent for millets and minus 8.2 per cent for wheat and minus 3 per cent for barley.

I do not think that we should feel happy or satisfied with our better performance as compared to Pakistan. What is more important is to feel disturbed at our comparatively slower progress, despite our Plans, in the country's agricultural productivity. We should also feel disturbed at our much lower level of productivity in absolute terms in regard to practically all our crops. Here again I feel that the time is now ripe for a complete re-examination of the whole problem of productivity in Indian agriculture not only in terms of inter-crop differences, but also in terms of absolute levels of productivity as also the insufficient degree of increase during the two Plan periods.

It will be seen from Table 20.3 that the highest percentage increases in acreage were recorded by mesta, potato, wheat, maize, gram, cotton, rape and mustard, groundnut and sugar cane. The lowest percentage increases were recorded by *tur*, barley, rice, *ragi*, *jowar* and *bajra*. Linseed and sesamum actually showed a decline. The disquieting feature revealed by Table 20.3 is the lower percentage rise of the area under rice. It is, however, encouraging that then low yield cereals like *jowar* and *bajra* showed a comparatively low percentage rise in cultivated area but the increase itself is not a matter for satisfaction.

Table 20.3

Area Under Cultivation: All-India Figures

(Figures in '000 acres)

Crop	*Annual Average of Three Years 1949-50 to 1951-52*	*Annual Average of Three Years 1958-59 to 1960-61*	*Rise in Acreage*	*Percentage Rise in Acreage*
Rice	75,106	82,534	7,428	9.9
Jowar	39,561	42,283	2,722	6.9
Bajra	25,524	27,604	2,080	8.1
Maize	8,067	10,640	2,573	31.9
Ragi	5,435	5,972	537	9.9
Small Millets	12,522	12,361	- 161	- 1.3
Wheat	24,083	31,811	7,728	32.1
Barley	7,838	8,168	330	4.2
Gram	18,802	24,573	5,771	30.7
Tur	5,664	5,883	219	3.9
Other pulses	23,959	28,608	4,649	19.4
Groundnut	11,934	14,965	3,031	25.4
Sesamum	5,782	5,260	- 522	- 9.0
Rape and Mustard	5,325	6,812	1,487	27.9
Linseed	3,563	4,334	771	21.6
Castor	1,422	1,169	- 253	- 17.8
Cotton	14,798	19,234	4,436	30.0
Jute	1,508	1,675	167	11.1
Mesta	462	741	279	60.4
Sugar Cane	4,211	5,252	1,041	24.7
Tobacco	819	933	114	13.9
Potato	615	862	247	40.2
Pepper (black)	198	232	34	17.2
Chillies (dry)	1,373	1,478	105	7.6
Ginger (dry)	47	40	- 7	- 14.9
All Crops	298,618	343,426	44,808	15.0

The crop pattern as such in the sense of the percentage distribution of the total cultivated area under different crops does not show much significant variation over the period. The major crops which have increased their share in the total cultivated area are wheat (1.2%), cotton (0.64%), groundnut (0.36%) and gram (0.86%), while those that showed a decline are rice (-1.12%), *jowar* (-0.94%), *bajra* (-0.51%) and barley. Details are given in Table 20.4.

On the whole, it can be stated that the crop pattern has changed in the right direction so far as it has increased the share of the higher yielding crops with one significant exception, namely rice. And it appears that one of the factors contributing to increased agricultural productivity during this period has been this change in the crop pattern. Changes in the crop pattern, however, need to be studied in more detail and by regions and sub-regions with reference to both technical and economic factors, if proper relationship is to be established between changes in crop pattern and changes in agricultural productivity.

An even more important aspect of agricultural production during the 10-year period is the change, if any, that has taken place in the comparative position of the different states in the Union. An oft-repeated consideration in economic planning in the case of a large country like India is the need for balanced regional development and the minimisation of differences in respective inter-state economic levels. Too often the factor to which one turns for redress of regional inequalities is industry; and even when one turns to agriculture, one does not take an overall comprehensive view of differences in agricultural productivity and yields per acre. And yet, agriculture is the largest single determinant of state levels of income except perhaps in one or two states like Bombay and Bengal that claim such a large share of India's organised industry. Differences in agricultural productivity constitute the largest single reason for differences in inter-state economic levels, and it is high time that we looked at Indian agriculture not only from the overall Indian angle but also from the individual state angles.

Table 20.4

*Percentage Distribution of Total Cultivated Area at Different Crops**

Crop	*1951-52*	*1960-61*	*Difference Col. 3–Col. 2)*
(1)	(2)	(3)	(4)
Rice	25.15	24.03	- 1.12
Jowar	13.25	12.31	- 0.94
Bajra	8.55	8.04	- 0.51
Maize	2.70	3.10	+ 0.40
Ragi	1.82	1.74	- 0.08
Small Millets	4.19	3.60	- 0.59
Wheat	8.06	9.26	+ 1.20
Barley	2.62	2.38	- 0.24
Gram	6.30	7.16	+ 0.86
Tur	1.90	1.71	- 0.19
Other pulses	8.02	8.33	+ 0.31
Groundnut	4.00	4.36	+ 0.36
Sesamum	1.94	1.53	- 0.41
Rape and Mustard	1.78	1.98	+ 0.20
Linseed	1.19	1.26	+ 0.07
Castor	0.48	0.34	- 0.14
Cotton	4.96	5.60	+ 0.64
Jute	0.50	0.49	- 0.01
Mesta	0.15	0.22	+ 0.07
Sugar Cane	1.41	1.53	+ 0.12
Tobacco	0.27	0.27	0.00
Potato	0.21	0.25	+ 0.04
Pepper (black)	0.07	0.07	0.00
Chillies (dry)	0.46	0.43	- 0.03
Ginger (dry)	0.02	0.01	- 0.01
All crops	100.00	100.00	

Note: The percentages are calculated on the basis of triennial averages, i.e., annual average of 1949-50 to 1951-52 for col. 2 and that of 1958-59 to 1960-61 for col. 3.

Inter-state comparisons in improvement in productivity during the Plan period is, however, not possible in the absence of adjustment of the figures of area and production for changes in size of reporting areas and methods of estimation of crop yields. Adjusted figures of production are not available for the majority of Indian states. It is therefore somewhat risky to attempt conclusions on comparisons of inter-state progress in agricultural production on the basis of published data. I may perhaps illustrate this by referring to one or two glaring examples of the difference that is made to production data in some states by the application of improved methods of crop estimation and more complete coverage of area under cultivation. Thus, in the case of Andhra, the traditional method of estimation was replaced by crop sampling for rice, *jowar*, *bajra* and maize in 1952-53, for *ragi* in 1955-56, for sugar cane in 1956-57, and for oilseeds in 1959-60; this was for the Telangana region, while in Andhra region, improvement in estimated yield was effected in later years. The result was a significant element of exaggeration in the recorded figures of production as an index of increase in agricultural productivity during the period, this being particularly so in regard to rice, *jowar*, maize and groundnut. Old Bombay (i.e., Maharashtra and Gujarat) and Madhya Pradesh show even larger variations between adjusted and unadjusted yield partly because of the different years during which this was done to their component regions, while the largest differences are found in the case of Rajasthan where improved methods of estimation were applied practically to every crop during this period. Large differences are also noticeable in the case of the Punjab, UP, Bengal and Bihar show the least differences, improved methods of estimation having been effected in their case before the advent of the Plan period.

It is very difficult to work out figures of adjusted production by states in the absence of the kind of data that can be available only to the Ministry. Nevertheless, I have attempted a heroic exercise in perhaps what may be called guesswork—I hope intelligent guesswork —by attempting to work out figures of adjusted production by states for their different crops. The difficulties in doing this are many and complex, and some may be mentioned briefly here. One is the different years in which the improved estimation is applied for the same crop in the different regions of what are now reconstituted states. Another is that of linking up improvement effected in later years of the period with the unadjusted figures of its earlier years of the period with the unadjusted figures of its earlier years in view of the differences made by the seasonal factor and a possible secular trend in productivity. I have tried to deal with the seasonal factor by taking the annual average of a three-year period following the introduction of improved estimation and that immediately preceding it. But for the genuine increase in productivity that may have taken place during the period, I have had to make arbitrary assumptions and take recourse to value judgements that cannot stand any close scrutiny. With all these limitations, however, the results arrived at are extremely interesting and certainly make for some revision in the estimates of comparison of inter-state progress in agricultural production during the 10 years of the two Plan periods. Thus, for example, Rajasthan and Madhya Pradesh have certainly not done as well in agricultural progress over this period as may appear from an uncritical acceptance of the published figures of their agricultural production; and the same is true in varying degrees of Andhra, Maharashtra and Gujarat (the Bombay of 1957), Punjab and Orissa. I have, however, not ventured to use in this address the results of this period, and possibly foolhardy, exercise in adjustment of the change in yields per acre for different crops by the different states and do not therefore say anything about the trends in inter-state productivity in agriculture.

There is, however, no reason why we should not deal with current position in regard to differences in inter-state productivity by different crops. If we take the latest available data for the states, namely around average of the three years ending with 1960-61, they are least likely to suffer from merely statistical increases in area and output, as by 1958-59, better coverage of area and better estimating of output had been completed in all the states for most of the crops dealt within this paper. Relevant figures are given in Table 20.5.

The data are given only for those states which have 4 per cent or more of the all-India acreage under each crop. It is seen from Table 20.5 that in the case of rice, the yield per acre is well below the national average in the case of four states with more than 10 per cent of the all-India acreage, *viz.*, Bihar, Madhya Pradesh, UP and Orissa. Orissa and UP which between them have 23.6 per cent of the all-India acreage have a yield per acre which is less than 73 per cent of the

Table 20.5

Annual Yield Per Acre in Lbs. for the Period 1958-59 to 1960-61

	Rice	Wheat	Jowar	Bajra	Ragi	Maize	Barley	Gram	All Foodgrain	Sugar Cane	Groundnut	Total Oilseed	Cotton	Tobacco
Andhra Pradesh	1130		471	445	730	671			641		702	518	57	692
Assam	850								824					
Bihar	747	554			473	680	440	402	615	3344		272		
Maharashtra and Gujarat	850	465	456	266	700	840		299	561	6534	579	509	86	554
Kerala										1137				
Madhya Pradesh	731	619	585			824	742	543	566		600	318	71	
Madras	1323		682	543	863				954		1081	955	136	
Mysore			350	200	659				464		539	484	67	499
Orissa	627				393				594			274		
Punjab		978		286		1024	755	716	738	3295			217	
Rajasthan		792	268	196		830	1020	550	392			210		
Uttar Pradesh	616	787	571	441		629	741	574	669	2541		448		698
West Bengal	925								845			202		
All India	850	715	455	282	663	771	737	565	606	3330	671	456	94	645

national average and less than half of the yield per acre in Madras. In the case of wheat in Maharashtra and Gujarat with nearly 11 per cent of the national acreage as a yield of 35 per cent less than the national average. In the case of *jowar*, Mysore and Rajasthan have yields well below the national average with nearly 20 per cent of the acreage between them. As for *bajra*, this is true of Mysore and Rajasthan, with 38 per cent of the acreage between them. For maize, UP has an yield well below the national average, though it is the largest single state under that crop (26.4%). Bihar occupies a similar position with respect to barley, though occupying the third place in area with an acreage of only 10.6 per cent, while with regard to *ragi*, yields well below the national average are shown both by Bihar and Orissa. Taking foodgrains as a whole, the difference in yield are startling. Madras, Kerala, Bengal and Assam stand well above the national average, with Punjab trailing at some distance; UP, Andhra, Bihar and Orissa are about national average, while Madhya Pradesh, Maharashtra and Gujarat, Mysore and Rajasthan are all below the average. In the case of sugar cane, it is well known that yields in UP are well below the national average, even though it accounts for nearly 60 per cent of the acreage. The highest place in term of yield per acre is occupied by Madras, for 5 crops out of the 7 for which it is listed in the table, with the second place for the remaining crop. Punjab occupies the highest place for 4 out of the 7 crops for which it is listed in the table, with the second place for one of the remaining crops. Thus, both north and south lead in agricultural productivity. All these differences in inter-state productivity cannot be explained merely in terms of natural factors. Capital, labour and agricultural practices must have something to do with these differences; and it is high time that the whole question gets the benefit of an authoritative and comprehensive examination, mainly with a view to determining an agricultural policy designed to reduce inter-state differences in agricultural yield and at the same time maximise the national yields per acre.

What I have said above gains added importance when we remember that the states with the highest agricultural productivity, *viz.*, Madras, Punjab and Kerala account between them for only 11.4 per cent of the total area under cultivation, while the states with the largest area under cultivation like Maharashtra and Gujarat (18%), Madhya Pradesh (12.7%), Rajasthan (8.8%), Bihar (7.8%), and Mysore (6.6%)—all fare rather badly in terms of their agricultural productivity. If, as is quite certain, the future of agricultural production in India is going to turn on increase in productivity, it follows that maximum attempts should be made to raise the yield per acre in these studies, thereby not only making the most effective contribution to increasing the national average, but also in the process, minimising inter-state difference in levels of agricultural productivity.

I may now turn in conclusion to a few observations on the future. The Third Plan contemplates an increase in agricultural production on the order of 30 per cent, as against an increase of 36 per cent during the 10 years preceding it. Projection made in the Third Plan report appear to contemplate a rise of another 25 per cent during the Fourth Plan period, so that over the decade 1961-70, an increase of about 60 per cent is contemplated in agricultural production. This means that, broadly speaking, it is hoped to step up the increase in agricultural production during the 10 years covering the Third and Fourth Plan periods to about twice the rate achieved during the 10 years covered by the First and Second Plan periods. This must be further interpreted in the light of the fact that whereas nearly half of the increase in agricultural production during the 10 years ending with 1960-61 was the result of an increase in the area under cultivation, it would be very unwise to expect a similar situation in the 10 years following 1960-61. In fact, we are practically reaching the limits of extensive cultivation in India; and the scope that remains for increasing output by increasing the area under cultivation is only marginal. The main brunt of the task of increasing agricultural production and reaching the Plan targets in the future is bound to fall therefore on increased productivity. Increasing the yield per acre by about 50 to 60 per cent on an average in the next 10 years is the supreme task that confronts the agricultural planner in this country. The Third Plan lists in some detail the steps it contemplates for increasing agricultural production by about 30 per cent in the next 5 years; and, for the first time, details are given by states for more than irrigation.

It is a matter for some satisfaction that an attempt is being made during the Third Plan period to minimise some of the existing differences in productivity among the different states. This is seen especially in regard to the programme of extending the area under improved seed, where the proportion of cultivated area sought to be covered is practically the same in all the states. This is well illustrated from Table 20.6 compiled from the Third Plan report.

An attempt has also been made in the same direction, though to a much smaller extent, in regard to the programme for extending the area under irrigation. Table 20.7 gives the relevant figures.

Table 20.6

Third Plan Programme for Extending Area Under Improved Seed

(in '000 acres)

State	*Area in 1960-61*	*Area Under Improved Seed*			
		1960-61	*Percentage*	*1965-66*	*Percentage*
Andhra Pradesh	21,091	1,230	5.8	12,780	60.6
Assam	4,553	438	9.6	3,000	65.9
Bihar	25,002	2,618	10.5	11,800	47.2
Maharashtra and Gujarat	41,467	3,678	8.9	17,563	42.4
Kerala	2,064	500	24.2	1,200	58.1
Madhya Pradesh	37,258	6,300	16.9	15,298	41.0
Madras	12,152	7,250	59.7	9,450	77.8
Mysore	17,321	4,869	28.1	8,876	51.2
Orissa	10,976	1,200	10.9	6,200	56.5
Punjab	17,863	3,000	16.8	9,000	50.5
Rajasthan	27,136	4,140	15.2	17,000	63.0
Uttar Pradesh	45,215	18,961	41.9	29,301	64.8
West Bengal	13,685	1,000	7.3	6,000	43.8
All India	279,613	55,441	19.8	148,253	53.1

Table 20.7

Irrigation Factor by States

State	*Percentage of Area Sown (1960-61)*	*Percentage of Area Irrigated (1956-57)*	*Col. 3 Expressed as a Percentage of Col. 2*	*Percentage of Area to be Irrigated Under Third Plan*	*Col. 5 as Percentage of Col. 2*
Andhra	7.6	12.7	167	11.7	154
Assam	1.6	2.8	175	1.8	113
Bihar	7.8	7.9	101	12.0	154
Maharashtra and Gujarat	18.0	6.5	36	15.0	83
Kerala	0.6	1.5	250	1.2	200
Madhya Pradesh	12.7	3.7	29	6.1	48
Madras	4.7	9.9	211	3.2	68
Mysore	6.6	3.3	50	4.1	62
Orissa	3.4	4.3	127	4.8	141
Punjab	6.1	13.4	203	9.1	149
Rajasthan	8.8	6.3	72	6.4	73
Uttar Pradesh	16.5	20.5	124	16.0	97
West Bengal	4.4	5.4	123	7.0	159
All India	100.0	100.0	100	100.0	100

Thus, it can be seen from Table 20.7 that the shares of Maharashtra and Gujarat in the irrigation programme under the third Plan is 15 per cent as against its existing share of irrigated area in the country as a whole of only 6.5 per cent. Similarly, in the case of Madhya Pradesh, the figures are 6.1 per cent and 3.7 per cent respectively. On the other hand, the share allotted to Madras under the Third Plan programme is only 3.2 per cent as against its existing share in the national irrigated area of 9.9 per cent. Similarly, the corresponding figures for Punjab are 9.1 per cent, and for UP 16.0 per cent and 20.5 per cent respectively. While this is an attempt in the right direction, the fact still remain that there is wide variation between the different states in regard to the area under irrigation and that the states with the largest area under cultivation with the exception of UP are the very ones with the least area under irrigation. No doubt, this is the result of natural factors as far as major or even medium irrigation works are concerned. But this difficulty should not be equally great in the case of minor irrigation works and more imaginative and more vigorous planning is required to extend the area under minor irrigation, including tube-wells and storage of rain water by tanks not only for increasing agricultural productivity but also for minimising inter-state differences.

The programme contemplated for fertiliser distribution also takes into account the need for some correspondence between area cultivated and amount sought to be distributed, with the exception of Madhya Pradesh and Rajasthan that do not fare so well in the distribution. Details are given in Table 20.8.

Table 20.8

Chemical Fertilisers: State Shares

(in percentage)

State	*Percentage of Area Under Cultivation (1960-61)*	*Ammonium Sulphate*		*Superphosphate*		*Muriate of Potash*	
		1960-61	*1965-66*	*1960-61*	*1965-66*	*1960-61*	*1965-66*
Andhra Pradesh	7.6	21.3	10.0	22.3	14.3	–	–
Assam	1.6	0.8	1.1	1.2	1.0	–	–
Bihar	7.8	3.9	7.6	3.7	4.9	4.3	7.2
Maharashtra and Gujarat	18.0	15.2	16.8	25.0	17.0	–	4.7
Kerala	0.6	2.9	3.6	1.7	7.8	4.3	20.1
Madhya Pradesh	12.7	2.0	3.8	1.2	1.6	–	
Madras	4.7	11.7	11.1	14.8	13.1	52.2	32.4
Mysore	6.6	6.8	5.7	4.9	13.1	–	4.0
Orissa	3.4	2.3	3.8	2.0	2.0	–	0.8
Punjab	6.1	3.1	5.3	0.5	1.0	–	–
Rajasthan	8.8	1.2	1.9	1.0	1.3	–	–
Uttar Pradesh	16.5	23.2	18.9	14.8	12.3	34.6	28.8
West Bengal	4.4	3.1	9.5	6.2	10.2	–	–
All India	100.0	100.0	100.0	100.0	100.0	100.0	100.0

In spite of the attempt to take into account regional considerations in regard of the programme of improved seed, irrigation and fertilisers, the fact remains that the comparative position of the different states is not likely to be substantially altered at the end of the third Plan period as compared to its beginning. This is particularly so in the case of food grains, as can be seen from Table 20.9.

Table 20.9

Production of Foodgrains at the Beginning and End of the Third Plan Period

(By Percentage)

State	*(1960-61)*	*Addition During Third Plan*	*1965-66*
Andhra Pradesh	8.3	10.2	8.7
Assam	2.3	1.8	2.2
Bihar	8.1	8.6	8.2
Maharashtra and Gujarat	10.9	10.9	10.9
Kerala	1.4	1.3	1.4
Madhya Pradesh	11.8	7.1	10.7
Madras	6.7	7.0	6.8
Mysore	5.0	4.3	4.8
Orissa	5.2	6.8	5.6
Punjab	7.8	7.8	7.8
Rajasthan	6.5	6.8	6.6
Uttar Pradesh	17.6	20.1	18.2
West Bengal	6.8	6.1	6.6
All India	100.0	100.0	100.0

In fact, Table 20.9 shows that Madhya Pradesh and Mysore will be contributing a smaller share of national food production, while UP, Andhra and Orissa are expected to record a significant increase. The differences in the case of the other states are marginal. In view of the existing difference in inter-state yields per acre, this means that not much will be done during the Third Plan period to minimise these differences; and that is not a satisfactory conclusion from the point of view of balanced regional development, which is an accepted principle of Indian planning.

I am also not certain in my mind that the investment contemplated in agriculture during the Third Plan period would be sufficient to bring about the desired increase in agricultural production in view of the fact that there is no longer much scope for increasing the area under cultivation. I am even less sure in my mind that the investment in agriculture contemplated for the different states does adequate justice to the principle of balanced regional development and the need for minimising a part at least of the startling differences that now exist in inter-state agricultural productivity. Above all, I am convinced that the size of the targets contemplated during the Third and Fourth Plans, and, even more, that which will be necessary during subsequent Plan periods, cannot be achieved merely by attention to investment. Agricultural practices must improve beyond measure and the whole country brought up to levels which now prevails in some of its best agricultural areas. The human factor needs far more attention than it has received so far in Indian agriculture, in spite of community development programmes and village *panchayats*. Land reforms have presented a sad story during the first 10 years of planning; adult literacy is unconscionably low in the rural areas, and the scientific temper and technological outlook, about which the Prime Minister talks so much has still to permeate the countryside. The man behind the plough still remains a neglected figure as far as motivation, outlook and training for increased production are concerned. The problem is so serious that I would in all humility, suggest that the time has now come—in fact it is already with us for an authoritative and thorough examination of the entire problem of Indian agriculture. Such an examination should concentrate on the problem of agricultural productivity, the task of balanced regional development and minimisation of inter-state inequalities, and the vitalising and modernisation of the man behind the plough in India. Nearly 35 years ago, an agricultural-minded viceroy appointed a Royal Commission on Indian Agriculture. Would it be too much to hope that the peasant-conscious President of the Indian Republic will appoint another commission on Indian agriculture before he steps out of his presidential office!

Indian Journal of Agricultural Economics, Vol. 17, No. 1, January-March 1962.

21

TARLOK SINGH

Agricultural Policy and Rural Economic Progress

I

For more than a decade, continuous efforts have been made to transform India's economy through a series of Five Year Plans. The formulation of each plan has brought up prominently issues concerning the place of the rural sector in the economic development of the country, its rate of growth and investment requirements and the relations between the rural and the non-rural sectors of the economy. Large changes have taken place in many directions, new institutions have been established and the rural environment today is different in many ways from that of 10 or 12 years ago. Yet, a general impression remains that economic progress in rural areas has not been achieved in adequate measure, that partly is consequence of planned development, the gap between levels of income and opportunity in rural and non-rural areas is widening, and that the agricultural economy has not been strengthened sufficiently to meet the demands of rapid industrial and economic development. At this juncture, when a crisis in our affair has forced us to re-examine our assumptions and practices, it would be useful to explore how far the common beliefs concerning the progress of the rural economy are borne out by the available evidence, whether the framework of policies and institutions on which the Five Year Plans are based calls for fundamental changes in emphasis and direction and, finally, whether the instruments through which the various objectives are sought to be attained are adequate for the purpose. These questions, important as they are for an appraisal of the past, become even more significant in relation to the perspectives for the future.

In recent years, a vast amount of statistical, economic and other data regarding conditions in rural area has become available. One has only to recall the work of the Reserve Bank of India (RBI) in the Rural Credit Survey and its series of follow-up surveys, statistical and other information presented by the Ministry of Food and Agriculture, the village surveys and resurveys carried out by the Agro-Economic Research Centres, the Farm Management studies, reports of the Programme Evaluation Organisation, and the wide range of monographs prepared under the auspices of the Research Programmes Committee, the universities, Board of Economic Enquiry and other institutions, to realise the wealth of knowledge now available for gaining deeper understanding of the basic problems of our rural economy. Doubtless, there would still be many aspects on which answers might be difficult. Nevertheless, in the degree to which we formulate issues correctly, even if precise answers are not to be had immediately, means now exist for gaining greater insight and devising new policies on the basis of tested knowledge. However, to use these means to the best advantage, there is need for much closer interchange of experience and for a common frame of reference between those engaged in carrying out studies and surveys and the workers, both official and non-official, who are actively participating in the specific tasks of developing the rural economy.

In assessing the progress of a national economy, growth in national and per capita income, rates of growth in different sectors and the rates of investment and domestic savings are among the principal indicators. Statistical measures have considerable validity in the organised sector of an economy. In the unorganised sector and, more specially, in relation to the rural areas, such indicators are not always readily available. To the extent statistical data can be obtained, even apart from inherent limitations, they throw only partial light on the total process and degree of change that occurs over a period. Data which permit comparison with given bench marks exist at present only to a small extent. Moreover, many aspects of rural change are not sufficiently susceptible of measurement. Without, therefore, under-estimating the importance of statistical measures of economic progress in rural areas, one may legitimately suggest that for any just appraisal, it is essential to take a wider view of the changes through which the rural economy is passing under the impact of planned development and the economic, technical, social and administrative policies and measures associated with it.

II

Before proceeding to this appraisal, certain broad magnitudes and trends may be briefly stated. These are not intended to be a summing up of the total picture, but may at least point to some of the problems. The rural sector accounts for about 70 per cent of the total net national output. Its pace of advance is therefore, the most dominant factor in the overall growth of the economy. The growth of population is throwing an increasingly heavy burden on the rural areas. According to the final population totals for the 1961 census, during the decade 1951-61, as against an increase of 34 per cent in the total number of workers and of 35 per cent among non-agricultural workers, the addition in agriculture was of the order of 33 per cent. In terms of absolute figures, this has meant an increase in the number of agricultural workers from about 98 million to about 131 million, the proportion of agricultural workers to the total working force remaining at about 70 per cent,

This large increase in the pressure of population on land should have been balanced by corresponding increase in production in rural areas. Between 1950-51 and 1960-61, as against increase in national income of 42 per cent, the net output of agriculture, animal husbandry and ancillary activities rose by 33 per cent. It has been estimated that over the period 1950-51 to 1960-61 the average net output per worker increased by 21.5 per cent. Against this, the net output per worker in agriculture rose by about 15 per cent, while that outside agriculture increased by 25.5 per cent. Consequently, a small decline is observed in the proportion between the net output per agricultural worker and the net output per worker outside agriculture. At the same time, such statistical data as are available suggest that capital formation in rural areas has increased during the first two Plans at a much lower rate than outside the rural sector or in the economy as a whole.

Since agriculture accounts for most of the economic activity in the rural sector, the presumption is that in an economy which, as a whole, has been advancing only at a moderate pace, agriculture has tended to lag behind and, secondly, that with the growth of population, the gap between incomes in rural and urban areas may have actually widened. Over the period 1961-1976, the total increase in population may be of the order of 187 million, the increase in the labour force being about 70 million. With this prospect of population growth, it becomes all the more imperative that there should be the most concentrated attack on the development of agriculture and the rural economy. It is from this aspect that we have to give thought to changes in policy and, what is not less important, to changes in the machinery and methods of practical implementation which are called for in the light of experience. The trends which have been cited suggest that the task of lifting the rural economy from the stagnation which characterised it in the past and of enabling it to function in a truly dynamic manner is much bigger and more difficult and will make larger demands on our resources and our institutions than might have been allowed for.

III

As different policies and programmes embodied in the Five Year Plans are executed, it is important to keep in mind the character of the rural problem as a whole and of the strategy which has been evolved for dealing with it. Prior to the First Plan, much study was devoted to the problem of uneconomic holdings and of rural indebtedness, and the need for agrarian reform was widely felt. In the Five Year Plan, an attempt was made to approach the rural economy in an integrated manner. From a more or less static rural structure, which laboured under feudal tenures, the aim was to build up a progressive and rapidly growing economy founded on peasant proprietorship and organised increasingly on cooperative lines. The main features of this approach may be briefly stated.

To secure the development of the rural economy, the entire machinery of district administration had to be strengthened and identified with the welfare of the people. This involved steps in three directions—firstly, provision of technical skills through a network of extension services; secondly, a coordinated approach to the village community and the problems of the peasant; and thirdly, building up institutions by means of which the people could largely undertake development through their own initiative and participation. These steps led to the emergence of community development and the National Extension Service and, in due course, to the establishment of democratic panchayat raj institutions at the district and block levels in addition to the *panchayats* in the villages.

The Plans have sought to raise the productivity of land, above all, through irrigation, supplemented to an increasing extent by programmes for soil conservation, extended use of fertilisers, development of local manorial resources and adoption of improved agricultural practices.

The Five Year Plans visualise the rural economy being reorganised and developed basically along cooperative lines. The precise content of this cooperative approach has taken shape more slowly than had been hoped for. While large parts of what is broadly accepted in principle still remains to be carried out in practice, the objective clearly is that cooperation should become, as speedily as possible, the principal basis of organisation in many branches of economic life, notably in agriculture, small industry, processing, marketing, distribution, supplies, rural electrification, housing and construction, and the provision of essential amenities for local communities.

The scheme of land reform has been regarded from the beginning as being fundamental to development. Its main features have now been embodied in the legislation which has been enacted in the states. The primary aims were to eliminate intermediary tenures, establish security and reasonable rents for tenants, limit the size of agricultural holdings, progressively provide the rights of ownership to tenant-cultivators and create conditions for the growth of a cooperative rural economy. In the earlier stages, it was also hoped that substantial areas of surplus lands would become available for distribution to the landless.

The fact that a sizeable proportion of the rural population consists of landless or virtually landless agricultural labourer has been a source of anxiety in each of the Plans. In practice, the scheme of land reform has provided mainly for tenant cultivators. Even in its conception, it did not seek to provide to every agricultural labourer a piece of land, however small it might be, for the simple reason that such a step would have rendered peasant farming untenable and would have led to the breakdown of the agricultural economy. Measures for raising agricultural productivity could bring only gradual and limited benefits to agricultural labourer. The main hope of bringing agricultural workers on to a level of equality in status and opportunity with others in the village community, therefore, lay in the development of industry and in the diversification of the rural economy. This is one of the major compulsion for rural industrialisation, for closer integration of the rural with the industrial economy, and for accelerated development both in regions which have favourbale condition for growth and in those with considerable rural unemployment and under-employment. It should be added that in the social field, through community development and other programmes, there has been already a certain impact on the life and conditions of rural labourer.

Thus, in terms of policy, the Five Year Plans provided the machinery for rural development and set out the economic and social goals to be achieved. They were weak in dealing with the problem of landless labour, but it was thought that with the growth of the economy as a whole, increase in the productivity of land, development of cooperation and changes in the agrarian structure, this problem might become more manageable. In reviewing economic progress in rural areas, particular attention should be given both to the strengthening of policy and objectives and to eliminating the gaps between policy and implementation which have tended to reduce the social and economic impact of planned development. Such a review should encompass changes in the rural environment which are under way, changes in the agrarian structure which need urgent thought, directions in which the process of building up a cooperative rural economy must be strengthened, measures required for speeding up technological change in agriculture and for raising the productivity of land, and policies designed to achieve intensive development in each area and closer integration between the rural and the industrial economy.

IV

In the building up of the rural economy, changes of an environmental and social character have a significance which is not fully described in statistical terms. There is an intimate and growing connection between economic, social and technical change. The emergence of an environment, in many rural areas favourable to economic progress is a vital fact. In bringing about such an environment many factors are at work. The general development of the economy and expansion of industry and trade and of transport and power and increasing urbanisation would in any case exert a powerful and continuing influence on the rural environment. The main role of community development thus far has been to serve as an agent of change close to the rural population and as a link between national and local planning. The process of community organisation has not yet gone far enough to be able to tap sufficiently the local manpower and other resources of each area. Social research and improvements in organisation and planning could help in reducing this shortcoming. The existence of institutions through which the people of each area can largely determine their own course of development should be a factor of increasing importance in the coming years and one capable of imparting richness and vitality to the process of growth at the grass-roots. Already, the development of an expanding network of communication has brought the town and the village nearer to one another. The Benchmark surveys undertaken by the Programme Evaluation Organisation have shown that in most cases there is increasing readiness to adopt improved agricultural practices. Thus, in the 18 block which were studied, between the first survey in 1953-54 and the second survey five years later, the percentage of household using improved seeds increased by 70, of those using improved implements by about 50, and of those making use of fertiliser by about 150. In many areas the limits to the acceptance of improved practices are now set less by resistance on the part of farmers and far more by the ability of the extension services to make the necessary supplies and technical guidance promptly and readily available.

In the nature of things, changes in the rural environment are by no means evenly spread. They are more marked in areas in which new resources have come to be developed, be they by way of irrigation, or rural electrifications or the growth of processing and other industries, or new communications. However, these changes are being stimulated in all areas by the development of education, health and other social services. Although the re-planning of village sites and improvement of rural housing are necessarily long term undertakings, it is interesting to observe that in several areas the proportion of brick built houses is increasing fairly rapidly. Thus, in the Programme Evaluation Organisation's studies in one area in Punjab, the proportion rose from about 3 to 23 percent, in another from 7 to 34 per cent. Increases, somewhat smaller in dimensions, are also reported from a number of other states and are borne out by several re-surveys of villages which have been undertaken in recent years. Changes in the rural environment, which are now proceeding under the influence of developments outside as well as within each local area have to be harnessed purposefully as a positive manpower resources, specially for creating new community assets, improving living conditions for the weaker sections of the village community and expanding agricultural production.

V

The scheme of land reform was evolved in detail over a period of several years after Independence. As its main features come to be accepted at the level of policy, legislation followed in the states, in the process responding to local pressures and permitting variations which were often better avoided. The piecemeal character of land reform legislation and its excessive reliance on the machinery of revenue administration, without adequate education of rural opinion and support by way of credit and supplies and programmes for resettlement on land, has unfortunately tended to diminish the impact of land reforms on rural welfare and development. The main consideration at the present time is the need for early completion of the process which has been initiated and intensification of development in areas in which land reform entails substantial changes in the distribution of right in land.

Land reform comprises four sets of measure—abolition of intermediary tenures, tenancy reform, ceilings on agricultural holdings and conferment of rights of ownership on tenant cultivators. Once accomplished, these measures are intended to pave the way for the growth of a cooperative rural economy. At independence, intermediary tenures accounted for about 45 per cent of the total area of the country. Except for the payment of compensation—a process spread over 20 to 40 years—this phase of land reform has been completed. Similarly, despite occasional problems arising from attempts of owners to resume land for 'personal cultivation', and with certain variations between states, the programme for tenancy reform has been also substantially undertaken. Together, abolition of intermediary rights and tenancy reform involve considerable de facto redistribution of land. On the other hand, proposals for imposing ceilings on agricultural holdings have only slowly found their way into state legislation, and a high proportion of the 'surplus' lands has passed out of hand through 'transfers', many of them being essentially evasive of land reform. In due course, large holdings will disappear in fact as in law but, for the time being, an important public purpose sought through land reform has not been served. Nevertheless, whatever the limitations of the legislation, it is essential that it should be implemented as early as possible.

Apart from ceiling, the next major step in land reform is the grant of rights of ownership to the bulk of tenant cultivators, especially to those occupying non-resumable lands. The direction of policy is that this task should be accomplished before the end of the Third Plan. Some states have undertaken the necessary legislation and are giving effect to it, elsewhere the objective has not been given so far the priority due to it.

In theory land reform was urged and broadly accepted as part of the Plans. In its actual implementation, instead of being developed as a movement rooted in mass opinion, land reform has been generally treated as an isolated programme. In every area in which land reform is carried out, there is need, firstly, for an intensive agricultural drive and provision of supplies, credit and other services; secondly, for greater emphasis on the organisation of cooperative activity; and thirdly, for supporting rural employment programmes. Without such measures, the positive gains from land reform cannot be fully realised.

For the agricultural economy to be placed on a sound basis, the problem of small and uneconomic holdings still remains to be faced. To an extent, in most states, extreme fragmentation of holdings is now avoided, but the limits are necessarily very low. Consolidation of individual holdings has been undertaken on a notable scale only in three or four states. It is doubtful if the total area benefitted will at any time be much in excess of 50 to 60 million acres, the largest scope for consolidation being in areas which are irrigated or are likely to come under irrigation. Cooperative farming, rural work programmes, land settlement schemes and intensive development of rural industry are essential not only in themselves but as a means for securing efficient cultivating holdings are completing the reorganisation of the agrarian structure. These elements have to be brought into the Five Year Plans on a much larger scale than at present.

Over the past few years a distinctive cooperative sector has begun to take shape. The principal advances have been in the field of credit, specially short-term credit, in processing, in particular, of sugar cane, and in the handloom industry. At the end of the Second Plan, there were about 200,000 primary agricultural credit societies with a total membership of about 17 million. Of these nearly 8,000 with a membership of over 3 million were large-sized societies. The total short and medium-term credit advanced through cooperatives amounted to more than ₹200 crores, as against about ₹50 crores five years earlier and ₹20 crores at the beginning of the First Plan. The number of cooperative sugar factories increased steadily from 3 with a total production of 30,000 metric tonnes in 1955-56 to 30 with a total production of 440,000 metric tonnes. The number of industrial cooperatives increased from about 8,000 in 1951 to about 15,000 in 1956 and, at the end of the Second Plan to about 30,000, of which about 8,000 were cooperatives of handloom weavers. For the rest, despite some advances, the role of cooperative institutions is still quite small, and the main task of building up a strong and diversified cooperative movement still remains to be accomplished.

The magnitude of the task ahead will be seen from some of the goals which have been indicated even in the Third Five Year Plan. These goals will not be attained at the present rate of progress. The volume of agricultural business conducted by cooperative marketing societies was estimated at the end of the Second Plan at about ₹200 crores—itself a mere fraction of the total trade in agricultural produce —which it was hoped at least to double during the Third Plan. In the field of cooperatives of the existing units engaged in the processing of agricultural produce as well as for the allocation to the cooperative sector of the greater part of the expansion in processing industries in consequence of the increase in the production of agricultural raw materials and in the consumption of the final product. In cooperative farming, the broad aim is that over a period of ten years or so a substantial proportion of agricultural land should be cultivated on a cooperative basis. However, in two directions some advance is now expected, namely, consumer cooperatives and, under the impulse of the rural works programme, in the organisation of labour and construction cooperatives. In the reconstruction of the rural economy and in the harnessing of the resources of rural areas, there are no larger or more far-reaching objectives than those connected with the building up of a strong and dynamic cooperative sector. These objectives have not yet been fully incorporated into concept and practice of community development at the level of the block and the village. Their attainment is a vital part of the country's agricultural effort and indeed of the entire scheme of planned development.

VI

To bring about a sharp increase in agricultural production and a rise in the productivity of land have been among the central aims of planning for more than a decade. These tasks have been approached from several directions. In considering the impact of planned development on the rural economy, it is difficult to separate policy from administration and the technical from the economic and social factors. The scope of the agricultural effort under the Five Year Plans may be summed up in three broad propositions. Firstly, the scale of investment in agriculture has increased from about ₹800 crore in the

First Plan to about ₹1,200 crore in the Third Plan. Secondly, at the end of the Second Plan, some 14,000 agricultural graduates, 5,000 veterinary graduates and about 40,000 village level workers trained in agricultural extension were engaged largely in the task of providing technical guidance and improving the level of agriculture in different parts of the country. Thirdly, steadily increasing resources by way of water, improved seed and fertilisers have been made available, so that the essential technical foundations for raising the productivity of land have been strengthened. Thus, over the first two Plans the net area irrigated increased from about 51 million acres to about 70 million acres, the consumption of nitrogenous fertilisers rose in terms of nitrogen from 55,000 to 230,000 tonnes, and about 4,000 Government farms were set up for multiplying improved varieties of seeds produced in the research stations.

These measures have led to perceptible results, but certain important deficiencies have also been revealed. Between 1950-51 and 1960-61, the total agricultural area increased by about 19 per cent, the area under food grains by about 15 per cent and that under crops other than foodgrains by about 39 per cent. Agricultural production as a whole rose by about 40 per cent, production of foodgrains by 36 per cent and of crops other than foodgrains by 49 per cent. Despite year to year fluctuations, the output of cotton increased nearly two-fold and of sugar cane by about 74 per cent. The average agricultural productivity rose by about 18 per cent, the increase in respect of rice, wheat, *jowar* and cotton being distinctly higher than, for instance, for maize, pulses and sugar cane. In some states progress in agricultural production has been fairly marked, although the crops whose output has increased vary widely in different areas. Table 21.1 brings out these broad facts in greater detail.

Increase in irrigation and changes in demand have led in many areas to be substitution of more remunerative for less remunerative crops. This is borne out by special surveys which have been undertaken and is also reflected in changes which have taken place in the crop patterns of different states. Thus, while there are important variations between states, area under crops such as *jowar*, *bajra* small millets, barley and seasamum have tended to diminish, while those under

Table 21.1

Index Numbers of Agricultural Production, Area and Yield Per Acre

(Agricultural year 1949-50=100)

Crop	Agricultural Production			Area			Yield per Acre		
	1950-51	1955-60	1960-61	1950-51	1955-56	1960-61	1950-51	1955-60	1960-61
Rice	87.9	114.2	136.2	100.9	103.1	109.4	87.1	110.8	124.5
Wheat	101.1	131.3	162.7	99.9	126.7	133.0	101.2	103.6	122.3
Jowar	89.8	96.7	134.6	100.2	111.6	110.8	89.6	86.6	121.5
Bajra	83.8	108.3	101.2	97.4	122.4	123.3	86.0	88.5	82.1
Maize	84.4	112.3	144.5	96.4	112.7	132.9	87.6	99.6	108.7
Gram	98.0	138.9	162.3	91.2	118.0	113.6	107.5	117.7	142.9
Total pulses	91.7	118.4	128.6	91.9	116.8	116.6	99.8	101.4	110.3
Total foodgrains	90.5	115.3	135.6	97.9	111.9	114.5	92.4	103.0	118.4
Sugar Cane	113.7	119.8	17.9	116.4	125.1	158.7	97.7	95.7	109.6
Oilseeds	98.5	108.6	131.2	106.5	119.3	133.4	92.5	91.0	98.3
Cotton	110.7	153.9	203.3	119.2	164.0	154.8	92.9	93.8	131.3
Jute	106.3	135.8	121.7	121.3	149.4	129.7	87.6	90.9	93.8
Tobacco	97.3	112.9	116.4	102.7	117.8	115.1	94.7	95.8	101.1
Total Non-Foodgrain	105.9	119.9	148.5	110.8	130.7	138.9	95.6	91.7	106.9
All crops	95.6	`116.8	139.9	99.9	115.0	118.5	95.7	101.6	113.1

Source: Ministry of Food and Agriculture, Directorate of Economics and Statistics.

crops like rice, wheat, gram, groundnut, sugar cane, cotton and jute have shown an increase. The overall position is illustrated in Table 21.2.

Table 21.2

Changes in Cropping Patterns

('000 acres)

Crop	*1953-54*	*1961-62*	*Increase (+) or Decrease (-) (%)*
Rice	77,318	83,669	+ 8.2
Wheat	26,394	33,240	+25.9
Bajra	30,145	27,027	- 10.3
Jowar	43,882	43,074	- 1.8
Barley	8,719	18,255	- 5.3
Small Millets	14,028	11,714	- 16.5
Gram	19,689	24,078	+ 22.3
Groundnut	10,496	15,848	+ 51.0
Sesamum	6,351	5,561	- 12.4
Sugar Cane	3,485	5,942	+ 70.5
Cotton	17,265	18,710	+ 8.4
Jute	1,228	2,259	+ 84.0

Source: Ministry of Food and Agricultural, Directorate of Economics and Statistics.

In a planned economy, in those areas in which crop prospects are reasonably secured through irrigation and assured rainfall, cropping patterns should be a matter of systematic and studied planning and should not be subject to any marked extent to short-term shifts in prices and demand.

We may next attempt to identify certain weaknesses shown by the agricultural trends of the past decade or so. This will suggest the directions in which policy, planning and administration need to be strengthened. Under conditions of planned development in India, three important tests should be met in planning for agricultural development. The first is that the agricultural economy should be developed and stimulated to the extent of providing for increase in the requirements of food and raw materials in accordance with the estimates of the Plan. Variations on account of seasonal factors could be progressively provided against through buffer stocks and other regulatory measures. Apart from such variations, in keeping with the crop plan, the relative prices and other economic incentives should be as nearly equal as possible between different crops. Secondly, there should be steady increase in production and yields in all parts of the country and, in particular, in areas with irrigation and assured rainfall. In the third place, through increase in yields, adoption of improved techniques and improvements in the size of the unit of cultivation, cost per unit of production should be kept down and even reduced. The high level of costs in Indian agriculture at present is both an index of backwardness and a drag on the development of the economy as a whole. On these three tests, both the techniques of agricultural planning which have been actually followed and the performance of agriculture have on the whole fallen short, the gap being much larger in some sectors than in others. Thus, for oilseeds which are important for domestic consumption as well as for exports, over a period of ten years the yield per acre scarcely increased. Though the output of cotton has increased considerably, both on account of improvement in yields and increase in area, the output is still significantly short of requirements. In the case of an export crop like tobacco, the yield per acre has not materially improved. In respect of foodgrains, the goal of self-sufficiency, which was set for the First Plan and again for the Second, has yet to be reached.

The Benchmark survey undertaken by the Programme Evaluation Organisation show that except in a few areas there has been relatively small increase in the double-cropped area. A number of re-surveys of village which have been undertaken , for instance, in Maharashtra, Punjab, Uttar Pradesh (UP), Mysore and elsewhere, indicate that even where irrigation has led to changes in cropping patterns, in the methods of cultivation and in the tools and implements employed by cultivators, there have been few radical improvements. This points to a persistent inadequacy both in the organisation and in the techniques and objectives of extension work.

Lags in the utilisation of irrigation present a more complex problem reflecting, on the one hand, deficiencies in agricultural policy and administration and, on the other, existing weaknesses in the organisation of community effort and in extension methods. The problem has been examined in considerable detail in a number of inter-state conferences, in the reports of the Irrigation Team of the Committee of Plan Projects and in a study undertaken by the Programme Evaluation Organisation. On the available data it appears that at the end of the Second Plan, as against a total potential of about 12 million acres from major and medium irrigation projects, the actual utilisation was of the order of 9 million acres. The gap between the potential and the actual utilisation appears to be still of this order. The same problem also exists in respect of minor irrigation works such as tanks and tube-wells. It is obvious that a many-sided approach has to be adopted to achieve the speedy utilisation of irrigation facilities. This must include more precise knowledge of the water resources of an area through surveys and investigations, co-ordinated action by different agencies, evolution of suitable crop patterns, supported by measures to facilitate their adoption on the part of cultivators, acceptance of the obligation to construct and maintain field channels, not only by the beneficiaries individually but by the village community as a whole, and policies relating to the distribution of water and irrigation cesses which are better calculated to promote speedy utilisation. Since a variety of factors are responsible for what is doubtless a critical weakness in the present agricultural situation, in addition to steps which are already being initiated, it would be desirable to take up some areas on a pilot basis with the deliberate object of evolving technique under field conditions for eliminating the time lag between the creation of the irrigation potential and its utilisation and for promoting a higher level of agricultural technology simultaneously with the introduction or expansion of irrigation facilities.

This brief review points to four main conclusions. Firstly, in the planning of agriculture a much more central role should be assigned to price policy and the organisation of marketing than has been done so far. This involves both closer analysis of cost and returns for different crops as well as continuous study of the conditions of cultivation and marketing actually prevailing in different parts of the country. Secondly, there are definite indications that the existing agricultural administrations in the states, the numbers and quality of extension personnel and the techniques of extension do not answer adequately the requirements of rapid development in agriculture. The community development organisation in the blocks has to be greatly strengthened in terms of technical, agricultural and extension personnel. The earlier expectation that the village level worker, as the common agent of the various development agencies functioning in the Block, would also meet adequately the requirements of agricultural extension needs to be reviewed. For this there are two reasons. With the development of *panchayati raj*, even if agriculture is always given the first place, the more general administrative and organisational functions will frequently tend to take precedence over work of a routine and technical nature. Secondly, the progress already achieved in agriculture and the new environment and outlook which has emerged in the villages in many parts of the country has created a demand for higher levels of skill and knowledge on the part of the agricultural extension service than village level workers can at best be expected to provide. There is also need for a much more intimate connection between agricultural research and its application in the field. An expert committee has recently listed outstanding agricultural practices in a number of areas and has attempted to identify the practices to which the high yields of crops like rice, wheat, sugar cane and cotton obtained in these areas may be attributed. The variations in yields and in the general level of agriculture even between different parts of the same state are extraordinary wide. In the effort to bring up levels of cultivation and agricultural practices to similar standards, it would be desirable to spare resources neither in men nor in money. The requisite agricultural personnel can be provided only over a period, however, recent developments such as the accelerated programmes for cotton and oilseeds, for soil conservation and in the intensive agricultural districts emphasise the urgency of a new outlook on agricultural administration and the development of extension services.

The third main conclusion which may be drawn from current experience concerns the place of community action in the scheme of agricultural development, specially at the level of the village. This is an indispensable condition for realising substantial increase in agricultural production, widespread adoption of superior techniques and practices and the organisation of cooperatives, specially for marketing and production. The most important aspect of such community action must be the full utilisation of the manpower resources of each village and each area. In a system of peasant farming, the labour of each household may be expected to be well utilised. However, in the Indian situation the existence of small and uneconomic holdings and of large bodies of agricultural labourers accompanied by rapid growth of population leave a considerable reserve of man-power which could be mobilised for productive work. With a larger and better organised labour input, it is possible to extend irrigation, undertake soil conservation, dry farming, land reclamation and afforestation, and adopt labour-intensive agricultural practices on a scale far exceeding that yet attempted in any part of the country. Hiring of labour by individual households cannot provide for continuity of work, and is adequate as a means for harnessing all the manpower available in the village. Such a task can only be undertaken by the village community as a part of a wider area plan which ensures the requisite resources, technical guidance and other facilities. It is therefore necessary to redefine the relationship between the individual landholder or cultivator and the community in terms of the present imperative of rapid economic development and the fullest utilisation of manpower.

In many areas it should be possible to create conditions in which the demand for labour could be greatly increased, not merely in busy seasons but throughout the year. As is well known, in present day Europe, a number of countries receive a regular influx of unskilled workers from other countries which have larger labour reserves and are absorbing these workers permanently within their own economies. This is because their own economic development is rapid and renders them continually short of labour. In the same manner, agricultural activities could be intensified in areas with irrigation and assured rainfall on a scale which could, in time, not only assimilate the local labour force but also create conditions in which there would be additional demand for labour from other areas and, in some cases, even from other states. Thus, a policy aimed at the fullest utilisation of rural manpower resources will also bring out the possibilities of planning for the redistribution of the rural population within individual states and, progressively between different states.

A marked intensification of agricultural operations, supported by extension activities, community action and the full use of manpower and by cooperative organisations, will call for a bolder and more forward-looking approach to investment in agriculture. The agricultural growth potential of many areas cannot be realised without a much higher level of investment that the existing plans and schemes of community development provide for. The consequence is that planning becomes rigid, the local manpower resources are not fully harnessed to productive effort and the growth of the agricultural economy is in fact retarded to an extend that may not be always realised. There should be a special provision in the scheme of annual allocations under the National and state Plans for additional resources beyond the customary norms for areas in which intensive agricultural operations are likely to yield significant additions to output. In these areas, to begin with, the scheme of land utilisation, cropping patterns, the inputs of capital, labour, fertilisers, etc., as well as the marketing and processing of agricultural produce should be planned for the village and the block as a whole, and responsibility for the enforcement of standards and for the fullest utilisation of manpower and other available resources should be placed on the community.

VII

In the numbers it engages and in the contribution it makes to the national product, agriculture will always be India's leading industry. In the more advanced countries, on account of improvements in the techniques of agriculture, production has continually increased even as numbers on the land have diminished at the same time, with agriculture still the main occupation in rural areas, the distance between the village and the town has been broken down. In fact, the rural and the urban areas, which retaining their distinct character, yet form part of a single integrated national economy. In these countries, it is the urban areas which constantly pull the rural areas in their own direction. The nature of the problem of poverty in India, the size of the rural mass, the limited number of centres of major industrial activity and the very process of planning for development both in economic and human terms, to which we are committed, lead to the conclusion that the pattern of relationships between the rural and the urban and the industrial economy which we seek to build up over the next 15 to 20 years should be carefully thought out at this stage in our development. In the nature of things, many decisions have to be taken which involve choice of location, choice of technology, selection of industries and the building up of economic and social overheads. These decisions have profound long term effects and set in train further chains of action and reaction which will later leave the room for manoueuver to those concerned with planning or legislation or the making of policy. Sooner than may be perceived, the balance and momentum of a complex economic structure largely begin to dictate the course.

In what manner, then, should we plan for agriculture and the rural areas so that, in time, the rural and the urban economies will merge into one another? As mentioned earlier, the gap between the output per worker in the village and in the towns is already appreciable and in the absence of a larger strategy of development, the forces leading to a further widening of this gap will prove too powerful to resist. However intensive the process of industrialisation and urbanisation, by its very nature, its influence will be felt most fully in small and limited areas and the impact further afield will be much smaller. In other words, one of the essential aims of economic planning in India must be to find a way to avoid a marked and continuing dichotomy in the pace and direction of development in urban and rural areas. An impelling reason for this is the existence of a large labour surplus in the village which no amount of agricultural development can by itself absorb to the extent of providing full time gaining employment. Moreover, beyond a point, the growth of agricultural output will be held back unless there are strong incentives toward increase in agricultural production associated with new economic opportunities and improvements in levels of living.

Under the existing conditions of agricultural organisation, the rural economy is not well placed for mobilising its manpower and other resources and gaining adequately from the processes of industrial and economic development which have been initiated. There is a viscous circle in as much as an under-developed region cannot secure or sustain the economic and social overheads needed for rapid growth and, without these, the balance of economic advantage continues to go against it. It may need all the skills and leadership it can muster and more. Yet, these it must lose to regions which are developing more rapidly. There is, thus, a close connection between securing rapid increase in agricultural production, creating in the rural areas expanding opportunities for non-agricultural employment and building up an efficient rural economy along cooperative lines. It is in the measure in which these three aims are pursued as part of the same basic concept of agricultural policy that it will be possible to achieve the necessary integration between the rural and the industrial economy and to raise significantly the levels of income and living for the bulk of the people in the rural areas.

Indian Journal of Agricultural Economics, Vol. 18, No.1, Jan-March 1963.

22

B.N. GANGULI

Agricultural Development and the Price Mechanism

THE topic I have chosen for my address is 'agricultural development and the price mechanism'. To avoid needless controversy, let me assume that, whatever the socio-economic system, the price mechanism plays a functional role in the allocation of resources and the achievement of output targets. This is not an unwarranted assumption at least in respect of the agricultural sector of the economy. The price mechanism cannot be planned away if we have in mind a sensible planned economy. I am assuming that the innumerable participants in a planned economy-state as well as private, collective as well as individual-must indicate their needs and the relative urgency of these needs not only through the statistical estimates of the Planning Commission but through the direct pressure of supply and demand. I am assuming that a plan is checked by market indicators and is also realised through the operation of the market mechanism, that the regulation of the market, if necessary, depends upon the tendencies revealed through the market, and that the blue-prints of the Planning Commission must demonstrate their worth in terms of commercial calculation. I hope you will agree that this is at least the current Indian context. I suspect that this is also the current context of agricultural development in many socialist countries, not to speak of free enterprise economies, in which the price mechanism reflects as well as generates the basic economic impulses.

Once you grant my assumption, I can hope to develop the thesis that the perspective of agricultural development over time shows a surprising degree of approximation in spite of the differences in the character of the economic systems. One may indeed build up a general conceptual frame-work as a tool of analysis which may have a much wider application than we usually suspect.

The general perspective of agriculture seems to be its comparatively low man-hour productivity. Technological progress has been greatly reducing the number of persons required to produce a given amount of agricultural output. In the United States of America (US) 2 per cent largest farmers obtain no less than 26 per cent of the total farm income. This means that the majority of farms have low productivity per man-hour. The farm hand is the lowest paid wage earner. The southern share-cropper as well as the tenant-farmer who supplies a share of equipment and livestock and shares the annual farm return, has a comparatively precarious position. The low-productivity marginal farmers are a drag on the economy, and there is a shift of the rural population to the urban areas in which it seeks shorter hours, better wages and better social life. And the dice is heavily loaded against the small or even medium family farmer who decides to stick it out.

In socialist countries economists have recognised that indices of net production per person employed in industry and agriculture show differences in labour productivity. For example, Strumilin has shown that in 1925-26 Soviet productivity in industry was three times higher than in agriculture. In Poland today net production per worker has been estimated between 2 and 2½ times higher in industry than in agriculture. One must take into account in this context not man-hour productivity in the technical sense but the net return per person employed in agriculture. In the case of small and even medium-sized farms the 'indivisibility' of labour as a factor of production, deriving from the family character of farms, makes for large fluctuations in the number of the farmer's annual working days, depending upon the size of the holding and the intensity of agriculture.

It is against the background of such reasoning that an economist in a socialist country very aptly tells us that: "migration from agriculture and the more rapid tempo of development of industry are general economic truths. Smaller or greater differences in the rates of development of agriculture and industry may take place within this frame-work. In certain situations the difference is so great that it leads to a disproportionate development of the whole economy." The law of unequal development of industry and agriculture is thus as valid and significant as the law of unequal development of capitalism. In *An Outline of Political Economy*, written in USSR more than 30 years ago, I read: "By industrialisation we must, therefore, understand such economic development as would be accompanied by a more rapid development of industry than of agriculture."[1] In 1925 the Planning authorities in USSR had over-estimated the harvest of the year and had framed an over-ambitious industrial plan. The plan had to be cut down. There was then an insufficient supply of manufactured goods. The USSR economy was operating on the basis of minimum dependence on foreign trade and loans. The difficulties connected with the 'commodity famine' were supposed to be temporary at that time. But Soviet economic system also experiences difficulties of a more stable character. These difficulties arise from the disproportion in the development of the productive forces in state industry and agriculture."[2] Economic developments in the subsequent decades have indeed confirmed the basic character of these difficulties in USSR as also in other developing socialist countries which have faced, from time to time, the problem of disproportionate development of the entire economy due to lack of balance between agricultural and industrial development.

Here we come across a contradiction in the process of economic development, which transcends the differences in the nature of the economic systems. Lack of balance between industry and agriculture in terms of productivity is the precondition of industrialisation, the pace of which is accelerated by such imbalance threatening to become wider. The economic history of Czechoslovakia, Sweden and UK in the 19th century shows that the maximum gap in the rate of development between industry and agriculture and this was achieved through the favourable terms of trade between industry and agriculture operating in the context of international division of labour. The familiar aspects of the dynamics of such a process of economic development have

1. Lapidus and Ostrovityanov, *Outline of Political Economy*, Martin Lawrence, 1929, p.483.
2. Ibid.: 521-22.

been migration of agrarian surplus population, industrialisation, capital formation and levelling up of productivity as between agriculture and industry. But inherent in this process is an economic paradox which is like the famous paradox of trade and factor price equalisation. There is trade because of inequality of factor prices, but the effect of trade is to bring about factor price equalisation. Trade, however, goes on, because factor prices can never be equal for various reasons. Similarly, economic development is conditioned by differential levels of productivity in agriculture and in industry. The effect of economic development is to level up productivity in both sectors, but the levelling up process is never complete. Indeed, it is incomplete levelling up which supplies a momentum to development. In fact an economy can make a virtue of the necessity of an unequal rate of growth of productivity. While the new dynamic industries are continually lowering prices, the stagnant industries, including agriculture, may be compelled to raise prices in certain conceivable circumstances. The balance of these forces may thus tend to produce a fairly stable general level of prices.

Countries caught in the paradoxical situation I have just explained, have reacted to it in different ways depending upon circumstances. In free enterprise industrial economies while labour productivity rises as rapidly as money wage rates (W/E remaining constant) or more rapidly (W/E falling) in the industrial sector, labour productivity lags behind money wage rates in agriculture, which means that efficiency wages (W/E) rise resulting in rising farm costs and potentially higher prices of farm products. Since money wage rates are relatively too low in agriculture the state subsidises agriculture by means of price-support and non-price support programmes to make it possible for efficiency-wages to rise in agriculture without their being reflected in higher farm prices. But the perspective of agriculture development is different in different countries in this category. In US the highly favourable man/land ratio associated with the amazing increase in productivity per acre has created the problem of burden—some agricultural surpluses for which price and income support hardly offers a solution, and all devices designed to bring about a reduction in total output, have not yielded the expected result. In western

Europe where the man-land ratio is unfavourbale subsidisation not only buttresses up agriculture, but for security reasons also aims at both extensive and intensive agricultural development. In both cases, of course, there is the common restrictive attitude to agricultural imports and a tendency to subsidisation of exports.

Socialist countries of Europe and also Sweden have recently shown strikingly different trends. Sweden has maintained a constant level of agricultural output for 10 years, and, while encouraging mass migration from agriculture to industry has chosen to meet the rising demand for agricultural commodities by imports. A similar process is at work in Czechoslovakia. In both countries the marginal physical productivity of labour, expressed in world prices, is higher in industry than the marginal physical productivity in agriculture, expressed in world prices, of labour migrating from the rural areas to the industrial centres. Therefore, it pays to export industrial products in exchange for food imports than to rely on domestic production. In comparatively underdeveloped socialist economies the possibility of converting farmers into industrial workers has been limited by the scarcity of investment goods, while there is great scope for raising the productivity of labour in agriculture. These countries now increasingly rely on a greater degree of approximation between the rate of development of industry and that of agriculture through a proper readjustment of their plans of economic development. What has happened in Sweden or Czechoslovakia or Poland cannot be without its relevance for the objective situation in the Soviet Union. From the early days of economic planning, Soviet planners were aware of the problem of disproportionate development of industry and agriculture. The writers of the first official text-book on Political Economy in the Soviet Union, whom I have already quoted, stressed: "the necessity of starting a lasting process of remoulding the peasant on the basis of his own private interests so as gradually to pull him through the medium of the market, through the medium of buying and selling cooperation into collective, social forms of labour."[3] "Taxation and price policy", they said, "must be such as to give the

3. Ibid.: 516.

peasants a chance to develop and crate a market for expanding socialist industry."[4] There cannot be any clearer enunciation of the necessity of operating the price mechanism as an instrument of agricultural development. Since there was misplaced optimism about the quick response of the peasant who knew his strength and also about the levelling up of labour productivity in agriculture as in industry, and since the Soviet economy was a closed economy, the course of events was, however, different. The Soviet text-book writers had said that "Small agriculture cannot be driven with a stick into the socialist paradise."[5] But the big stick was used with devastating effect and socialist paradise was not in sight. Indeed, agriculture has proved to be the Achilles Heel of Soviet economic planning even after several decades of remarkable economic progress. If my analysis of the end-results on the assumption of the operation of the price mechanism is correct we should not be surprised if the Soviet Union were to seek a better balanced economy through agricultural imports flowing in the process of Soviet Union's participation in an international division of labour confined not merely to the Socialist Bloc of countries but a much wider world which has agricultural surpluses to sell for the products of Soviet industry and mining.

I have presented so far a perspective of agricultural development as a long-range process, which clearly shows certain basically common characteristics in spite of the differences in the economic structure and market forms. In the short run, there is the familiar problem of fluctuations of output and prices over time—the problem of short cycles. The conceptual framework within which we usually understand this problem is so familiar that it would be unpardonable to talk about it in a gathering of specialists. I am, however, tempted to place before them few lines of thought which have considerable theoretical as well as practical importance from the point of view of the perspective I am trying to present.

As regards the utility of the cob-web theorem as a tool of analysis, we must abandon the implicit assumption of discontinuous and

4. Ibid.: 515.

5. Ibid.: 516.

alternating reactions on the supply side. In real life there is a more rounded movement rather than sharp fluctuations so elegantly depicted on the curves. One must be sure whether the curves are in fact rectilinear. If they are not, if the supply curve is first comparatively flat and steeper than the demand curve, later on, the final outcome will be a kind of cyclical movement with constant amplitude, preceded successively by decreasing and increasing fluctuations. Are constantly increasing fluctuations possible? If the amplitude of output and price fluctuations goes on increasing, after a point there would be an 'explosion' of the system, but before this point is reached the curves would have changed through a transformation of the structure of the Society and of the market. I draw your attention to these theoretical refinements, because they provide a rational basis for avoiding the extremes of price policy in relation to the operation of the price mechanism. There is no doubt that continually increasing fluctuations of price and quantity in agriculture have to be avoided even before a point is reached where the 'explosion' of the system is prevented by social revolution. At the same time panicky and unwise intervention in the determination of price and quantity would clearly be neither necessary nor desirable. The 'cob-web effect' is most evident in the case of agricultural production with a long gestation period where extremely violent changes in output are inevitable if all producers, when starting new production, expect to realise the prices of the moment. The answer to this may be rationalisation of expectations by various means rather than panicky price and output controls long after the event. Nevertheless, it would be clearly necessary to keep constantly in view the amplitude of the cyclical movement in both plantation agriculture and arable farming and to ensure that the agricultural economy in the course of its development is moving steadily towards the narrowing of the range of fluctuations of output and prices and ultimately towards a cyclical movement with constant amplitude. The widening amplitude of the fluctuations of output and prices is, of course, a danger signal which a country can afford to ignore at its peril.

The short-range problem of fluctuations of output and prices presents itself to a particular country according to the clarity of its own perspective of agricultural development. The nature of the measures it adopts is also conditioned by its own long-range problem. An affluent society with a burdensome surplus in agriculture and looking for means of restricting acreage and output thinks of farm prices in the context of national income distribution. Other industrial countries with highly protected agriculture adapted to the requirements of national self-sufficiency take complicated measures to stimulate production on the basis of prices which are high enough above world prices to ensure high farm incomes. Taking wheat as an example, prices have been 25 to 100 per cent above the world price in a number of European countries. On the other hand, in relatively less industrial countries stabilisation measures have resulted in farmers receiving prices that are actually lower than the world price. Policies thus vary according to the levels of national incomes, according as agriculture is regarded as a peripheral or a strategic sector of the economy, and according as the planned target is an increase or decrease or stability of the total output of agriculture.

Let me now try to apply the analysis I have developed so far to the concrete Indian situation and to assess broadly the trend of agricultural development with special reference to the operation of the price mechanism. We can put India in the general perspective if we reason in terms of certain statistical indicators of the dynamics of agricultural development.

Per capita national income reflects the average labour productivity of the economy as a whole compounded of the average labour productivity in 'agriculture' and in 'non-agriculture'. The lower the proportion of persons engaged in agriculture the higher the per capita national income. Taking countries with a sizeable agricultural sector, in 1949, for example, US per capita income of $1,455 was associated with 16 per cent of persons (engaged in agriculture) at one end of the scale and the Indian per capita income of $57 associated with 65 per cent of persons (engaged in agriculture) at the other end of the scale. In between there were highly progressive agricultural countries with not a high degree of industrial development, like New

Zealand and Denmark, with dependence on agriculture of 18 and 27 per cent of the population respectively and per capita incomes of $856 and $689 respectively.

The gap in productivity between 'agriculture' and 'non-agriculture' is broadly indicated by the disparity between the percentage of persons engaged in agriculture and the percentage of gross national income shared by agriculture. For UK in 1949 the two percentages were equal (6%), indicating a broad equilibrium of productivity and income at the same level in 'agriculture' as well as 'non-agriculture'. The other striking case was that of New Zealand in which 18 per cent of persons (engaged in agriculture) shared a much higher proportion of the gross national income, *viz.*, 25 per cent. Otherwise, the income share was invariably less than the percentage of population engaged in agriculture: 10 per cent and 16 per cent for US; 22 per cent and 27 per cent for Denmark; 30 per cent and 40 per cent for Italy. In India in 1951 69.74 per cent of persons were engaged in agriculture proper; where the share of agriculture, forestry, fishery and animal husbandry was 48.8 per cent (estimate for agriculture alone is not available). This is a big enough gap which indicates: (i) the relatively low technological as well as social productivity of labour engaged in Indian agriculture, and (ii) the disparity in the inter sectoral income distribution.

Has agricultural development in India during the last decade or so made an impact in terms of narrowing this gap? The overall increase in agricultural production during the 11 year period, 1951-52 to 1961-62, has been 43.40 per cent, with an annual rate of increase, at the compound rate, of 3.34 per cent. The annual rate, however, declined from 3.68 per cent during the period of the First Plan to 2.38 per cent during the period of the Second Plan. The annual rate of increase of industrial production (judged by the index of industrial production) at the compound rate has been 6.22 per cent, i.e., almost double, with an overall increase of 106.16 per cent. The sharply differential rate of growth in the two sectors is quite evident. But this has made only a very small dent on the percentage of persons engaged in agriculture. This percentage has declined from 69.74 in

1951 to only 69.53 in 1961. An appreciable proportion of migrants must have been absorbed in 'non-agriculture'

Other than industry, accounted for by large investments in economic and social overheads during the last decade. Net value added by labour in this sector may well be as high or as low as that of unskilled labour in industry. The real social cost of the shift of people from agriculture to non-agriculture has been high. The increase in income and consumption levels as a result of the shift is greater than the rise in productivity so that capital formation is less than the increase of productivity. We have to add to this the cost of housing and other facilities of even a low level of community life.

In theory, reduction of concealed unemployment in agriculture brings about a rise in the levels of rural investment and consumption. We have had, however, evidence to show that the means of consumption released by those who were disguisedly unemployed before are absorbed by those who remain in agriculture and that this tends to reduce the volume of marketed surplus and to necessitate food imports, especially in bad seasons, because the volume of agricultural output has not increased substantially over time. Here we come across the hard core of the problem of agricultural development, *viz.*, how to raise the social productivity of labour engaged in agriculture. If the farmer is not sufficiently interested in investment the kind of situation I have just outlined will persist. Creating means of production in the rural areas by way of social and economic overheads is certainly a desirable policy, but as an offsetting factor the problem of lower social productivity of labour in agriculture is aggravated by unfavourable agrarian relations and wide marketing margins both of which weaken the incentive to save and invest and to produce more.

In macro-economic terms it is not so much the relative social productivity of labour (net national product per worker) in agriculture compared with industry, which is the problem in India as the absolute rate of increase of productivity over time. Net output per worker in agriculture and allied industries at 1948-49 prices has increased in India by only 8.8 per cent in 1959-60, compared with 1950-51. Taking a somewhat shorter period, 1953 to 1960, Taiwan (China) has shown, on a similar basis of calculation, an increase in agricultural labour productivity of as much as 20.8 per cent.

The process of economic development generally involves a decrease in the relative importance of agriculture. On the basis of the data for 34 countries, Egbert de Vries came to the conclusion that, for every 10 per cent increase in per capita real income, the proportion of national income arising from agriculture decreased by 1.5 per cent. In the case of India the per capita real income (per capita income at 1948-49 prices) has increased in 1961-62, compared with 1951-52, by 17.2 per cent. The share of agriculture, fishery, forestry and animal husbandry in the net national product (national income) has decreased during the same period by 3.8 per cent. This gives a rate of 2 per cent decrease, against every 10 per cent increase in per capita real income i.e., only 0.5 per cent above the average percentage found by de Vries. We may derive satisfaction from being above the average, but we must not forget that Taiwan (China) has shown a figure of 2.7 per cent above the average during the period 1953-1960.

Economic development is also normally associated with a shift of acreage from food crops to non-food crops. The cropped area under foodgrains has declined from 77.4 per cent in 1949-50 to 76.1 per cent in 1961-62, by only 1.3 per cent. The annual rate of increase of output of non-food crops since 1949-50 has been 4.19 per cent, while the rate has been 3.60 per cent in the case of food crops. In the category of food crops, the crops the output of which has shown a higher than the average rate of increase are rice (3.75%) and wheat (6.01%). Higher than average percentage increase of output has been recorded for groundnut (4.59%), sugar cane (5.61%) and cotton (6.56%) in the category of non-food crops. A decade is, of course, not a long enough period for trends to be established. There is, moreover, to erratic influence of seasonal fluctuations. Nevertheless it is necessary to examine the economic factors steadily at work many of which are traceable to human arrangements, which I propose to pass rapidly in review.

Before I turn to this let me carry the analysis a little farther. Output is a function of two variables—area and productivity. In the case of foodgrains there has been 1.58 per cent annual increase of area and 1.81 per cent annual increase of productivity during the period 1949-50 to 1961-62. For non-food crops the corresponding figures are 1.84 per cent and 1.68 per cent respectively. The differences are not appreciable, although the increase in both area and productivity and higher in the case of non-food crops. It is, however, significant that the higher than average annual rate of increase of output is explained by the higher than average annual rate of increase of area in the case of wheat, groundnut, sugar cane and cotton. It is only in the case of rice that the annual rate of increase of area (1.05%) is below the average for foodgrains (1.58%), but there has been an impressive annual increase of productivity of 2.50 per cent, as against the average of 1.81 for foodgrains. The increase in productivity has been below average in the case of wheat amongst food crops, and particularly groundnut and sugar cane amongst the non-food crops. It is only in respect of cotton that we find a high annual rate of increase of area (4.03%) and a higher than average increase of productivity (1.71%). It is necessary for specialists to turn their attention to these crops which have shown evidence of dynamism during the last decade. Is the increase in area due to the favourable price factor or the extension of facilities of irrigation? Is it marginal land, as understood by economists, in the context of population pressure? Is the annual increase in productivity in the case of cereals, in general, and rice (2.50%), wheat (1.64%) and *jowar* (1.47%), in particular, due to larger and better input? If so, what has been the incentive?

In terms of the title of my theme I would, however, like to briefly address myself to one aspect of the problem-the operation of the price mechanism, with particular reference to the agricultural price policy pursued by the state in recent years, which, as I shall try to show, follows a pattern conforming to the general conceptual framework I developed earlier. I shall confine my remarks to mainly these five crops which would excellently serve as a peg to hang my analysis on.

The responsiveness of farm output to prices is the very core of the dynamics of agricultural development and has rightly claimed the attention of competent analysts in recent years. Too often the official analysis runs, I am afraid, on unconvincing lines. Over the last decade the coefficients of variation of harvest prices have been as follows:- Sugar cane:8.1; Wheat: 10.9; Cotton: 11.6; Rice: 12.5; Jute: 14.3; *Gur*: 17.1; *Jowar*: 17.8; Groundnut: 20.3. In the case of sugar cane, wheat, cotton and rice the coefficient of variation has been low. There is minimum price regulation (on the basis of a high minimum price which has become traditional) in the case of sugar cane. The prices of rice and wheat have been kept down by a number of regulatory measures, and maximum and minimum prices are fixed for cotton. In the case of jute, *gur*, *jowar* and groundnut prices are more or less unregulated or there is less continuous regulation. Comparative stability of harvest prices of sugar cane, rice and wheat are supposed to be due to regulatory measures. You will agree that there are other complex factors in the situation. One realises their existence when one goes a little more deeply into the problem. How does one explain the dynamism of supply response in the case of these crops? It is not entirely convincing to say that output has increased because of relative price stability, although prices have been kept down in the case of rice and wheat by regulatory measures, and because the farmer values stability of price even at a comparatively low level of price. Groundnut prices have been more or less uncontrolled, and the coefficient of variation is the highest in the range, *viz*., 20.3; yet the annual rate of increase of output has been 4.59 per cent on the average, as against 4.19 per cent in the case of non-food crops, in general. The annual rate of increase of output of cotton, a controlled commodity, the price of which has been kept down by regulatory measures, is as high as 6.56 per cent. Dr Raj Krishna of the Institute of Economic Growth has made an excellent econometric study of farm response for pre-partition Punjab, which throws interesting light on these enigmas.[6] His study is a model of research which, in my opinion, will yield valuable results for other regions of India, by way of inter-regional comparisons of responsiveness of supply. He finds that price alone was the important factor in the determination of the acreage of maize and sugar cane. In the case of cotton price was more

6. Raj Krishna, "Farm Supply Response in India-Pakistan: A Case Study of the Punjab Region", *The Economic Journal*, September, 1963.

important than yield. Irrigation capacity was more important than price for cotton, *bajra* and irrigated wheat. Yield was more important than price in the case of rice. For *jowar* Dr Raj Krishna found a possible negative response to price movements. I wonder whether irrigation capacity rather than the price factor explains the recent increases in both the area and yield of cotton and wheat in India as a whole. Similarly, his finding that yield was more important than price in the case of rice is very significant in the light of the fact that in the case of rice during the last decade we notice the highest annual increase of productivity, *viz.*, 2.50 per cent. The importance of the price factor in the determination of the sugar cane acreage has been too familiar a phenomenon to leave any scope for speculation.

Farm response would seem to depend on the constellation of price and non-price factors which act and re-act on one another through an intricate chain of relations. The diagram given in the next page is designed to present a synoptic view of the operation of these factors.

Beginning with the point A on the circle, output of foodgrains is linked in a chain of mutual causation with the output of non-food crops, prices of input and consumer goods purchased by farmers, prices of non-food crops and food crops and the level of self-consumption. Output of non-food crops (B) is lined with the other factors mentioned except that self-consumption is either non-existent or is insignificant in the case of commercial crops.

The segment of the diagram including the chain C-D-E-F is of considerable importance in the Indian agricultural economy. The factors involved here are: (i) the parity in the sense of the ratio between the prices received and the prices said, and (ii) the inter-crop price-parity. It is quite evident that any disequilibrium in these parities has adverse reactions on agricultural development, at any rate in the long run. Since this question is linked with agricultural price policy and non-price regulatory measures it will be discussed a little later. It is clear from the diagram that prices of input and of consumer goods purchased by farmers (*gur*, for example) also depend on food prices and prices of non-food crops. This segment of the chain of relations becomes significant in an inflationary situation.

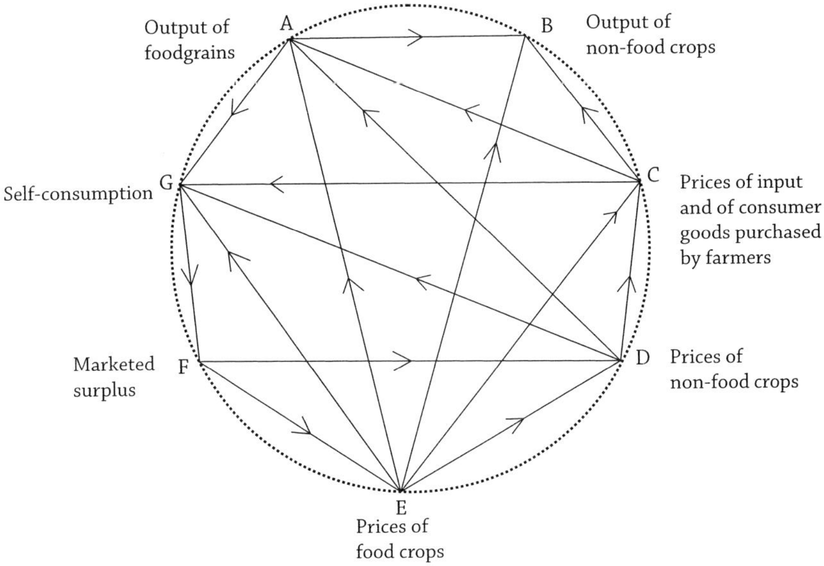

Self-consumption, marketed surplus, and de-stocking and re-stocking operations of the primary producers have assumed quite some importance in recent years. How they influence prices is easy to understand, but the reaction of prices and particularly any disturbance in inter-crop price-parity, on farm consumption, stocks and marketed surplus is an interesting phenomenon revealed by current research. The marketed surplus is 31 per cent and 37 per cent of production in the case of superior foodgrains, like rice and wheat respectively, for which demand has been increasing steadily, and varies between 19 and 26 per cent in the case of inferior grains. For non-food crops self-consumption is naturally an insignificant factor. A dynamic element in the situation is that even small holdings below 10 acres contribute appreciably to the marketed surplus not only in the case of a heavy-yielding crop like rice but even in the case of wheat. There is clear evidence to show that the price factor does influence the size of the surplus. Small cultivators have a deficit of foodgrains. Their distress sales result in their contribution to the market surplus, but that is more than offset by their purchases of foodgrains later in the season provided the prices of commercial crops they sell are high enough. But to the extent that small farmers produce more food, although their self-consumption increases, since their demand on supplies flowing into the market perhaps later in the year will be less, there would be a net increase in marketed surplus. Of course, in the case of medium and large farmers the marginal rate of marketable surplus would be high as well as their capacity to hoard and not hoard according to the market situation. In this case the higher prices of non-food crops which they may get will increase not their self-consumption but their capacity to hoard supplies for a rising market. These trends should be studied, if I may suggest, in the context of seasonal price fluctuations which are such a great headache for harassed food ministers. The increase in the area under more paying non-food crops is not an unmixed blessing from the point of the flow of the marketed surplus of foodgrains. The planners here are really on the horns of a dilemma.

This brings me to the problem of inter-crop price-parity in respect of which there is a good deal of ambivalent thinking and policy making in the name of planning. For the five crops which, as I have said, have shown some evidence of the farmer's dynamism and enterprise the inter-crop price-parity is apt to be disturbed, because while, on the one hand, the prices of wheat and rice are kept down, the price of cotton is subject to maximum and minimum limits and the price of sugar cane is kept comparatively high; on the other hand, the prices of substitutes like jute, *gur*, *jowar* and groundnuts are more or less unregulated. This creates a distortion in the price-mechanism. Ad hoc price-fixing by government is a de-stabilising factor. In 1958-59 relative cheapness of rice in the open market and the high black market prices resulted in withholding of supplies. Similarly, relative cheapness of wheat in comparison with gram, peas, oilseeds and other pulses in UP, Madhya Pradesh and Punjab induced farmers to sell more of these crops and to hoard wheat. Similarly the disparities between the price of jute and the price of rice in West Bengal, between the price of sugar cane and the price of wheat or rice in UP and elsewhere, between the price of cotton and those of its substitutes have created imbalances for which government's unwise interference with the price mechanism is mainly responsible. The assumption underlying the 'cob-web' theorem that farmers are influenced by their immediate experience rather than knowledge of the basic trend does hold good when one studies the de-stocking and re-stocking movements in the countryside. Their bad experience with government's regulative measures has been an important disincentive.

Another disincentive, particularly from a long-range perspective, is the unfavourable level of parity prices or the terms of trade of agriculture as a whole. I shall content myself with offering only two comments on this problem. As my earlier analysis has shown, this is a kind of evil paradox inherent in the process of economic development. The answer to it is to subsidise agriculture through transfer payments of various kinds. It is, at bottom, a problem of better distribution of national income not only through direct subsidies, but also through increasing investment in economic and social over-heads in the agricultural sector. Maintaining parity prices by government price-fixing in order to obtain a better distribution of national income between the agricultural and the non-agricultural sectors, has serious limitations because of the contradiction between the objective of influencing national income formation and that of the optimum use of production factors in agriculture. In practice, agricultural price policy has different aims which show contradictions. To resolve these contradictions the state has recourse to other methods of influencing agricultural development—adjusting the taxation system, providing cheap credit and other kinds of aids to production as well as developing the 'infra-structure'. What I have said is generally true of states with different social systems.

Let me carry my analysis a little farther to deal briefly, in conclusion, with the vexed question of agricultural price policy. Many of us may be inclined to question whether in India the state has an agricultural price policy and to think that whatever has happened by way of agricultural development has happened not because of any price policy but in spite of it. As I have tried to show, there are a number of non-price factors in the situation, which have influenced the direction of our agricultural development. In terms of the simple diagram I have used, it is possible to imagine that the state may exercise controls at all points-C to G-and even A and B directly. This would be an instance of an 'overdetermined' system. The very nature of Indian planning rules out an 'overdetermined' system of this kind. What has been attempted in India is fixation of maximum prices of certain commodities, combined with regulation of the movement of supplies. This does not make for an 'overdetermined' system. Instead of two variables: (i) actual price, and (ii) actual quantity, there are in our system two additional variables that are determined, (1) the excess quantity sellers have to supply, and (2) the deficit in the quantity buyers want to buy, at the price and quantity fixed. These hypothetical quantities represent 'tensions' which also become variables in the system. These tensions are periodically released through rules being set aside or altered or reduced in number, e.g., by political compromise between groups and sectional interests. This also happens even in a totalitarian, 'overdetermined' system.

Indeed, in India as well as in some of the socialist countries which still maintain family farms, price regulation in the agricultural sector shows a high degree of ambivalence and flexibility. This is explained not necessarily by slipshod thinking, but by the peculiar complexity of price formation in agriculture. Prices may be fixed before farmers take their production decisions. They are supposed to be guaranteed, but there may be no guarantee as regards the amount to be bought by the state. Purchase prices later in the season may, in fact, be elastic. This kind of wariness is due to a number of reasons. Depending on the operation of imponderable psychological factors, price-fixing may operate to stabilise the market as well as to intensify speculation. As our own experience has shown, fixing a high price may restrict the flow of supplies. If the objects are intensification of agriculture and a high rate of capital formation, particularly high prices may stimulate self-consumption and frittering away of resources or otherwise dispersion of purchasing power, which, even if it leads to investment will not materialise in large investments needed for agricultural development. High prices may only benefit speculators and not the primary products, specially the small farmers. Fixing all prices faces almost insurmountable difficulties. Supply elasticities for individual crops are not known. Planning authorities, in fact, cannot weigh properly the factors bearing on the direction of agricultural production, such as changes in crop rotation and in the intensity of agriculture and the farmer's estimation of the long-run price advantage in respect of particular crops. The problem of agricultural price policy, as it poses itself to planning authorities, is not that prices fixed by the state should deviate from equilibrium prices, but to create conditions for market equilibrium in the case of selected crops. Buffer stocks are supposed to be a very handy instrument for this purpose. But buffer stock operations have been bedevilled in India by the influence if demand variations getting dovetailed into the effect of harvest variations which buffer stocks are designed to counteract. The problem becomes serious enough when buffer stocks are largely dependent on costly imports of supplies of 'strategic' agricultural commodities like wheat and rice.

If there is an agricultural price policy in India, it follows not an unfamiliar pattern. Highly industrialised countries subsidise farm income by means of 'deficiency payment' offers a weaker incentive to increase sales than if the subsidy were paid on the price of the product sold. Subsidies have also taken the form of remission of Central and state taxes, or subsidisation of the prices of fertilisers and better varieties of seeds or provision of credit at special concession rates, etc. The object is two-fold: to lower cost and to encourage better agricultural practices.

Poor countries cannot afford deficiency payments, and have to rely on public investment in the overheads of agricultural development as well as on subsidisation of better farming practices, as India has done. A Polish economist believes that a proper agricultural policy should be to rely less on price-fixing and more on a combination of state investment and farm investment, reduction of costs of production and various kinds of aids to production through subsidisation of the prices of means of production.

The striking fact to which I wish to draw your attention is that India, like a number of less industrialised countries, has been, to a certain degree, following the policy of agricultural protection in the reverse. In Australia agricultural marketing arrangements have been such as to hold down the price of wheat to the domestic consumer below the world price of wheat and to pay the Australian wheat producer a price which is an average of the world price and the domestic price, or which is less than the world price. In Argentina the incidence of taxes on exports of wheat and meat, the proceeds of which financed industrialisation, was thrown back on the domestic farmer and the consumer. In West Africa the marketing arrangements have resulted in the domestic producer of cocoa getting less than the world market price. In Burma the difference between the price of rice paid by the state to the primary producer and the high export price has been the main source of state finance. In the case of India, to quote the Haberler *Report on Trends in International Trade*, "the price of wheat and so the reward offered to domestic wheat producers has been controlled at so low a price that even the imports of US wheat provided under special surplus-disposal arrangements have had to be subsidised in the Indian market to bring the price down to the low level earned by the domestic producers."[7] Thus while industrial countries have been pursuing generally protectionist policies to stimulate agricultural development, "agricultural anti-protection devices", as the Haberler Reports puts it, "are to be found mainly in some of the less industrialised countries where they tend to discourage agricultural production."[8] Is this not a paradoxical policy for a country which aims at agricultural development and self-sufficiency in foodgrains? So I conclude, as I began, with a paradox.

---

7. *Report on Trends in International Trade*, GATT, Geneva, October,1958.

8. Ibid.

Indian Journal of Agricultural Economics, Vol. 19, No. 1, Jan-March 1964.

23

ASOKA MEHTA

Towards an Agricultural Revolution

A stagnant agriculture is the main feature of a traditional society. In such a society one finds stereotyped agricultural practices, inadequate use of water and fertilisers, widespread land fatigue and soil erosion, weak and unprofitable strains of seeds and antiquated agricultural tools and implements. Modernisation of the economy and accelerated economic growth are possible only if this stagnation is broken for good. The first 25-year cycle of Indian planning covering the five five-year plan periods from 1950-51 to 1975-76 is based on this premise, although there would appear to be many doubters and dissenters today of this fundamental policy.

To those who doubt and dissent, my only plea will be to disengage their vision from one or two specific five-year plans where emphasis may have been placed on this or that particular aspect of the economy and to look at this pulsating cycle of growth in the round. If they do so they will find that the true implication of this 25-year cycle lies in the heroic attempt of the Indian people to modernise themselves, not in industry or in agriculture or in transport and communications alone, but in every sphere of life, economic, social, psychological and cultural. It is an attempt to bring India up-to-date, to make her participate in the stream of modern life with all its existing amenities and promises for future rewards. In the economic sphere proper this means simultaneous, large-scale, integrated industrialisation and the development of agriculture and the rural community. It is not agriculture at the cost of industry or industry at the cost of agriculture which is sought to be developed. It is the development of both in a harmonious, rising crescendo of activity.

I may be permitted to refer at this stage to a pet hobby-horse of some economists. It is pointed out by scholars that the industrial take-off in England and Western Europe between 1750 and 1850 was really made possible by the agricultural surpluses of the countries concerned preceding this period. It is also pointed out sometimes that the economic development of the USSR turned on the creation of agricultural surpluses, whatever might be the means adopted. It is further pointed out that the spectacular industrial development of Japan in this century has been largely due to the agricultural surpluses that were generated in her economy during the last two decades of the 19th and the first two decades of the 20th century. Nearer our own times, Communist China's or even Nationalist China's experience is drawn upon to make the point that agricultural surpluses are the true foundation of industrial development. Arguing from these historical examples, many economists have stated that India's emphasis on heavy industry has been too premature, that we should have first made agricultural surpluses secure and that consequently India is caught today in a crisis of ambition.

I do not deny the importance of agricultural surpluses in industrial growth. It is a commonplace of economics that people who are put in industry, particularly heavy industry which has a long period of gestation, have to be fed, clothed and housed before the fruits of their labour begin to come forth. For this purpose an agricultural surplus within the country is certainly handy. But what if the country has a vast population pressing on its soil, rising at a high rate, and trapped in the vice of outmoded agricultural practices? What if the country chooses the democratic path of action of providing political freedom and economic liberty to its humblest citizen, and what if the humblest citizen, in this case the poorest farmer, is moved by a thousand hunger for a fuller life, for himself and his family? If we admit these initial conditions into our system of analysis, then the facile generalisation of mopping up agricultural surplus through some form of authoritarian or autocratic fiat for investment in industry is no more acceptable. Then we have to be on the look-out for a system which combines three factors, the creation of agricultural surpluses wherever possible, but also of industrial surpluses wherever conceivable in order to modernise agriculture, and finally, a continuum of democratic economic institutions which make for this inter flow between agriculture and industry. Such a system, even in its bare conception, is a much more intricate, delicate, and I believe, civilised, mechanism than the one which existed in England and Western Europe in the 18th and 19th centuries or anywhere else nearer our own times. As far as I am able to see, India's economic planning is dedicated to the task of devising this mechanism.

Few people pause to think that if the performance of agriculture has been limited in the Second Plan and is going to believe our original expectations in the Third, it is not for lack of emphasis on agriculture. It is to a considerable extent due to the failure of industry, power and irrigation to supply the growing needs of agriculture. In concrete terms, it is the inadequacy of fertilisers, pesticides, farm equipment and other material inputs produced by industry for use in agriculture, insufficient and delayed utilisation of water resources and our failure to take the fullest advantage of the increasing supply of power for agricultural uses which are really at fault. This is not to call away attention from some of the lapses which have occurred in agriculture itself, for instance, the slow pace of implementing the land reforms, inadequate deployment of extension services and the lack of a proper price policy. This is meant only to restore the sense of perspective which should govern analysis in this difficult field. In fact, placed as we are a revolution in agriculture is unthinkable in terms of something happening in agriculture alone. It is to be thought of as a series of interchanges between agriculture and industry with rising intensity, industry supplying the basic needs of material inputs for agriculture, agriculture feeding back its surpluses for the development of industry and industry supplying back the various consumer goods on which the agricultural surpluses can be spent. Any other kind of development can only be partial and will not be in conformity with our view of modernisation of the country's economy.

It is, thus in this double harness of simultaneous industrial and agricultural advance with a democratic mode of functioning as our guide that we are moving towards an agricultural revolution in India. Such a revolution is rather simple to outline. It means a revolution in the yield per acre of our farm lands. But this would involve a very great use of commercial fertilisers, made possible by vastly improved methods of water and pest control. It would involve better cultivation, sowing and weeding of the growing plants. It would involve selective breeding, propagation and distribution of those strains of rice, wheat and other crops grown in this country which respond to heavy dosage of fertilisers. Though simple in outline, such a revolution is extremely difficult to bring about. It impinges on centuries old agricultural practices which die hard. It calls for demonstrations on a gigantic scale of the results of the new techniques of agricultural production. It demands intensive botanical and biological research into untried fields. It requires organisational and managerial innovations in the field of agriculture on a scale large enough to affect the six crore farmers in this country. It is not going to be an easy task.

And yet it is precisely this which happened in Japan during the four decades from 1881-1890 to 1911-1920. As a result of these basic changes in Japanese agriculture, its agricultural output increased by more than 75 per cent during this period. The area under cultivation increasing by over 20 per cent and yield per acre by over 45 per cent. During this period in Japan the population increased by 44 per cent but the agricultural labour force fell away by 14 per cent. The per capita food supplies increased by over 70 per cent, the output per farm worker increased by more than 100 per cent. The starting point of this agricultural revolution in Japan was also more or less the same as ours in India today: over-population, low per capita incomes and an average farm of two to three acres of fragmented land.

It is however necessary to remember that what was possible for a small country like Japan with a fair amount of uniformity among its people is not going to be so easy for us with our vast land, a very large population and diverse culture. It is scientific research in the sense of agricultural, chemical and biological research on the most intensive scale and the application of the results of this research on the widest possible scale which can deliver the goods in future. Once these are organised and brought under control, the other fringe improvements should more or less automatically come in their wake. I mean greater mechanical aids to agriculture and so forth.

That we are moving towards such a revolution can be seen from the tentative targets envisaged for the Fourth Plan. The total plan outlay in agriculture and in public and private sectors taken together is likely to be of the order of ₹3,100 crore and in agriculture and irrigation taken together, of the order of ₹4,100 crore. The total investment in agriculture in the public sector alone estimated at ₹2,400 crore is going to be nearly two and a half times that in the Third Plan, and that of agriculture and irrigation taken together estimated at over ₹3,400 crore for the public sector twice that in the Third Plan. In terms of specific targets, the area under improved seeds and under green manures would be doubled, the consumption of fertilisers would be trebled, and the area under plant protection measures would be quadrupled. There would be an increase of nearly 50 per cent in area under major irrigation, of nearly 25 per cent under minor irrigation and about 70 per cent under soil conservation measures. The coverage under plant protection measures would be nearly half and that under improved seeds nearly 80 per cent. While the coverage under chemical fertilisers is difficult to calculate at this stage, it is estimated that about 40 per cent of our total requirements of nitrogenous fertilisers will be met, this showing the leeway to be made up in respect of fertilisers in the subsequent plans. The tentative picture outlined above would indicate the enormity of the tasks before the agricultural scientists and the administrators. In short, it means that the results of scientific research on the widest possible scale should be made available by administrative machinery which is equal to the task to every nook and corner of our countryside.

That this task is going to be truly herculean can be realised if we remember that the performance of agriculture in the Third Plan is likely to fall substantially short of our original targets. As against our original targets of 100 million tonnes for foodgrains production, we are likely to end up with only 90 or 92 million tonnes. Similarly, production of oil seeds is now anticipated to be only 7 to 8 million tonnes against a target of 9.8 million tonnes, and that of cotton 6 million bales as against 7 million bales as originally targeted.

This unsatisfactory performance is primarily due, as I have said earlier, to the failure of inputs for agriculture. As against the target creation of cumulative potential for irrigating 29.5 million gross acres under major and medium schemes, we now expect to create a potential for only 20 million gross acres, of which only 16 million gross acres may be actually utilised. Against a target production of 1 million tonnes in terms of nitrogen of chemical fertilisers, we are likely to produce less than 500 thousand tonnes at the end of the current plan. As against a target of 166 thousand units of diesel engines for use in agriculture, we now hope to produce only 78 thousand units.

It is the view of our experts that with the gigantic effort in the field of agriculture in the Fourth Plan outlined before we can hope to increase agricultural output at least at the rate of 5 per cent per annum. Even this modest rate of growth would require over 3 million tonnes of chemical fertilisers, 2.5 million tonnes of iron and steel, 2.5 million tonnes of coal, 5.5 million tonnes of cement and 2.5 million tonnes of pesticides besides a large number of small machines, equipment and other industrial products. According to a recent analysis of the growth of agricultural production during the period 1951-54 to 1958-61, 33 out of 184 districts for which the statistics were reviewed are seen to have achieved an average annual increase of 7.5 per cent or more per annum and 69 districts out of the total number studied, to have done better than 5 per cent a year. These findings confirm the view that it should be possible to attain a rate of growth of over 5 per cent per annum in agricultural production. What is important is to formulate targets for each one of the 300 districts in our country and execute these plans with the utmost determination. In view of the massive nature of the inputs proposed, the actual results may even be better provided these inputs are timely. The real problem is to translate technological feasibility into an administrative certainty.

For this purpose we shall have to depend in the years to come more and more heavily on the extension services of the community development blocks on the one hand and on rural cooperatives on the other. The central task of the community development organisation in the blocks is now recognised to be to mobilise the entire rural community and the extension services for sustained and concentrated efforts in order to increase agricultural production. It is true that the community development organisation in the blocks includes extension officers not only for agriculture, cooperation and animal husbandry, but also for *panchayats*, women's welfare and rural engineering. But unless agricultural improvements are made the focal point of all these various services, the real task of dynamising agriculture will never be performed. This does not, however, mean that community development activities other than in the field of agriculture are not going to be important in future. In fact, with greater and greater availability of power in the rural areas, many more units for processing agricultural products and their wastes will be set up. The community development organisation will naturally have to foster these and bear the responsibility of consciously bringing about the long-awaited socio-cultural transformation of the countryside in their wake.

Rural cooperatives will also have to play their part in a more efficient manner in supplying the needs of the farmer. It is not enough to say that the membership of the primary agricultural credit and services societies exceeded two crores at the end of 1962-63. As it is, this compares unfavourably with the figure of the total number of agricultural families which is six crores. What is more important than numbers is that their work should be improved in terms of management of business, administration and supervision of units. Large over dues and diversion of loans to non-productive uses still continue to plague us. It is a notorious fact of the current pattern of development that farmers having sub-marginal holdings who are preponderant in number are not in a position to avail themselves of the facilities of the cooperative movement to any significant degree. Besides, the weaker and the relatively more vulnerable sections of the community are also not able to take the fullest advantage of the cooperative apparatus. To solve this last mentioned problem two alternatives are being considered: earmarking of separate societies for this section of the community and earmarking of separate funds for special programme for them. Unless these defects are removed without delay one important aspect of organisation for agricultural advance will remain ineffectual.

Another important organisational parameter is the pace at which land reform measures are completed. The progress in this field has not been spectacular. Administrative arrangements for supervising our land reform enactments have not been adequate and even complete and up-to-date records of rights of tenants do not seem to exist in several states. The consequence of all this has been that the poorer farmers and the landless labourers are fast becoming the single weakest spot in Indian agriculture, a source of serious disaffection. Efforts are, however, being made to implement the reforms in full before the Fourth Plan is launched. States are taking expeditious measures to complete the work in time while the Centre is supervising the progress of the work.

The people who are responsible for the formulation and execution of agricultural policy in this country obviously are not unaware of the many tricky problems. In fact, a number of interesting departures from the past policies have been made in recent times. One such innovation is the emphasis now placed on regional specialisation and area approach, utilising to the full comparative advantages of different regions in the production of different agricultural commodities. Another is the enlargement of the intensive agricultural district programme in an increasing number of districts, the two essential features of this programme being: (i) the development of a package of improved agricultural practices for each important crop based on the latest research findings, and (ii) assistance to the cultivator to develop farm production plans. A third is the crash programme in agriculture now under way and the intensification of the production of subsidiary foods, growing of vegetables and poultry farming etc. a beginning has recently been made in building up a structure of reasonable prices for agricultural commodities, particularly foodgrains, which make agricultural investments worthwhile. A Food Trading Corporation has been set up which along with the Agricultural Prices Commission, which is in the offing, state warehousing and credit facilities is bound to have an impact on production, through a rational structure of prices for foodgrains, lowering of their transportation cost and provision of better storage facilities for them. Finally, the plan for technical training is also being tailored to the increasing needs of agriculture in the Fourth Plan.

All this, however, means unremitting, back-breaking toil day in and day out for the next few years to come. It is time for us to realise that there is no short-cut to the agricultural revolution. The agricultural revolution of which we conceive today will depend for its success on our ability to adapt ourselves to the different techniques best suited for different regions and different crops, techniques either already devised or to be devised through scientific research and technical advances in different fields. This adaptation would call for the reorientation of the human factor in agriculture through continuous education, demonstration and training as well as through the provision of the essential inputs and services to the farmers. It is only if we succeed in making this reorientation shall we be in a position to bring the small farmers and the landless agricultural labourers into the vortex of the revolution. Without them, however, the revolution will die at its birth. I have no doubt that the members of the Indian Society of Agricultural Economics will look upon themselves as the carriers of this revolution throughout the length and breadth of this country.

Indian Journal of Agricultural Economics, Vol. 20, No. 1, Jan.-March 1965.

24

J.P. BHATTACHARJEE

Agricultural Economics and the Agriculturist

HAVING been an evaluator for quite some time, I could not help taking an evaluative look at myself and, in the process, settled for a similar look at the present state of agricultural economics, especially with reference to its ability to explain the behaviour—economic behaviour if it pleases you—of our agriculturists. What I have to say will probably sound like loud thinking, besides being more negative than positive in its implications. If it is so, I seek your indulgence as I go through this exercise. In any case, in the year 1965, one can hardly say much that is absolutely original. However, my intention is an honest one, namely, to provide you with a change of fare, a relief from another analysis of the urgent problems of our agricultural economy.

In making this last comment, my intention is not to underrate the gravity of the food and agricultural situation in the country. It may be one of the worst we have had to face, and is likely to figure in the conference discussions on one of the topics. As a body, we should do everything possible to help the government and the country tide over it. Fortunately, the problem has been anticipated in advance, and we hope that the different state governments will take a national view of the emergent crisis. We also hope that friendly nations will take a sympathetic attitude and generously extend needed aid and assistance.

To get the theme, my main problem is that the available economic theories do not explain sufficiently usefully the practice of agriculture in the so-called traditional societies and do not offer meaningful guidelines for its development. When I say this, I am referring not only to macroeconomic but also microeconomic theories. My reference is essentially to theory, not to measurement. There have been important and significant advances in econometrics, which have immensely improved our ability to specify measure and solve structural and even functional problems. But these have not changed, except in a few sidelines, the framework of economic theory. In fact, in some ways, they have diluted some of the principles. As an illustration, I may mention the convenient way in which principles like diminishing returns have been bypassed in all the models based on the linear system. So, I shall be concerned only with the theory and principles of economics and agricultural economics.

I would, first of all, refer to the general problem many of us have often faced in our attempts to apply the analytical principles of agricultural economics to the solution of the economic problems of development of agriculture and especially the agriculturists. As we all know, the proposition that is most often questioned is whether farmers in traditional societies such as India's do actually and invariably seek to maximise profit or net return of some sort. This question comes up awkwardly in nearly all attempts to apply the principles of farm planning and management and of production economics in general. What is doubted is the behavioural motivation assumed in economic logic; and the ground for doubt is provided by observed discrepancies between the actual and the logically rational behaviour. However, what we do not usually try to differentiate in this argument is whether this assumption does not appear to be realistic only in dynamic situations, or both in static and dynamic circumstances. I shall come to this later on.

Once this doubt is raised, the questioning should not end there, for the whole of the economic logic gets implicated. The more important elements of this logic as far as microeconomic theory is concerned may be summarised in the following four propositions, if we leave out the one already questioned as well as the most general one's concerning wants, ends and means. One of these may be stated as: production is also a techno-economic activity in which inputs combine and produce outputs according to certain predictable relationships. Secondly, the input factors as well as output products have markets in which these can be, and do get exchanged through the medium of money and in this process acquire a valuation of price through the operations of the forces of supply and demand or some variant of these. Thirdly, this market valuation process operates smoothly or perfectly enough in a competitive environment to provide guidelines for a rational allocation of economic resources. Finally, there are many variables such as technology, population, uncertainty, preferences, etc., which are recognised to be of crucial importance but have been traditionally treated as external ones, outside the system for most purposes. To understand the economising behaviour of the farmers, it is necessary to evaluate these basic ingredients of economic theory. For the purpose of this evaluation, I shall be having in mind the type of traditional agriculture which is found in a dual (or plural) economy such as India's. In other words, I shall not assume the extreme or the very early type in which money economy hardly exists, or if at all, on the fringe, and there is almost nothing but the subsistence sector.

How far is production a techno-economic activity under these circumstances? By and large, the question boils down the nature of the crops of products grown, the so-called 'state of the arts,' the availability and elasticity of substitution of different inputs, and the nature and realm of decision-making. A broad general description of the type of agricultural economy I have in mind will be somewhat as follows: The crops or products grown are largely for the farm household sector; but a certain portion of the area is used for growing commercial crops. The 'art of farming' is widely diffused among farmers at a level which has been stabilised over more than one generation of trial and adaptation. The level of technology is low; and technological change or advance is not a continuing process generated within the local community. Differences in skill do exist among the farmers, but these are within a given technological horizon which has evolved in relation to the natural hazards and environmental risks. Among the input factors, land is the most important for a variety of reasons. It also, in the context of India, suffers from a relative inelasticity of supply. There is no sizeable segment of capital in the form of reproducible goods used as a means of production. Those generally used are the conventional tools and implements and draught animals. This does not mean that there has been some improvements made in land and irrigation sources constructed in some cases; but these are all embodied in land. Most of the current inputs are home-grown (seeds, fodder, farmyard manure, etc.) and generally tend to be applied in quantities or does that, subject to availability, are considered optimal by the local community. It is obvious that under the circumstances, these inputs tend to combine in given proportions, the only constraint being availability.

Labour is however, in a separate category. The available supply is determined by the size and composition of the family as modified by the rather limited scope for alternative employment opportunities in the locality. Since the supply is almost pre-determined, there is an attempt to utilise it as fully as possible. The daily wage cost is not a deciding factor (except in respect of hired labour), since the annual subsistence bill has to be borne in any case. It is not surprising, therefore, that in families with small farms there is a tendency, if not a premium, on intensifying the family labour input on the limited land resources so that the available labour gets as much work as can be created. One can even say that it is to them immaterial as to whether all of this labour is technically needed for the volume of production handled by them. The imputed return per unit of such labour thus becomes low, often lower than the conventional wage rate. But what is important for our analysis is that any measurement of this labour use shows not so much the magnitude of the labour input variable in the technical production functions, as that of labour availability. There is always a difference between works for the sake of keeping busy and labour for production in the techno-economic sense. To a certain extent, the same holds also for family bullock labour.

For these and some other reasons production activity in agriculture cannot exactly be described as a techno-economic process. Not only are there far too many constraints, but also the technical relationships get damped because of the nature of inputs. Further, the nature and scope for independent decision-making is limited. It is not surprising, therefore, that production functions derived for such farming activity are not very reliable or meaningful for prescribing methods of improving allocative efficiency in the use of resources.

This point emerges more clearly when we look at the second and third of the hypotheses enumerated earlier. Basically, these relate to the existence of monetisation, market mechanism and competition. That the household sector is dominant in such an economy both in production and consumption is an integral aspect of its specification. The nature and composition of the output is determined up to a certain limit by the domestic or subsistence needs. This forms a solid hard core not easily amenable to change. On the outer perimeter of this core can be placed the so-called cash or market oriented output. This follows from the dual or plural nature of the economy I have assumed. The volume of the latter is influenced by the market pricing mechanism. And to the extent the latter changes, the former also has to change by reason of substitutability. The influence of market prices on the subsistence part of the output may thus be said to be at best indirect, and any observed relationship should be interpreted with caution before any inference of a causal nature is drawn. In any case, if the bulk of the production of the majority of the farmers is for domestic consumption, the volume produced, as distinct from marketed, and is more likely to be influenced by prices if purchased inputs amount to a significant portion of the costs.

That brings me to the market for input factors. Here we find an even more heterogeneous situation. Until the time that fertilisers and pesticides begin to be used, most of the current items of input such as seed, fodder and farmyard manure are home grown or obtained from neighbours on barter. Payments for repair and servicing of tools and implements are even now made in kind after harvest, the basis often being a quantity of grain per implement per year. Land revenue and irrigation charges are in cash; but payment for the use of land by tenants is in kind, generally as a share of the produce. Wages of hired labour are paid partly in cash and partly in kind, the kind part being usually a quantity of grain conventionally considered adequate to meet the labourer's consumption needs. The extent of monetisation in the factor markets is thus very low indeed. What is, however, more important is that valuation of sales, purchases and hire is often socially determined and the terms are not subject to negotiation, bargain or contract for each deal. The element of competition is functionally restricted by social codes; and in the case of land it is also structurally restricted.

When the markets suffer from such serious social constraints and structural rigidities, most of the new-classical theoretic assumptions do not or cannot hold. Thus, the markets are not competitive; and the usual demand-supply-price interactions either operate imperfectly or are not strong. Capital in the form of means of production is not in any way meaningful sense differentiable from land; and if investments on land improvement (including irrigation) with their unknown age-composition are excluded, there is precious little left over to form a meaningful category of capital. In short, capital, in whatever way this factor is defined, is not amenable to meaningful measurement. Finally, the scope for factor substitution is severely restricted; the elasticity is very low indeed. These conditions make a nonsense of the neo-classical theory of distribution with its underlying basis in the calculus of marginal productivity. Even Marshall would probably turn in his grave if somebody tries to tell him—in a séance session—that the impersonal theory of distribution based on assumptions regarding competitive market forces and possibility of continuous substitution along the production function, applies to such an economy or society.

In such dual societies, the 'economics' of production in the non-monetised or imperfectly monetised sector cannot neatly fit into the framework of the conventional equilibrium analysis. The confusion becomes more confounded if an attempt is made to introduce into this theoretical analysis such dynamic elements as uncertainty and technological change. These are elements taken as given in the conventional microeconomic theory; but their importance, especially that of uncertainty, is basic to the economising behaviour of individuals in the type of society we are considering. The incidence of natural hazards may or may not be any higher in these economies that in the developed ones. But the facilities and arrangements—technical or infra-structural, social or governmental—that help the people to bear them, and in the event of their incidence, get over them without economic ruin, are extremely underdeveloped in these countries as compared to the developed ones. Consequently, the feeling of insecurity and uncertainty in the minds of individuals in the former countries is much greater in relation to natural hazards, environmental risks, etc. innovations, modernisation and encroachment of the market economy add further elements of uncertainty except when these changes are designed to reduce hazards of risks that were obtaining earlier. In general, however, there are very few improvements in cultivation methods or practices and production techniques that satisfy the famous *ceteris paribus*—other things, remaining the same. Consequently, the general attitude tends to be one of 'playing safe'. The decision rule is not necessarily to maximise, but something less than that—mini-max or some such criterion that is now being played around by the Game theories.

Unfortunately, we still do not know how best to incorporate this uncertainty element into economic theory. A convenient way for purposes of operational use is to adapt a discounting approach to the net return or the variable (vector) assumed to be maximised by the individual. This procedure does not, however, explain how uncertainty affects the motivation of the individual, not does it specify how the discount rate is to be arrived at. Another possible approach to measurement is to evaluate uncertainty on a scale of values and then add it to the cost of production. This is the procedure adopted for risks, but many not be easy or even feasible is respect of uncertainty which by definition is different from risks and, as such, not amenable to actuarial assessment. In any case, there is long deductive exercise inviting us almost as a challenge, before we can hope to grapple with the uncertainty factor and understand its implication for the economising behaviour of the individual. The game-theoretic approach seems to me a worthwhile one to pursue, provided psychologists are willing to help us in the process of conceptualisation and abstraction of observed behaviour.

What I have said so far is mainly in the nature of an analytical description of the economic system in which the vast majority of the cultivators in a dual society's such as India's have to behave and perform. Many of the aspects of the situation I have been mentioning as specifications without any explanation of the whys or hows behind them. Economists, however, have not been interested very much in these, but other social scientists have been; and I would like to make a brief reference to the relevance of a few of their findings for our purpose. I shall confine myself only to the main sociological findings. These are usually couched in the sociologists' terminology which has developed into quite a jargon, thanks to the limitations of the English language. I sometimes feel that communication between them and us would improve considerably if their works were rewritten by persons like H.J. Fowler, the master of English usage. I hope my sociologist friends will not mind this jab.

Institutional structure has played only an incidental role in economic theory proper, and social structure has received even lesser attention. To the sociologists, however, these have been among the most important areas of study. According to them, the Indian society is rigidly structured on the basis of the system of caste and kinship; and its functioning has been influenced or guided by the *Yajmani* system. The former has defined the hierarchy in terms of status and roles of the family and the individuals, as well as their economic activity; while the latter has prescribed the socio-economic relations among the members of different caste groups in terms of service, activities and payments. This is, of course, a very broad and simple generalisation of the very detailed study of the social structure and relationships that sociologists have been concerned with. It is against this background that they have tried to analyse and explain phenomena like exercise of power and authority, leadership and conformity, innovation and diffusion, achievement and ascription, tension and personally and attitude and behaviour. It is not possible to go into all these aspects in the course of this address. I shall merely refer, somewhat illustratively, to the implication for economics of some of their propositions.

The social system in India affects economic relations, among other things by restricting the mobility of individuals in respect of occupational pursuits in agriculture, limiting the play of competitive forces in intra-caste activities, inhibiting the growth of contractual institutions and impersonal rules of behaviour, putting a premium on allegiance to primary group members and interests and discouraging departments from traditional forms of socio-economic behaviour especially in respect of individual achievement. In retrospect, the system appears to have been designed more for survival and security than for growth and development. It had its good points, but in the changed political economic environment it is faced with a very difficult job of adjustment. All these features of the social system need not necessarily be obstacles that have to be removed for the sake of development. Some of them like dominance of kinship ties or action allegiance may even be turned into forces favourable to development. There will, however, be a considerable degree of creative disruption in the transitional period.

So far as the core of economic theory is concerned, the findings of the sociologists more or less corroborate the weakness or invalidity of the assumptions that I have discussed earlier. The assumptions specially challenged relate to the competitive nature of the markets, particularly the factor markets, smoothness of the price-quantity curves describing the supply and demand functions, elasticity of factor substitution and the sovereignty of the profit motive. Cooperative research between economists and sociologists have not advanced to the point where the needed modifications can be precisely quantified. There is no doubt that the theory of distribution will have to be taken away from its functional base and linked to the institutional set-up, and the production function approach, to be meaningful, will have to take into account a number of social constraints on the substitutability of inputs.

In a traditional dual economy, production and distribution are not mere economic activities. They are also socially influenced, if not determined. It is only a socio-economic view of production that seems meaningful for an analysis of the production behaviour of the agents engaged in it. Since distribution of income among the agents is often predetermined by social norms such as in the case of payments to artisans or hired labourers, the scope for economising decisions in respect of these is limited. In short, production and distribution will have to be viewed together as aspects of the same process.

When we take this point of view, we shall find that the 'economic'—socio-economic in my scheme—behaviour of the traditional farmer in a dual economy stands the test of rationality.

This rationality has the logic of the world he lives in. Professor T.W. Schultz in his book *Transforming Traditional Agriculture*, has also characterised the behaviour of the traditional farmers as rational; but his argument and approach are different, based as they are on the assumption of a low level of marginal productivity of capital, lack of so-called surplus element in manpower, (low level of skill and training of the human elements in agriculture) and the unattractiveness of the available technological improvements. In my way of thinking, the concept of the marginal product of capital is not clearly distinguishable in traditional agriculture, not is the marginal productivity of labour a determining consideration. Given the product distribution pattern in the society and the nature and extent of the uncertainty element that the farmer has to contend with, his allocation of land which is the most important asset entering into his calculations, among enterprises or production lines, is likely to be rational as long as technological or market opportunities do not change the static assumptions.

This brings me to a brief consideration of some macroeconomic aspects have a bearing on the development of such a society. I would start off with a bold statement that in a typically traditional society, there is nothing like a national economy which as an entity can be distinguished from an arithmetic aggregation of the large number of local, sometimes even village economies. What I mean is that in its extreme form, the economy of a traditional society does not have a separate integrated structure other than a sum total of the structure of the local economies. This is, however, an extreme situation which I shall not try to develop or analyse. By the time such an economy acquires a dual character as a result of either colonial development or foreign enterprise or even urban-based native enterprise, an organised foreign trade and urban consumption sector has developed along with a limited industrial, commercial and transportation base. This organised sector manifests the characteristics of a national economy having control levers which can be operated through fiscal, monetary and market mechanisms. But the dispersed rural sector continues in its disaggregative state of organisation, not any longer as distinct area economies, but in respect of the subsistence sector.

A state of functional dualism is thus introduced into the rural agricultural economy. The central problem of development of such a dual economy, as I see it, is one of 'nationalising' the local economies based on subsistent self-dependence. As and obiter dicta, I may add that the main problem besetting the community development movement is also the same, namely, how to 'nationalise' the village community which is cut across by partially overlapping social, political and economic systems.

Two sets of economists have tried to tackle this problem, one as practitioners and advisors, and the other as empiricists or theorists. The practitioners in the profession have been concerned with the problem of what is to be done to move such an economy, while the theorists have tried to find out how it does or can move. The pressing needs of planning and policy formulation have attracted to the former group of economists like Tinbergen, Frisch, Nurkse, Rosenstein-Rodan, Hirschman, Leibenstein, Chenery, Chakrabarty and others. They have made important contributions to the planning and development strategy with special reference to resource allocation, choice of techniques, etc. we would, however, have been in a much better state of enlightenment if they had started from a theory of development of a dual economy. Unfortunately, they have taken such a society as a datum and imposed growth conditions on these. Since my concern is primarily with the behaviouristic aspects of such an underdeveloped economy, I would have to bypass this group.

Surprisingly, there are very few economists who have been interested in a theoretical formulation of the behaviour of the agricultural economy in its passage from a dual to an integrated stage. Arthur Lewis,[1] who was the first economist in the post-war era to draw attention to the vast army of underemployed agricultural labour creating a situation of 'unlimited supply of labour' at real wage rates fixed institutionally, emphasised, among the conditions of growth, the transference of this labour to the industrial sector. His assumptions were realistic; but he did not pay much attention

1. Lewis, Arthur (1954). *Development with Unlimited Supplies of Labour*, The Manchester School, May.

to the agricultural sector. It is this line of explanation that has been further developed by the joint team of G. Ranis and J.C.H. Fei,[2] who have presented a more comprehensive analysis of the behaviour of the two sectors and their terms of trade. According to them, the rate of labour reallocation between the two sectors is the crucial variable to manipulate for converting a dual economy into a fully commercialised, market economy. The absorptive capacity of the non-agricultural sector should set the pace of labour release from the agricultural sector for achieving a balanced growth. As long as dualism exists, there are two different wage schedules, the one for the traditional sector determined by the subsistence level. All this sounds like a return to the lines of classical economics, though with a difference.

One of the weaknesses of the Ranis-Fei approach lies in the treatment of technological improvement and innovation. Another weakness lies in the assumption that the rate of growth of employment in the industrial sector will outpace the rate of growth of the labour force in agriculture. The recent experience in the developing countries shows that in reality, the rate of growth of net industrial employment is likely to be rather low in the initial stages of industrialisation and development. If the real wages in the agricultural sector are to be maintained even at the current low levels, the productivity of labour in agriculture has to be raised faster than population growth, as a *sine qua non* of development. Otherwise, the system is likely to break down.

The other economist who tried to explain the behaviour of the traditional agricultural economy is T.W. Schultz, to whose contribution I have already referred. Schultz's main concern has been the explanation of the long run equilibrium of such economies at the subsistence level. His analysis brings out, as prime movers, the right type of technology which Rani-Fei assume to be remaining stacked up on the shelf, and the functional education of the cultivators.

2. Ranis, G. and J.C.H. Fei (1964). *Development of Labour Surplus Economy.*

An economist who has attempted to tackle the challenging job of synthesising economics, sociology and psychology is E.E. Hagen.[3] His is an admirable work that starts with the inadequacies of the economic theories of growth and then goes into social structure and personality formation conducive to entrepreneurial and innovational activity. This is the first inter-disciplinary work that I have seen and can commend to other. Economists, however, will probably with that he had attempted in the end a reconstruction of economic theory proper, by introducing the personality factor and its capacity to live with tension and sustain change, among the relevant variables.

This brief survey of the recent lines of macroeconomic theorising on the development of the agricultural sector shows that all the loose ends have not yet been tied together. I should now conclude this part by presenting some of the lines of analysis that appeal to me. I stated earlier that at the macro-level, the problem of development of agriculture can be viewed as one of nationalising the local economy of the non-commercialised sector within it. This commercialisation has to proceed *pari passu* with a large enough growth of productivity, especially since large scale transference of the so-called 'surplus' labour can be ruled out in the short and intermediate run. As the traditional local economy has been stagnant, its structure and operation can be assumed to be not conducive to quick, regular and sustained absorption of technological improvement and innovation which alone can raise productivity. Among the factors which are relevant for consideration in this connection, the more important ones are capital *vis-à-vis* technology, uncertainty, growth of the labour force, quality and skill of the labour force, enterprise and innovational propensity of the decision-makers, the role of the Government in research, extension and organisation, and the nature and tempo of development of the non-agricultural sector. These are the operational variables that need to be kept in view in formulating a theory of development of traditional agriculture.

The first pre-requisite for such a theory is an explanation of the relationships between technological improvement and capital labour

3. Hagen, E.E. (1962). *On the Theory of Social Change*, Illinois.

requirement, innovation and capital accumulation, technological change, its impact on uncertainty and the education and skill required for its adoption. This may not be an easy area for theorising, even for developed economies. As for the developing ones, there are very little experience and data to go by. There is no other way but to conceptualise hypothetical situations.

The second area of inquiry relates to what has been traditionally called distribution theory. There is here, a need for a thorough overhaul, not merely a nut-and-bolt job. The starting point should be the definition and identification of land and capital as items of factor inputs; and the inquiry should proceed from there to valuation and measurement of these factors against the background of the social valuation attached to these. Then there is the whole field of labour supply—wage labour as well as family labour—wage determination for hired labour and returns for family labour. Both institutional and educational-cultural factors are bound to play a significant role in this area of study. Otherwise, it may not be possible to explain situations such as in Kerala (probably the best example of a 'labour surplus' economy) where wage rates necessary to attract rural labourers are about the highest among the states in India, though the conventional wage rates in the traditional jobs are rather low. There is some evidence to show that once new jobs outside the traditional socio-economic pursuits are created, something like an entirely different social or group norm about the minimum level begins to operate, in spite of a statistically large enough supply. Finally the influence of the caste system operates to strengthen the non-competing groups.

Lastly, the production relationships should have social constraints built into them, in addition to technology and uncertainty. It is with these building boxes that a general theory of development of traditional agriculture can be attempted. It would take me too far out—and I do not have the time also—to attempt even a preliminary outline of such a theory. Obviously, there have to be at least two sectors, each with its behaviouristic relationships and areas of interaction with the other. All this may sound like a highly ambitious specification and a tall order; but I feel safe, for nobody can be elected President of this Conference for a second time.

What I have presented so far is an analysis of the theory on the basis of which agricultural economics is taught, practiced or enquired into. Such an analysis does not immediately throw up findings of relevance to policy, though the address should, I understand, offer some suggestions on policy issues. In order not to go against this tradition, what I can at best do is to offer some views, purely as obiter dicta.

The analysis in the earlier part of the address indicates the need for a cautious and understanding approach to farm management research and extension. Workers in these areas would probably do better, if they expressly take into account the uncertainty situation facing farmers and the latter's assessment of these, in formulating resource allocation decisions. Secondly, input-output and other production relationships derived from non-experimental data should not be looked at as sacrosanct or invariant techno-economic measures. They are variable and depend on the socio-economic background of the farmers to whom these relate. It is, indeed, very necessary in any search for these relationships to decide initially on the groups from whom such data could be collected, for meaningful analysis and future application. Thirdly, I would be interested to see someone try out an accounting method under which identifiable improvements on land are evaluated separately and included under capital, and then work out the relative returns to factors. Another accounting procedure for which there is considerable justification is to evaluate the total input of family labour in a year at the total subsistence bill for the family for the whole year. These are a few of the changes suggested by the earlier analysis.

On the macroeconomic plane, there is a very strong case for evaluating all the technological improvements that are often stated to be waiting for use. Such an evaluation may, in the end, show that there are indeed not too many available to choose from, if the uncertainty and skill aspects are fully taken into account. After all, such improvements are expected generally to result in a reduction not only of input per unit of output, but also of the uncertainty element. It has been a common experience that the latter type of improvements gets accepted quickly and extended widely. Rust-resistant varieties of wheat provide a good example. I hope that in our search and research for improved technology, we will always keep one eye on the uncertainty element so that we do not blindly fall for things which promise very high output but only under the research station conditions of care and control.

My last point is on education, training and skill. By now, the value of education as an instrument of agricultural development is fairly well recognised; the question is one, of what type of education. There has to be, or course, elementary education corelated as far as possible with agriculture. For immediate impact, however, this may not be sufficient. Is it not possible to supplement it with a quick training scheme that would have the effect of imparting new skills? What I have in mind are training arrangements for the masses of petty cultivators, tenants and agricultural labourers, in the use of improved tools, implements and eventually of machinery. It is they who have to use them if these are going to be extended on a large scale. Once they get trained and familiar with these, there will be a desire for using them. Even if they cannot buy them immediately, there will be a force or pressure injected into the society and the landowners will eventually think of investing in them. Is it not possible for the community development agency to undertake such a training programme in all blocks?

I have now come to the end of my address. I do not know if I have been able to present you with a fare rich enough for this occasion. But if what I have said enkindles in you a desire to take a new look at the economic theory underlying agricultural economics, I shall be very happy indeed; and if I have succeeded in evoking in you a challenge for a battle with hypothesis and theory, I shall consider my effort more than amply rewarded.

Indian Journal of Agricultural Economics, Vol. 21, No. 1, January-March 1966.